DICTIONARY OF
ARCHITECTURAL SCIENCE

DICTIONARY OF
ARCHITECTURAL SCIENCE

HENRY J. COWAN

Professor of Architectural Science, University of Sydney, Australia

A HALSTED PRESS BOOK

JOHN WILEY & SONS
New York—Toronto

PUBLISHED IN THE U.S.A. AND CANADA BY
HALSTED PRESS
A DIVISION OF JOHN WILEY & SONS, INC., NEW YORK

Library of Congress Cataloging in Publication Data

Cowan, Henry J.
A dictionary of architectural science.

Bibliography: p.
1. Architecture—Dictionaries. 2. Building—Dictionaries.
I. Title.
NA31.C64   721'.03   73–15839
ISBN 0–470–18070–6

WITH 105 ILLUSTRATIONS

© APPLIED SCIENCE PUBLISHERS LTD 1973

Printed in Great Britain by Galliard (Printers) Ltd, Great Yarmouth, Norfolk, England

To Dr Kate Salisch

Preface

A selection of words for a dictionary inevitably involves a personal choice, and one has to balance a desire to be comprehensive against the need to be concise. The terms were compiled, in the first instance, from the subject indexes of about two hundred standard textbooks. I assumed that readers would range from architects, who might find some quite elementary scientific and mathematical terms helpful, to scientists with little practical building experience. Hence I have included simple terms, such as eaves, specific gravity and tangent.

I have endeavoured to cover words in neighbouring fields which architectural scientists might wish to have defined, since jargon is a great obstacle to reading outside one's speciality. An old-established subject, such as geometry or classical architecture, inevitably acquires a wide range of specialist terms. Their justification in a newly established 'in' subject is more debatable. Sports writers use jargon without explanation, not to show how much they know, but because it gives their readers a warm glow of satisfaction to be on the inside. The same motive may explain the extensive use of technical terms in many articles on modern art and on computer applications.

Evidently one must draw the line somewhere. From classical architecture I have included 'pediment', partly because it is sometimes confused with 'pedestal', but excluded 'metope', which describes a non-structural decoration. Among computer abbreviations, I have explained Fortran which has come to stay, but not Watfor which may be of only passing significance. Constructivism is mentioned because of its obscurity, but expressionism is known well enough to require no definition. Some of the terminology of differential geometry is explained, but tensors are beyond the scope of this book. Although strictly a term in architectural science, kitchen sink is omitted, because everybody knows what it is; on the other hand, there is an entry for ton, because there are three different kinds of ton.

The dictionary aims to be comprehensive within the field of architectural science proper, *i.e.* structures, materials, acoustics, lighting, thermal environment and building services. If there are any gaps, they are due to oversight. Timber classification is an exception, because the terminology is not internationally standardised, and a complete explanation may only add to the confusion. Except for a few timbers, like Douglas fir, the same name often denotes a different material in a different part of the world, as for example in the case of oak, maple or mahogany. May I refer the reader to the glossary which the local timber industry organisation usually provides with its compliments.

I have left out terms of marginal significance which posed similar problems. Thus invariant, paradoxically, has definitions which vary with the branch of science.

The illustrations may give the impression of a particular leaning towards structures and geometry. This is due to the fact that definitions in these areas particularly benefit from explanatory sketches, while they are of little help in elucidating thermal problems or material properties. No sketches have been included on decorative architectural features. I have mentioned Corinthian columns and Perpendicular Gothic for the benefit of readers who could not, on the spur of the moment, remember the difference between them and Ionic columns or Decorated Gothic. However, this dictionary is not intended to cover architecture as such. Some entries in pure science are very brief, for the same reason.

Names have been mentioned only incidentally. This is partly because I do not believe that either the history of architecture or the history of science is the history of great men, and partly because adequate biographical coverage would require at least a volume on its own.

On the other hand, I have deliberately included a few records. 'Taller than the Eiffel Tower' has a definite meaning, if one knows its height.

Words in SMALL CAPITALS indicate that further relevant information is given in a separate entry under that word. However, the absence of small capitals does not necessarily imply the absence of a separate explanation; for example the mention of fluorescent tube under 'electroluminescent panel' is not in small capitals, because readers who look up 'electroluminescent panel' are likely to be familiar with the meaning of 'fluorescent tube'.

Words are printed in *italics* for general emphasis. For example, the reader is directed under 'conoid' to consult 'surface of translation'. Under the latter heading, the word *conoid* appears in italics to attract the eye.

Metric units are used in the scientific entries. Quantitative information of a more applied character is given in the old British units with conversion in brackets to the SI system (the new metric system now being introduced in Great Britain). There are also conversion tables in Appendix G.

Reference is made especially to Britain, America and Australia. This does not imply any lesser regard for other English-speaking countries; it merely indicates that my acquaintance with them has been too fleeting to set myself up as an authority on their specialist terms. Spelling is in accordance with the Oxford Dictionary, and cross-references are given where American spelling differs significantly, particularly in the first three letters.

Most entries are short, and deal only with one word. However, in a few cases, where terminology is confusing, longer entries are used which list all the terms for one subject together.

It would be impossible to acknowledge the several hundred authors to whom I am indebted, although a significant proportion is mentioned in Appendix A. However, I would like to make special mention of the Encyclopaedia Britannica (9th (1875) and 1965 editions), the Shorter Oxford English Dictionary (third edition), Chambers' Technical Dictionary, the Penguin Dictionaries of Architecture, Building and Civil Engineering, and Knaurs Lexikon der Modernen Architektur.

I am particularly grateful to Mrs Rita Arthurson for typing the manuscript, and to Mr Paul Frame and Mr John Howard for preparing the drawings. Finally, I should like to express the hope that readers who find the mistakes, omissions and lacking cross-references which undoubtedly remain, will be so kind as to let me know, so that they can be corrected.

H.J.C.

Sydney

Contents

xi

A

ABACUS (*a*) A simple calculating device, particularly one consisting of beads sliding on bars. *See also* DIGITAL COMPUTER. (*b*) A slab forming the uppermost member or division of a column capital; it supports the ARCHITRAVE.

ABBEY Originally a church associated with a monastery, which contains the seat of the abbot. *See also* CATHEDRAL.

ABNEY LEVEL *See* CLINOMETER.

ABRAMS' LAW Experimental rule enunciated by D. A. Abrams in 1919;

'With given concrete materials and conditions of tests, the quantity of mixing water determines the strength of concrete, as long as the mix is of workable plasticity.' *See* Fig. 1 and WATER–CEMENT RATIO.

ABRASION The wearing away of the surface of a material by the cutting action of solids. There are numerous abrasion resistance tests. Comparing similar materials by the same test gives satisfactory results; however, the correlation between different tests is difficult, and the method of testing must always be specified. Abrasion tests may be rolling or sliding in nature, with or without abrasive. Usually the test is run for a definite number of strokes with a definite pressure, and the loss of weight is measured.

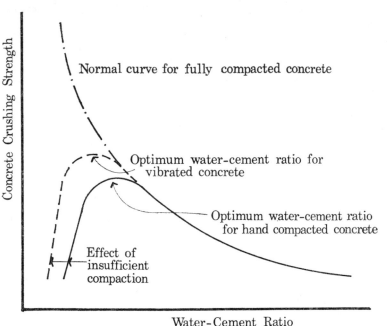

FIG. 1.   Abrams' law.

ABS PLASTIC Acrylonitrile–butadiene–styrene, a thermoplastic material of low cost and good chemical resistance, used largely for drain, waste and vent pipes.

ABSCISSA The *x*-axis, or horizontal axis, of a co-ordinate system (Fig. 13).

ABSOLUTE HUMIDITY *See* HUMIDITY.

ABSOLUTE TEMPERATURE Temperature measured from absolute zero, *i.e.* the lowest temperature which can be reached in theory, when the molecules of a perfect gas would possess no kinetic energy, and its volume would become zero. This occurs at $-273 \cdot 16°C$ and $-459 \cdot 67°F$. The absolute temperature scale, measured in Celsius (Centigrade), is called the *Kelvin* scale (K). Measured in Fahrenheit, it is called the *Rankine* scale, but this is rarely used.

ABSOLUTE VOLUME The actual volume of the particles of sand, concrete aggregate, etc. It is determined by immersing the aggregate in water, and measuring the volume displaced.

ABSOLUTE ZERO *See* ABSOLUTE TEMPERATURE.

ABSORBER, SUSPENDED *See* SUSPENDED ABSORBER.

ABSORPTION (*a*) The process whereby a liquid is drawn into the permeable pores of a porous solid. (*b*) Transformation of radiant energy to a different form of energy by the intervention of matter, as opposed to *transmission* which is the passage of radiation through matter without change of its frequency. (*c*) Absorption of sound. *See* SOUND ABSORPTION.

ABSORPTION COEFFICIENT *See* SOUND ABSORPTION COEFFICIENT.

ABSORPTION CYCLE A refrigeration cycle. It utilises two phenomena: (*a*) The absorption solution (absorbent plus refrigerant) can absorb refrigerant vapour; and (*b*) the refrigerant boils (flash cools itself) when subjected to a lower pressure. These two phenomena are used to obtain refrigeration. In the *lithium bromide* absorption machine, the bromide is used as an absorbent, and the water as a refrigerant. *See also* AIR CONDITIONING.

ABSORPTION FACTOR In lighting, the ratio of the LUMINOUS FLUX absorbed by a body to the flux which it received.

ABSORPTION RATE *See* SUCTION RATE.

AC Alternating current.

ACANTHUS A plant native to the Mediterranean. Its leaves are used as ornamentation, particularly on Corinthian column capitals.

ACCELERATED WEATHERING *See* WEATHEROMETER.

ACCELERATION The rate of change of velocity. The *acceleration due to the earth's gravity* has a mean value of 32·2 ft/sec/sec or 980·7 cm/sec/sec. *See also* NEWTON, KILOPOND, AND LBF.

ACCELERATOR A substance which speeds up a chemical reaction, as opposed to a RETARDER. (*a*) In concrete, an additive which increases the rate of hydration of the cement, and thus shortens the time of setting, or increases the rate of hardening or of strength development. (*b*) In synthetic resins or glues, a CATALYST which increases the hardening rate. The accelerator is mixed with the resin immediately before use.

ACCESS TIME The time required to accomplish the transfer of information in and out of a COMPUTER STORE.

ACCIDENTAL ERROR *See* COMPENSATING ERROR.

ACCLIMATISATION The process of adaptation by persons accustomed to a different climate, whereby the strain resulting from exposure to environmental stress is diminished. *See also* COMFORT ZONE.

ACETYLENE A highly inflammable gas (C_2H_2) which is colourless and highly poisonous. It can be generated by the action of water on calcium carbide; however, it is more commonly used in bottled form. It is occasionally employed for heating and lighting. Combined with bottled oxygen (*oxy-acetylene*) it is used for cutting and FUSION welding of steel.

ACHROMATIC LENS A lens designed to minimise chromatic aberration. *See* GLASS.

ACI American Concrete Institute, Detroit.

ACID A chemical compound containing hydrogen which can be replaced by metallic elements, and which produces hydrogen ions in solution. It neutralises *bases* to form *salts*, has a *pH value* of less than 7, and turns blue litmus red.

ACID ROCK An igneous rock with a preponderant silica content, *e.g. granite.*

ACID STEEL A steel made in a furnace lined with an acid refractory, such as silica, and under an acid slag. Under these conditions neither sulphur nor phosphorus is removed.

ACOUSTIC BOARD A low density fibre board with good SOUND ABSORPTION. It is often perforated to provide RESONANT ABSORPTION. *See also* PEG BOARD, INSULATING BOARD, and HELMHOLTZ ABSORBER.

ACOUSTIC CONSTRUCTION *See* DISCONTINUOUS CONSTRUCTION.

ACOUSTIC IMPEDANCE *See* IMPEDANCE TUBE and RESONANCE.

ACOUSTIC INSULATION *See* INSULATION, ACOUSTIC BOARD and DISCONTINUOUS CONSTRUCTION.

ACOUSTIC MODELLING Analysis of auditoria by means of models. As the scale is reduced, the frequency of the sound has to be increased in accordance with DIMENSIONAL ANALYSIS. If it is only required to check the direct sound lines, a much simpler and cheaper form of model analysis can be carried out by means of the *light analogy*. The sound source is replaced by a light source, and the amount of light received in various parts of the auditorium is determined by inspection or by photography.

ACOUSTIC PLASTER Plaster with a high sound absorption. If normally contains a substantial number of closed pores, which may be provided by gas bubbles resulting from aluminium powder or detergent (*see* CELLULAR CONCRETE) or by the use of VERMICULITE aggregate.

ACOUSTIC QUALITIES OF AN AUDITORIUM As defined by L. L. Beranek. *See* BRILLIANCE, DEFINITION, DYNAMIC RANGE, ENSEMBLE, INTIMACY, LIVENESS, TIMBRE, and WARMTH. *See also* ECHO and REVERBERATION TIME.

ACOUSTIC STRAIN GAUGE A *strain gauge* for measuring surface strains by vibrating a stretched wire which is matched against a note from a similar wire vibrating in a reference gauge. The tension in the wire changes with the strain, and the note changes accordingly. The instrument is very sensitive and can measure strains of 1×10^{-6}. Like the electric resistance strain gauge, it may be used in confined spaces, and controlled from a distance. Unlike the electric resistance gauge, it is not sensitive to moisture; but its minimum gauge length is greater.

ACOUSTIC TILE Square tile, normally placed on the ceiling, which has the capacity of absorbing sound. It is made of absorbent material, such as ACOUSTIC BOARD, and frequently perforated.

ACOUSTICAL CLOUD A *reflector* suspended from the ceiling of an auditorium, usually above the orchestra. *See also* SUSPENDED ABSORBER (whose purpose is to *absorb*) and SOUNDING BOARD.

ACOUSTICAL TRANSMISSION FACTOR The reciprocal of the SOUND REDUCTION FACTOR.

ACOUSTICS *See* DECIBEL and REVERBERATION TIME. *See also under* NOISE and SOUND.

ACROPOLIS The elevated part or citadel of an ancient Greek city; the best known is the one of Athens.

ACRYLIC RESINS Thermoplastic materials produced by the polymerisation of the monomeric derivatives of acrylic acid. They are obtainable in perfectly transparent and also in opaque form, and they have the best resistance to sunlight and outdoor weathering of all the transparent plastics; however, they are attacked by many organic solvents. The smoothness of surface and clarity of the cast sheets is superior to that of window glass, but they lack hardness and are easily scratched. Being thermoplastic, they are not fire-resistant. Acrylic resins are available as sheets, rods, tubes, castings and moulding powder. They are available white or tinted. They have PHOTO-ELASTIC properties. The ease with which they can be formed and joined, and their high elastic deformation, make them particularly suitable for experimental stress analysis, despite a high rate of creep. Acrylic resins are best known by the trade names *Plexiglas* (Rohm and Haas), *Perspex* (Imperial Chemical Industries) and *Lucite* (E. I. du Pont de Nemours).

ACSA Association of Collegiate Schools of Architecture, which has members in the USA and Canada.

ACTIVATION ENERGY Energy required to initiate a chemical reaction.

ACTIVE EARTH PRESSURE *See* EARTH PRESSURE.

ACTIVITY The term used in NET-WORK programming to denote a basic component of the work required for a building project. *Critical activities* are those which lie on the critical path. *Non-critical* activities are those which do not lie on the critical path. *Near-critical activities* are those which come to lie on the critical path if circumstances alter, *e.g.* if CRASHED TIME is used.

ACUITY *See* VISUAL ACUITY.

ADAPTATION (*a*) The process taking place as the eye becomes accustomed to the luminance or the colour of the field in view; or to its darkness. (*b*) The final state of the process.

ADDING MACHINE *See* DIGITAL COMPUTER.

ADDITIVE *See* CONCRETE ADMIXTURE.

ADDRESS A name or number identifying a particular location in the store of a computer.

ADHESION *See* BOND (*b*).

ADHESIVE *See* GLUE.

ADIABATIC A change in the condition of a body without any exchange of heat with the surroundings. An adiabatic change cannot normally be ISOTHERMAL.

ADMIXTURE *See* CONCRETE ADMIXTURE.

ADOBE Construction from large sunbaked, unburnt bricks, used in the south west USA, parts of Latin America and in other semi-arid countries. *See also* COB and PISÉ.

ADSORPTION Condensation of a gas on the surface of a solid. For example, SILICA GEL has the ability to collect water vapour by adsorption, and thus keeps dry the cases of instruments sensitive to moisture.

AELOTROPIC Having physical properties which vary according to the direction in which they are measured, as opposed to ISOTROPIC. *See also* ORTHOTROPIC.

AEOLIAN Wind-blown, *e.g.* aeolian soil (*loess*).

AERATED CONCRETE *See* CELLULAR CONCRETE.

AERODYNAMICS That part of the mechanics of fluids which deals with the dynamics of gases, particularly the study of forces acting on bodies in moving air. *See also* BOUNDARY LAYER, MACH NUMBER, and WIND TUNNEL.

AFNOR Association Française de Normalisation, Paris.

Ag Chemical symbol for *silver* (argentum).

AGATE A natural aggregate of crystalline and colloidal silica (SiO_2), coloured by metallic oxide. It is sometimes translucent or attractively banded; these varieties are classed as semi-precious stones, and they have been used in sculpture and architecture. It is also extremely hard, and for this reason used for the bearings of scientific instruments. *See* QUARTZ.

AGE HARDENING The hardening of an ALLOY which results from the formation of tiny particles of a new PHASE within the existing solid solution. *Also called precipitation hardening* or *ageing*.

AGEING OF CONCRETE The final stage in the chemical reaction between cement and water during which the concrete continues to gain strength slowly. It continues for many years. *See also* SETTING and HARDENING.

AGEING OF METALS The process of AGE HARDENING.

AGGLOMERATE Small particles bonded together into an integrated mass.

AGGREGATE *See* CONCRETE AGGREGATE.

AGING *See* AGEING.

AGORA An open space in an ancient Greek town, used as a market and general meeting place. The *forum* had the same function in Roman architecture.

AGRÉMENT SYSTEM A system established in France for certifying new materials and methods of construction, and later introduced in other countries. The *agréments*, or certificates of approval, are issued in France by the *Centre Scientifique et Technique du Bâtiment*. There is also a European Union for Agrément, to which France, The Netherlands, Portugal, Spain, Belgium and Italy belong, and a British Agrément Board.

AIA American Institute of Architects, Washington.

AIEE American Institute of Electrical Engineers, New York.

AIR-BAG LOADING A load applied in a model test or a full-scale test by pumping up a bag, pressing against a firm foundation, with an air compressor. Also called *pneumatic loading*.

AIR BRICK A brick perforated for ventilation purposes.

AIR CHANGES PER HOUR A unit of ventilation, defined as the volume of air passed through the ventilation system, per hour, divided by the volume of the room ventilated. The number of air changes required varies from 60 for laundries to 1 per hour for store rooms. The range for normal occupancies lies between 2 and 6. A more accurate method is to estimate the supply of air on the basis of persons occupying the space.

AIR CONDITIONING Artificial ventilation with air at a controlled temperature and humidity. Heating air, and moistening it if necessary, is relatively inexpensive. Air conditioning normally implies cooling and DEHUMIDIFICATION of the air.

AIR CONDITIONING DUCT *See* DUCT LINING.

AIR DRYING Drying a material, such as timber, in the air, instead of seasoning it in a kiln. *Air dry timber* has a moisture content which is approximately in equilibrium with that in the surrounding atmosphere. *See also* NATURAL SEASONING.

6

AIR-ENTRAINING AGENT An additive to cement or an admixture to concrete which causes minute air bubbles to be incorporated in concrete or mortar during mixing. It is claimed that this increases workability and frost resistance. *See also* ENTRAPPED AIR.

AIR FELTING Forming a mat from an air suspension of fibres. It is used for certain types of particle board to prevent the absorption of moisture which necessarily occurs in *wet felting*.

AIR OUTLET *See* CEILING DIFFUSER, PUNKAH LOUVRE and REGISTER.

AIR SEASONING *See* NATURAL SEASONING.

AIR SLAKING The process of exposing QUICKLIME to the air. It gradually absorbs moisture and breaks down into a powder. *See also* HYDRATED LIME.

AIR-TO-AIR HEAT-TRANSMISSION COEFFICIENT *See* THERMAL TRANSMITTANCE.

AIR TRAP A water-sealed trap which prevents foul air rising from the soil drain through the outlet of a wash basin, wc, etc.

AIR WASHER *See* HUMIDIFIER.

AIRBORNE SOUND Sound vibrations transmitted to a part of the building by airborne pressure waves, as opposed to *impact sound*. They may be caused by the original noise source, by reflections from another wall, or by the airborne vibrations of a part of the building, produced by sound waves propagated in it (*e.g.*

by a continuous steel structure). Insulation against airborne sound conforms to the MASS LAW.

AISC American Institute of Steel Construction, New York.

AISLE (*a*) A wing (Latin *ala*) attached to the NAVE of a church, usually separated from it by a line of columns. (*b*) Hence, any division in a church, such as a passage between pews. (*c*) Hence, a passage between seats in any building, such as a theatre or concert hall.

AITC American Institute of Timber Construction, Washington.

Al Chemical symbol for *aluminium*.

ALABASTER Pure GYPSUM in densely crystalline form. Due to its softness it is easily carved and polished.

ALBERTI'S TEN BOOKS *De re aedificatoria*, published by Leone Battista Alberti in 1452, and the first printed architectural treatise of the RENAISSANCE (Florence, 1485). It was translated into English by J. Leoni under the title *Ten Books of Architecture*, printed in 1755. *See also* FILARETE, VITRUVIUS and PALLADIANISM.

ALCLAD *See* CLADDING.

ALCOVE A recess in a room, originally vaulted or separated by an arch (from the Arab *al qobbah* a vault). The term now denotes a recess extending to the floor, while a *niche* does not extend to the floor.

ALGEBRAIC MEAN *See* MEAN.

ALGOL *ALGOrithmic Language,* an internationally agreed PROBLEM-ORIENTED computer programming language for scientific applications; however, FORTRAN is more widely used.

ALGORITHM Corruption of al-Khowarazmi (a ninth century Arab mathematician); the term originally used to denote arithmetic using the Indian–Arabic (*i.e.* decimal) numerals. It now generally implies a sequence of logical processing rules, set up to solve a problem.

ALITE *See* TRICALCIUM SILICATE.

ALKALI (*a*) A synonym for BASE. (*b*) More correctly, and in a more limited sense, a generic term for the hydroxides of sodium, potassium, lithium, rubidium, and caesium, which are called the *alkali metals.*

ALKALI–AGGREGATE REACTION Chemical reaction between certain aggregates and the sodium and potassium compounds contained in Portland cement. Aggregates liable to this type of reaction may cause deleterious expansion in mortar or concrete.

ALKYD PAINTS Paints using alkyd resin (derived from an alcohol, such as glycerol, and an organic acid, such as phthalic acid), as the vehicle for the pigment. There are interior, exterior and fire-retardant types. *See also* OLEORESINOUS PAINT.

ALLOTROPY The ability to exist in more than one state, *e.g.* carbon has three allotropic varieties which are diamond, graphite and amorphous carbon.

ALLOWABLE STRESS *See* MAXIMUM PERMISSIBLE STRESS.

ALLOY A substance with metallic properties, composed of two or more elements, which after mixing in the molten state do not separate into distinct layers on solidifying. Normally alloys are mixtures of metals. Structural steel (which is a mixture of iron and carbon) is a notable exception. Alloys may be composed of chemical compounds, solid solutions, eutectics, eutectoids, or of aggregations of these with each other and with pure metal.

ALLOY DIAGRAM *See* PHASE DIAGRAM.

ALLOY STEEL A steel to which one or more elements, other than carbon, have been added, as opposed to *carbon steel* (Fig. 71); although, strictly speaking, carbon steel is an alloy.

ALLUVIAL SOIL Soil deposited by flowing water (generally during a flood) in recent times (geologically speaking) on land which is not permanently submerged.

ALPHA BRASS A copper–zinc alloy which contains more than 64 per cent copper. It has good tensile strength and considerable ductility.

ALPHA IRON Unalloyed iron below 1670°F (910°C). It has a body-centred cubic space lattice, and it is magnetic below the magnetic change point which, in pure iron, is 1414°F (767°C). *See* Fig. 71 and also GAMMA IRON.

ALPHA PARTICLE A helium nucleus, *i.e.* a helium atom, which has lost two electrons and consequently has a double positive charge. It contains two protons and two neutrons. The *alpha radiation* emitted by radium etc. consists of a stream of alpha particles.

ALPHANUMERIC CHARACTER A term used in conjunction with digital computer printouts to denote both the letters of the alphabet, the ten numerical digits, and any special characters used. Generally 0 denotes zero, and the symbol Ø is used for the letter O.

ALTERNATING CURRENT An electric current which reverses its direction of flow at regular intervals (commonly at 50 cycles per second (50 Hz)), as opposed to a *direct current*.

ALTERNATOR An alternating-current generator. It is usually driven by a heat engine or a water turbine. *See also* DYNAMO.

ALTITUDE *See* AZIMUTH.

ALUMINA Aluminium oxide (Al_2O_3).

ALUMINIUM A white metal of atomic weight 26·98. Its chemical symbol is Al, its atomic number is 13, it has a valency of 3, a specific gravity of 2·70, a melting point of 660·1°C, and a coefficient of thermal expansion of $23·5 \times 10^{-6}$ per °C. Most aluminium is produced from BAUXITE by an electrolytic process, which depends on a cheap source of electric power for its economy. Aluminium metal is silvery-white in colour. It oxidises very readily, but the oxide skin formed provides a protective coating which inhibits

further oxidation. Aluminium is readily alloyed with copper, silicon, nickel, manganese, magnesium and other metals, and a very wide range of alloys are available for casting, forging, stamping, rolling and extruding. *Also called aluminum.*

ALUMINIUM FOIL *See* REFLECTIVE INSULATION.

ALUMINOUS CEMENT *See* HIGH-ALUMINA CEMENT.

ALUMINUM *See* ALUMINIUM.

AMALGAM An alloy of mercury and some other metal.

AMBIENT TEMPERATURE, HUMIDITY Temperature, humidity of the surrounding air.

AMBULATORY A place for walking in, particularly a covered passage adjacent to a building.

AMERICAN BOND *See* BOND (*a*).

AMERICAN EPHEMERIS *See* NAUTICAL ALMANAC.

AMINO-PLASTICS A generic term for urea FORMALDEHYDE and melamine formaldehyde resin.

AMMETER An instrument for measuring electric current, graduated in *amperes*.

AMMONIA A colourless, alkaline gas of composition NH_3. It has a freezing point of $-77°C$ and a boiling point of $-33°C$. Liquid ammonia is widely used in refrigeration plants.

AMORPHOUS Not crystalline.

AMPERE (A)   The unit of electrical current, named after the nineteenth century French physicist A. M. Ampère. One ampere is the constant current which, if maintained in two straight parallel conductors of infinite length, of negligible circular cross-section, and placed at a distance of 1 metre apart in a vacuum, produces between them a force equal to 2×10^{-7} NEWTONS per metre length.

AMPHITHEATRE   An oval, circular, or semicircular building with seats rising in ascending rows around an ARENA.

AMPLIFIER   A device for increasing the power level of a signal.

AMPLITUDE   The maximum value of a periodically varying quantity during a cycle, *i.e.* the crest or bottom of a wave, measured above or below the mean value.

ANALOGUE COMPUTER   A calculating device based on an analogy, *i.e.* a comparison of a lesser-known phenomenon to a well-known phenomenon. A SLIDE RULE is a very simple analogue computer. More complex types employ electrical currents, *e.g.* BUSH'S analogue for the slope-deflection equation, DANTER'S thermal analogue, and an analogue for the seepage of water through soil. The MEMBRANE ANALOGY and the SAND HEAP ANALOGY are used for solving the torsion problem, and the COLUMN ANALOGY for solving rigid frames. Analogue computers are generally designed to solve a specific problem, and they are therefore less versatile than DIGITAL COMPUTERS. They are also usually slower and less accurate; however, their cost may be quite small. *See also* INDIRECT MODEL ANALYSIS.

ANALYSIS   *See* STRUCTURAL ANALYSIS.

ANCHORAGE   Device, frequently patented, for permanently anchoring the tendons at the ends of a POST-TENSIONED member, or for temporarily anchoring the tendons of PRE-TENSIONED members during hardening of the concrete.

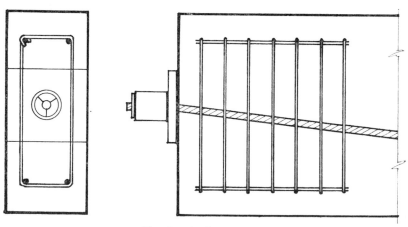

FIG. 2.   Anchorage zone.

ANCHORAGE ZONE In POST-TENSIONED concrete, the region adjacent to the anchorage of the tendon, which is subjected to secondary stresses resulting from the distribution of the prestressing force. Unless suitably reinforced (Fig. 2) the concrete may split due to secondary tension. In PRE-TENSIONED CONCRETE, the region in which the transfer bond stresses are developed.

ANECHOIC CHAMBER A room where all boundaries are highly absorbent, as opposed to a REVERBERATION CHAMBER. Also called a *free-field room.*

ANEMOMETER Instrument for measuring wind speed. High air velocities can be measured with the *deflecting vane anemometer* which gives a direct reading, and the *rotating vane anemometer*, which counts the rotations. The latter can also be used for low air speeds. Other low-speed instruments are the KATA-THERMO-METER (now largely obsolete), the HOT-WIRE ANEMOMETER, and the IONISATION ANEMOMETER.

ANEROID BAROMETER *See* BAROMETER.

ANGIOSPERMAE *See* HARDWOOD.

ANGLE CLEAT A small bracket formed of ANGLE IRON, which is used to locate or support a member of a structural framework.

ANGLE IRON A steel section, either hot-rolled or cold-formed, consisting of two legs at an angle (which is almost invariably a right angle). An angle iron may be *equal* (both legs the same width) or *unequal.*

ANGLE OF FRICTION The angle φ between the force due to the weight of a body resting on a surface, and the resultant force when the body begins to slide. The coefficient of friction $\mu = \tan \varphi$. *See also* FRICTION.

ANGLE OF INCIDENCE The angle which a ray or wave makes with the normal to the surface which is reflecting it.

ANGLE OF INTERNAL FRICTION The angle of friction (φ) in granular soils. It is defined by the equation shearing resistance = normal force on surface of sliding $\times \tan \varphi$. For perfectly dry or fully submerged granular soil (clean sand or gravel) it equals the *angle of repose, i.e.* the steepest angle at which a heap of dry soil will stand. *See also* RANKINE THEORY and BULKING.

ANGLE OF REPOSE *See* ANGLE OF INTERNAL FRICTION.

ANGLE OF SHEARING RESISTANCE *See* COULOMB'S EQUATION.

ANGLO-SAXON ARCHITECTURE *See* ROMANESQUE ARCHITECTURE.

ÅNGSTRÖM Unit of measurement for very small lengths, equal to 1×10^{-10} metre, 0·1 nanometre, or 0·1 millimicron; abbreviated as Å. Wavelengths of light etc. are commonly given in ångströms.

ANHYDRIDE An oxide which, if combined with water, produces an acid.

ANHYDRITE Anhydrous calcium sulphate ($CaSO_4$). It is found naturally as a mineral, or it may be made from GYPSUM by removing the water of crystallisation, usually by heating above 325°F (163°C). Anhydrite produced from gypsum is more reactive than the naturally occurring mineral. *See also* KEENE'S CEMENT.

ANHYDROUS A term applied to minerals which do not contain water of crystallisation.

ANHYDROUS LIME Quicklime.

ANIMAL GLUE *See* GLUE.

ANION A negative ion, the opposite of a *cation* which is positive.

ANISOTROPIC Having different properties in all directions, as opposed to ISOTROPIC.

ANNEALING Heating an alloy (such as steel) at a temperature about 50°C above the upper limit of the transformation temperature range. By contrast, TEMPERING is carried out below this range. The object is to remove stresses induced by previous treatment and improve ductility. The annealing temperature is held for at least an hour (more for thick pieces), and cooling is slow.

ANNUAL RING *See* GROWTH RING.

ANNULUS A figure bounded by two concentric circles, *i.e.* a thick ring.

ANODE The positive electrode of an electrolytic cell, or battery. The *cathode* is the negative electrode.

ANODISING A process of coating aluminium or aluminium alloys with a film of aluminium oxide, to protect the surface from corrosion. The film is slightly porous, and it is usually sealed, *e.g.* with lanoline dissolved in spirit. Colours can be introduced before sealing. However, aniline dyes are liable to fade if used externally.

ANT, WHITE *See* TERMITE.

ANTE-SOLARIUM A balcony which faces the sun.

ANTHROPOMETRY The measurement of the human body with a view to determining its average dimensions. *See* ERGONOMICS and MODULOR.

ANTI-ACTINIC GLASS A heat absorbing glass which absorbs about 85 per cent of the INFRA-RED radiation, but also about 50 per cent of the visible light; it is faintly green. Care must be taken that not too much of the heat absorbed is radiated into the room to be protected, and that due allowance is made for the thermal expansion of the glass.

ANTICLASTIC *See* GAUSSIAN CURVATURE and SADDLE.

ANTI-CORROSIVE PAINT Paint which delays corrosion, particularly of steel. It is used as a *primer*, rarely as a finishing coat. The best-known anti-corrosive primer is *red lead*.

ANTI-FRICTION METAL *See* WHITE METAL.

ANTI-LOGARITHM *See* LOGARITHM.

ANTIMONY A bluish white metal. Its chemical symbol is Sb, its atomic number is 51, its valency is 3 or 5, its atomic weight is 121·76, its specific gravity is 6·62, and its melting point is 630·5°C. Antimony and its alloys expand on solidification, thus reproducing the fine details of the mould.

ANTISIPHON TRAP An AIR TRAP which contains an additional volume of water in an enlarged pipe, to increase the resealing quality of the trap.

APARTMENT A dwelling unit, usually for rent, in a multi-storey building. The building is known as an *apartment house. See also* FLAT, MAISONETTE, TENEMENT, CONDOMINIUM and HOME UNIT.

APPARENT BRIGHTNESS *See* LUMINOSITY.

APPARENT VOLUME OF A POROUS SUBSTANCE The BULK VOLUME minus the open pores, or the true volume plus the closed pores. The *apparent density* is the mass, divided by the apparent volume.

APPROXIMATE NUMBER A number which cannot be represented exactly, such as 0·333 33 . . .. An *exact number* can be represented by one or more INTEGERS. *See also* IRRATIONAL NUMBER.

APRON A relatively wide vertical flashing, or a flashing that surrounds or partly surrounds a projecting construction.

AQL *Acceptable quality level*, the average quality at which the producer should work to satisfy the customer. It is a commonly used means of choosing a *sampling* scheme.

AQUA FORTIS Concentrated nitric acid.

AQUA REGIA A mixture of nitric acid and hydrochloric acid in the proportion of 1:3, so called by alchemists because it dissolves gold.

AQUEDUCT An artificial channel for the conveyance of water, usually an ancient structure carried on masonry arches. *See also* VIADUCT.

ARC OF A CIRCLE A portion of its circumference. The length of an arc, which subtends an angle θ degrees at the centre, is $\theta\pi D/360$, where D is the diameter.

ARC WELDING FUSION welding in which the heat is derived from an electric arc formed either between two electrodes, or between the parent metal and one electrode. *See also* ACETYLENE, and ARGON-ARC WELDING.

ARCADE Strictly a line of arches carried on columns, either free-standing or attached to a wall. The term was applied during the nineteenth century to glass-roofed shopping areas modelled on the CRYSTAL PALACE, and is now used for passages with shops on one or both sides, irrespective of construction.

ARCH A structure designed to carry a load across a gap mainly by compression, the opposite of a suspension cable (*see* Fig. 73). *See* CATENARY ARCH, CIRCULAR ARCH, CROWN, POINTED ARCH, PORTAL, and SPRINGINGS.

ARCH BAR A support for a FLAT ARCH.

ARCH BRICK A brick VOUSSOIR.

13

ARCH STONE   A stone VOUSSOIR.

ARCHIMEDES   *See* BUOYANCY and LEVER PRINCIPLE.

ARCHITECTURE   *See* SCIENCE OF ARCHITECTURE.

ARCHITRAVE   (*a*) The lowest of the three parts of the entablature of the classical orders. It is beneath the frieze and rests on the capital of the column. (*b*) Hence the moulded frame surrounding a door or window. (*c*) Hence the trim which covers the joint between an opening in a wall and the wall finish, particularly for a wooden door or window in a plastered wall.

ARCUATED   Spanning with arches, as opposed to TRABEATED.

ARE   A metric unit for area, equal to 100 m². 1 hectare = 100 ares = 10 000 m².

ARENA   (*a*) The sandy area forming the stage of an *amphitheatre*. (*b*) Any public place where contests are held.

ARENACEOUS   Sandy. Composed largely of sand, as opposed to *argillaceous*.

ARGILLACEOUS   Clayey. Composed largely of clay or shale, as opposed to *arenaceous*.

ARGON-ARC WELDING   ARC WELDING in an inert atmosphere of argon (which is an inert gas akin to, but heavier than helium). The argon is directed into the weld area through a sheath surrounding the electrode, thus preventing oxidation of both the electrode and the weld pool. The process is used for stainless and heat-resistant steels, and for aluminium and magnesium alloys which oxidise very readily. *See also* HELIARC WELDING.

ARITHMETIC MEAN   *See* MEAN.

ARM   *See* LEVER ARM.

ARMOURED CABLE   An insulated electric cable wrapped with a flexible steel covering.

ARMOURED CONCRETE   An obsolete term for reinforced concrete.

ARMOURED WOOD   Wood covered with metal.

ARRAY   A table of related numbers.

ARRIS   A sharp edge formed by the meeting of two surfaces, particularly two mouldings.

ARRIS GUTTER   A V-shaped gutter.

ARROW   The graphic representation of an ACTIVITY in a CPM NETWORK. One arrow represents one activity; however, it is not a vector quantity, and not normally drawn to scale (Fig. 67). The intersection of two arrows marks an EVENT.

ART NOUVEAU   A movement away from an imitation of the past, characterised by undulating curves, particularly in the form of flowing hair, flames, and flower stalks. It flourished at the beginning of the twentieth century, and took its name from a shop in Paris, opened in 1895. The German *Jugendstil* and the Italian *Stile Liberty* are essentially similar.

ART OF ARCHITECTURE *See* SCIENCE OF ARCHITECTURE.

ARTICULATED STRUCTURE A structure constructed with PIN JOINTS.

ARTIFICIAL LIGHT *See* FILAMENT LAMP, FLUORESCENT TUBE, LIGHTING, LUMINAIRE, MAINTENANCE FACTOR, and PSALI.

ARTIFICIAL SEASONING Seasoning timber by some means other than NATURAL SEASONING.

ARTIFICIAL SKY A hemisphere, 20 to 25 ft (6–7$\frac{1}{2}$ m) in diameter, lit inside to imitate the natural sky. A model, scaled in accordance with the principles of dimensional analysis, is placed at the centre of the hemisphere, and measurement of internal lighting conditions are made with photoelectric cells. Most artificial skies are calibrated for the overcast condition represented by the MOON-SPENCER SKY. A simpler type of artificial sky, which can only be used to investigate windows in a wall, consists of a mirror-lined box lit from above; this creates lighting through the windows similar to that from an infinitely distant source.

ARTIFICIAL STONE Precast concrete made with careful attention to appearance, frequently in imitation of natural stone.

ARTS AND CRAFTS The name given to a nineteenth century group of artists, of whom *William Morris* and the *Pre-Raphaelites* are the best known. It takes its name from the *Arts and Crafts Exhibition Society*, formed by C. R. Ashbee in 1888. The movement emphasised the importance of craftsmanship, and drew attention to the merits of medieval art. *See also* DEUTSCHER WERKBUND.

ASBESTOS Mineral occurring in fibrous form, which consists of various silicates. The fibres are thin, tough and flexible, and they can be woven like textiles. They do not change in high temperatures, and asbestos in sprayed form, as paper, and as cloth is consequently used for fireproofing. It is also a fairly good thermal insulator, particularly when sprayed with a gun, and an electrical insulator at low voltages. Asbestos is completely rot-proof and vermin-proof.

ASBESTOS CEMENT SHEET Building sheet made from cement with an admixture of asbestos fibres. It can be cut with a saw. *See* FIBRO.

ASBESTOS CURTAIN A fireproof curtain made of asbestos and other fire-resisting materials. It is frequently placed at the PROSCENIUM opening of a theatre stage, and closed in case of a fire on the stage to protect the auditorium.

ASCE American Society of Civil Engineers, New York.

ASEE American Society for Engineering Education, Washington.

ASHLAR Squared stonework as opposed to rubble. *Coursed ashlar* is laid in regular horizontal courses. *Random ashlar* consists of square blocks whose horizontal and vertical joints do not line up. *Ashlar brick* is brick which has been rough-hacked on the face to make it resemble stone.

ASHRAE American Society of

Heating, Refrigerating and Air-Conditioning Engineers, New York.

ASHVE American Society of Heating and Ventilating Engineers, now ASHRAE.

ASME American Society of Mechanical Engineers, New York.

ASPDEN, J. *See* PORTLAND CEMENT.

ASPHALT A black sticky mixture of hydrocarbons used for waterproofing basements and flat roofs. In America the term is used both for the naturally occurring product and that obtained from the distillation of petroleum. Elsewhere the artificial asphalt is usually called *bitumen*.

ASPHALT MASTIC *See* MASTIC.

ASPHALT ROOFING Waterproof roof laid either with bituminous felt or with mastic asphalt.

ASPHALT SHINGLES Roof shingles, widely used in America, made of felt saturated with asphalt or tar, and surfaced with mineral granules.

ASSAY Estimation of metal content in an ore by chemical analysis or heat treatment.

ASSMAN PSYCHROMETER A pair of WET AND DRY-BULB THERMOMETERS mounted in a nickel-plated cover to shield them from radiation. To achieve uniform ventilation, the air is drawn mechanically over the thermometers with a small fan.

ASTERISK A little star (*). In many ALPHANUMERIC printouts * is used for multiplication ($x*y$ means x multiplied by y), and ** for exponential ($x**y$ means x^y).

ASTM American Society for Testing Materials, Philadelphia.

ASTRAGAL A small semi-circular moulding, often decorated with a bead. In the classical orders it was used for the light mouldings separating the shaft from the necking of the capital, whereas the heavy moulding at the base of the Ionic and Corinthian columns was called a *torus*. The term astragal is sometimes used for mouldings covering a joint around a door or window, and for glazing bars.

ASTYLAR A façade without columns or pilasters. This is now common practice, but it was a design worthy of comment in the Renaissance and its revivals.

ATLANTES *See* CARYATIDS.

atm *Atmosphere*, a unit of pressure. The pressure exerted by the weight of the air at the surface of the earth is 14·7 psi, and this equals one atmosphere. 1 atm = 101·325 kN/m². *See also* MILLIBAR and BAROMETER.

ATOMIC HEAT The quantity of heat required to raise the temperature of one gram-atom of an element through 1°C.

ATOMIC NUMBER The number of a chemical element when arranged with the others in increasing order of atomic weight, as in the PERIODIC TABLE. It is equal to the total number of positive charges in the nucleus, or the number of orbital electrons in an atom of the element.

ATOMIC WEIGHT The relative weight of an atom of an element, taking the weight of an atom of oxygen as 16. The classification was originally based on Hydrogen as unity; hydrogen has now an atomic weight of 1·008. *See also* ATOMIC NUMBER and ISOTOPE.

ATOMISE To break up a liquid into very fine drops.

ATRIUM The main courtyard of a Roman house. In modern architecture an atrium house is one designed around a private court.

ATTENUATION Diminution or weakening, particularly of sound.

ATTERBERG LIMITS *See* LIQUID LIMIT, PLASTIC LIMIT and PLASTICITY INDEX.

ATTIC Roof space between the top-floor ceiling and the roof.

Au Chemical symbol for *gold* (aurum).

AUDITORIUM (*a*) Room especially designed to have satisfactory acoustics for speech, music or both. (*b*) In a theatre, the space assigned to the audience, as opposed to the stage. Fire regulations frequently require that the two spaces be capable of separation by means of a fireproof curtain.

AUGER (*a*) A tool for boring holes in wood. (*b*) A tool for boring holes in soil. The *post-hole auger* is a hand-operated tool which can be used to obtain BOREHOLE SAMPLES. Larger holes must be made with a power *earth auger*, or by conventional excavation methods.

AUSTENITE An allotropic form of iron (γ-iron), which has a face-centred lattice. It is not stable in carbon steels at room temperature. However, it exists in stable form in certain alloy steels containing nickel and chrome.

AUTOCLAVE A pressure vessel in which materials are exposed to high-pressure steam. In the building industry, autoclaving is used for the rapid curing of precast concrete products, sand-lime bricks, asbestos cement products, and hydrous calcium silicate insulation products.

AUTOGENOUS HEALING The closing of fine cracks in concrete and mortar through chemical action. It occurs naturally if the concrete or mortar is kept damp and undisturbed.

AUTOMATION Automatic handling of work during production and in transit between machines. The word was coined by the Ford Motor Company.

AVERAGE The arithmetic MEAN.

AVIARY A birdhouse.

AVOGADRO'S HYPOTHESIS 'Equal volumes of different gases at the same pressure and temperature contain the same number of molecules'; named after the Italian physicist who enunciated it in 1811. *Avogadro's number* is the number of molecules contained in the gram molecule weight of any gas; it is about $6·1 \times 10^{23}$.

AVOIRDUPOIS WEIGHT The normal system of weights used in FPS units. Other units are still used for weighing precious stones and precious metals.

AXIAL-FLOW FAN *See* CENTRI-FUGAL FAN.

AXIAL LOAD *See* CONCENTRIC COLUMN LOAD.

AXIOM *See* POSTULATE.

AXIS, MAJOR AND MINOR *See* ELLIPSE.

AXIS, NEUTRAL *See* NEUTRAL AXIS.

AXONOMETRIC PROJECTION *See* PROJECTION.

AZIMUTH The angle which the vertical plane through a point makes with the meridian plane. It is thus the horizontal co-ordinate for locating a point on the surface of a sphere. The vertical co-ordinate is the *altitude*. These two co-ordinates are required for determining sight lines, and also for defining the position of the sun, the moon or a star at any time of the day and year. The problem arises, for example, in the design of a *sun-shading device*. The altitude and azimuth are calculated by spherical triangles from the *right ascension* and the *declination* of the sun (its co-ordinates in the sky, obtainable from a NAUTICAL ALMA-NAC), from the geographical location of the building, and from the time.

B

BACK PRESSURE In plumbing, air pressure in a pipe which is above atmospheric pressure.

BACK SAWING *See* BOXED HEART.

BACKGROUND NOISE Blurred noise caused by air-conditioning machinery, the rumble of distant traffic, the rustle of wind in trees, the low sound of a large number of voices, etc. It is often regarded as a desirable 'sound-perfume' to blot out disturbing noise at close quarters, *e.g.* speech by other persons. It may also help to ensure the privacy of a conversation. On the other hand, a high level of background noise is considered a nuisance. To determine whether background noise is a significant factor, the sound source under investigation is turned off, and sound level measurements are taken before and afterwards.

BACKHAND WELDING Welding with the blowpipe directed towards the completed weld.

BACKING STORE The part of the computer store beyond the central processor unit.

BACKUP MATERIAL A material placed at the back of a curtain wall, particularly for fireproofing.

BAG OF CEMENT The bag of cement is frequently used as the unit for batching the aggregates and the water. It is commonly assumed that each bag contains the standard weight; however, some codes forbid that, and require that the cement be weighed. The standard weight of a bag of cement in the United Kingdom is 112 lb. In the USA and Australia it is 94 lb, and in Canada it is 87·5 lb. Most metric countries place 50 kg of cement into each bag.

BAGASSE Fibre obtained from sugar cane. It is used as raw material for both hardboard and particle board.

BAKELITE One of the earliest synthetic resins, named after its inventor L. H. Baekeland. It is produced by the condensation of phenol with FORMALDEHYDE. Many other plastics are now marketed by the Bakelite Company, which have different compositions.

BALANCED DESIGN In reinforced concrete, a design which produces simultaneous overstressing of the steel in tension and the concrete in compression; *i.e.* neither material reaches its limit before the other. It occurs only when a *balanced percentage of reinforcement* is used. A beam which has more steel is *over-reinforced*, and one which has less steel is *under-reinforced*.

BALANCED PERCENTAGE OF REINFORCEMENT The percentage which produces a *balanced design*.

BALANCED REINFORCEMENT A balanced percentage of reinforcement.

BALANCED SASH *See* SASH WINDOW.

BALCONY A platform projecting either from an inside or an outside wall of a building.

BALCONY BEAM A beam which supports a balcony. It usually has a horizontal projection, and is therefore subject to combined bending and torsion. *See also* BOW GIRDER.

BALK A large squared timber, usually of softwood. Also spelt *baulk*.

BALL CATCH A type of door fastening in which a spring-controlled metal ball in the door engages a hole in the door frame.

BALL COCK An automatic float valve for controlling the flush in a water closet served by an individual cistern. An empty ball of copper or plastic floats up as the water level in the cistern rises, and shuts the valve when a predetermined level is reached. It is thus only possible to use a definite quantity of water, and no more, for flushing. *See also* FLUSHING VALVE.

BALL MILL A mill in which material is finely ground by rotation in a steel drum with steel balls or pebbles. It is used for grinding cement.

BALL-PEEN HAMMER The engineer's hammer, used for metalwork and stone-masonry. It has a hemispherical peen (or small end) and a flat face. *See also* CLAW HAMMER.

BALL TEST A test for determining the WORKABILITY of freshly mixed concrete. A cylindrical metal weight with a hemispherical bottom is dropped from a standard height, and the depth of penetration is measured. *See also* SLUMP TEST.

BALL VALVE *See* BALL COCK.

BALLOON FRAME Timber frame in which the studs run in one piece to the roof plate. The floor joists are nailed to the studs. The development of the balloon frame in America in the 1830s led to a great reduction in the labour content of timber houses, by replacing the complex timber joints (such as *mortise and tenon, tongue and fork, dovetailing*), by simple nailed joints. It became possible only after suitable nails were mass produced.

BALSA WOOD   A very soft wood (although technically a hardwood) weighing only 7 lb/ft³ (110 kg/m³). It is widely used for models, and it can also be used as insulation in coreboard.

BALUSTER   Wooden post (in the past often elaborately carved or turned) holding up the handrail of a staircase, balcony, etc. *Also called banister.* The entire assembly of balusters is a *balustrade.*

BAND SAW   A saw consisting of an endless belt, one edge being cut to form the teeth, as opposed to a CIRCULAR SAW.

BAND SHELL   A curved sounding board placed over a bandstand.

BANDWIDTH   The range of FRE-QUENCIES in a wave-band. *See* SOUND-FREQUENCY ANALYSER.

BANISTER   *See* BALUSTER.

BAR CHAIR   A rigid device, usually of steel or plastic, which supports the reinforcement in its proper position and prevents its displacement both before and during concreting.

BAR CHART   *See* GANTT CHART and HISTOGRAM.

BAR, DEFORMED   *See* DEFORMED BAR.

BAR SUPPORT   *See* BAR CHAIR.

BARBICAN   An outwork defending the entrance to a castle.

BARGE BOARD   A sloping board covering the projecting portion of the timbers of a gable roof. In the nineteenth century it was often elaborately decorated. *Also called verge board.*

BARITE   Barium sulphate (BaSO₄). It is used as aggregate for high density concrete, *e.g.* for concrete used in radiation shields for atomic reactors. Also spelt *baryte.*

BARK POCKET   A patch of bark, partially or wholly enclosed in a wooden board. It is a source of weakness.

BAROMETER   An instrument used for the measurement of *atmospheric* pressure. The most accurate instrument is the *mercury barometer*, which consists essentially of a vertical tube, about 800 mm long, closed at the end, which is filled with mercury and placed in a pool of mercury. The atmospheric pressure on the pool of mercury balances the mercury column in the tube, and a vacuum forms above it. At standard temperature and pressure the height of the mercury column is 760 mm. A more convenient, but less precise form, is the *aneroid barometer*, which consists of a hermetically sealed metal box, exhausted of air so that the ends of the box approach or recede from one another with change in air pressure. The movement is magnified by levers. *See also* MILLIBAR.

BAROQUE   An architectural style which succeeded Mannerism in the seventeenth century, and was also widely used in the eighteenth; it strongly influenced Latin American architecture. It is characterised by exuberant decoration and curvaceous forms, which may degenerate into grotesque contortions. The term has been applied to modern architecture which shows these characteristics.

BARREL BOLT A round bolt for locking a door or a window. It slides in a cylindrical barrel, usually formed from sheet metal.

BARREL VAULT A semi-cylindrical or partly cylindrical roof structure of constant cross-section (Figs. 32 and 68). It was widely used in masonry construction, particularly in Romanesque architecture. A much thinner form is today built in reinforced or prestressed concrete as a *shell roof. See also* BEAM THEORY OF SHELLS.

BARYTE *See* BARITE.

BAS RELIEF Sculpture carved in low relief, *i.e.* the figures project only slightly from the face of the background. *See also* INTAGLIO.

BASAL METABOLISM *See* METABOLISM.

BASALT A fine-grained, dark-coloured IGNEOUS rock of basic composition.

BASE (*a*) A substance which neutralises acids, producing salt and water. It has a pH VALUE greater than 7, and turns red litmus blue. (*b*) The lowest part of a monument, column, wall or pier.

BASE LINE (*a*) In perspective

PROJECTIONS, the intersection between the ground plane and the picture plane. (*b*) In construction work, a definitely established line from which other measurements for laying out a building are taken.

BASEBOARD *See* SKIRTING.

BASILICA Originally a large, oblong hall built in the Roman empire for the administration of justice. Later, a Christian church of the same shape.

BAT A piece of brick used for closing a gap. One end is whole and the other is cut or broken. *See* HALF BAT.

BATCH MIXER A machine which mixes one batch of concrete or mortar at a time, as opposed to a *continuous mixer.*

BATCH OF CONCRETE Quantity mixed at one time, usually containing one BAG of cement.

BATTER Inclination from the vertical or horizontal.

BAUHAUS Literally the House of Building. *Walter Gropius* was appointed in 1911 to succeed Henri van de Velde as head of the School of Arts and Crafts at Weimar, and changed its name to Bauhaus. The school moved in 1925 to Dessau, where a glass-walled building, of the same name, was designed by Gropius to house it. Gropius left the Bauhaus in 1928, and the school was closed by his successor, *Mies van der Rohe,* in 1933 following the rise of the National Socialists to power. In spite of its short existence, the curriculum established at the Bauhaus had world-wide influence on architectural education, and on the development of the INTERNATIONAL STYLE. *See also* DEUTSCHER WERKBUND.

BAULK *See* BALK.

BAUXITE Hydrous aluminium oxide, named after Les Baux in France. It is the principal raw material for the manufacture of aluminium metal. It is also used in the manufacture of high alumina cement. Also spelt *beauxite*.

BAY An internal division of a building, marked by the column spacing, or an external division marked by the fenestration.

BCD Binary coded decimal.

BEAD *See* GLAZING BEAD and BEADING.

BEADING A small cylindrical moulding enriched with ornament resembling a string of beads used, for example, in Romanesque and neo-Romanesque architecture. Hence, an undecorated semi-circular timber moulding used to mask a joint.

BEAM A structural member which supports loads across a horizontal opening by flexure. *See* NAVIER'S

THEOREM. JOIST and GIRDER are synonyms for beam.

BEAM-AND-SLAB FLOOR A floor system, particularly in reinforced concrete, whose floor slab is supported by beams. *See also* FLAT PLATE and FLAT SLAB.

BEAM TEST *See* MODULUS OF RUPTURE.

BEAM THEORY OF SHELLS The MEMBRANE THEORY of shells is a reasonable approximation to the behaviour of *short* cylindrical shells, but it is unsatisfactory for *long* cylindrical shells. Although solutions exist for a general bending theory of cylindrical shells, and these have been partly tabulated, the simple beam theory is often used as an approximation. The shell is treated as a beam of curved cross-section (Figs. 3 and 32).

BEARING CAPACITY The load per unit area which a foundation can safely carry.

BEARING PILE A PILE which carries a vertical load, as compared with a SHEET PILE which resists earth pressure. The load may be carried on a load-bearing layer, such as rock (*end-bearing pile*) or by friction between the surface of the pile and

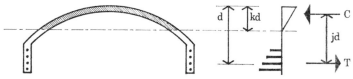

FIG. 3. Beam theory of long cylindrical shells. The concrete above the *neutral axis* (which is at a depth kd) provides the compressive component C of the bending moment, and the steel the tensile component T. The length of the lever arm is jd, and the bending moment $M = Cjd = Tjd$.

the surrounding soil (*friction pile*). *See also* BORED and PRECAST PILE.

BEARING PLATE A plate placed under a heavily loaded support of a beam, column, frame etc. to distribute the load over a wider area.

BEARING WALL A wall or partition which supports the portion of the building above it in addition to its own weight, as opposed to a *nonbearing* wall or partition which supports only its own weight.

BEAUFORT SCALE A scale for wind speed, which ranges from 0 for complete calm to 12 for a cyclone with a speed above 75 mile/h (120 km/h). The wind speed (in miles per hour) equals $1·875\ B^{1·5}$, where B is the Beaufort number of the wind ($3\ B^{1·5}$, if the wind speed is in kilometres per hour).

BEAUX-ARTS *See* ECOLE DES BEAUX-ARTS.

BEAUXITE *See* BAUXITE.

BEDFORD, T. *See* STUFFINESS.

BEDROCK Solid rock underlying superficial formations.

BEESWAX The natural secretion of the bees, from which the honeycomb is formed. It is soluble in turpentine and used in WAX polishes and stains.

BEGGS' DEFORMETER The original device for indirect model analysis, developed by G. E. Beggs at Princeton University in 1922. It applies vertical, horizontal or rotational deformation at one end of a scale model of the structure, and the corresponding thrust, shear or bending moment along the structure is then deduced from its deflected shape by the RECIPROCAL THEOREM.

BEL The basic unit of sound measurement, named after A. G. Bell, the inventor of the telephone. The practical unit is the DECIBEL, which is 0·1 bels. The reason normally given is that the bel is too large a unit; however, since the most common decibel readings are between 20 and 90, this appears illogical.

BELFAST TRUSS *See* BOWSTRING GIRDER.

BELFRY A tower in which a bell is hung.

BELGIAN TRUSS *See* FINK TRUSS.

BELITE *See* DICALCIUM SILICATE.

BELL-AND-SPIGOT JOINT *See* SPIGOT.

BELL PUSH A button which rings a bell when pushed.

BELL TRANSFORMER A small transformer which changes the mains voltage to that used by a bell or chime, usually 10 volts.

BELVEDERE A building, upper storey, lantern or turret, especially built because it commands a fine view. *See also* GAZEBO.

BEMA A platform built for public speaking or for ceremonies, named after the raised step from which orators spoke in the Pynx in Ancient Athens. *See also* ROSTRUM.

BENCH MARK A permanent reference mark, fixed to a building or to the ground, whose height above mean sea level has been accurately determined by a survey.

BENDING FORMULA *See* NAVIER'S THEOREM.

BENDING MOMENT The moment at any section of a beam of all the forces that act on the beam on one side of that section. There are two opposite sign conventions. The one normally employed in architectural text books and in building codes states that a bending moment is positive when the beam is bent concave or downwards ('*sagging*') and negative when it is bent convex or upwards ('*hogging*'). Thus the bending moment in a simply supported beam is positive, and that in a cantilever is negative (Fig. 66). *See also* SHEAR FORCE.

BENDING STRESSES IN SHELLS *See* MEMBRANE THEORY.

BENT A two-dimensional frame which is capable of supporting horizontal as well as vertical loads.

BENTONITE A clay composed principally of MONTMORILLONITE. Its expansion and contraction on wetting and drying are exceptionally high.

BENZENE A highly inflammable hydrocarbon of the aromatic series (C_6H_6), obtained from the distillation of coal tar. It is used as a solvent and a paint remover.

BENZINE A highly inflammable mixture of hydrocarbons obtained from the fractional distillation of petroleum. Apart from its use as a

motor fuel, it is employed as a solvent in quick-drying finishes.

BERANEK, L. L. *See* ACOUSTIC QUALITIES.

BERNOULLI'S ASSUMPTION 'In a beam, sections which were plane and parallel before bending (Fig. 4) remain plane after bending, but converge on a centre of curvature'. Consequently the relation between strain and distance from the neutral axis is linear. The assumption, which is easily proved in the laboratory, is the experimental basis of NAVIER'S THEOREM.

BERNOULLI'S THEOREM 'Along any one streamline in a moving liquid, the total energy per unit mass is constant'; it consists of the pressure energy, the kinetic energy, and the potential energy:

$$\frac{p}{\rho} + \tfrac{1}{2}v^2 + z = \text{constant}$$

where p is the pressure, ρ is the density, v the velocity and z the head of the liquid.

BERRY STRAIN GAUGE *See* DEMEC STRAIN GAUGE.

BESPOKE BUILDING A term coined by P. A. Stone to contrast industrialised or 'off-the-peg' building with traditional, tailor-made building.

BESSEMER PROCESS A method, developed by Sir Henry Bessemer in 1856, of producing steel. Air is blown through molten pig iron contained in a refractory-lined, pear-shaped cylindrical vessel, open at the upper end for the escape of gases,

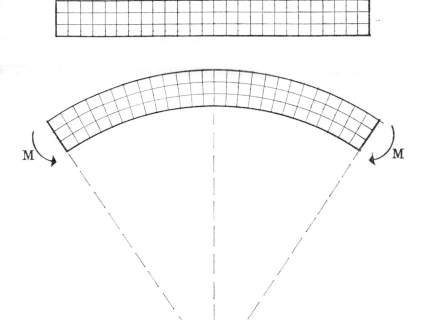

FIG. 4.   Bernoulli's assumption.

called the *Bessemer converter*. The heat is produced by the oxidation of impurities. The converter can be lined with acid or basic refractories; but only the latter can be used with pig iron which has a high phosphorus content. *See* ACID STEEL. *See also* OPEN-HEARTH PROCESS.

BETA BRASS   A copper–zinc alloy whose zinc content is between 46 and 50 per cent.

BETA PARTICLE   An electron. The *beta radiation* emitted by the atomic nuclei of radioactive substances during their spontaneous disintegration consists of a stream of electrons.

BÉTON   The French word for concrete; the same word is also used in German and Russian.

BÉTON BRUT   Literally concrete in the raw, *i.e.* the concrete surface left after removal of the formwork, without subsequent finishing. The term is used specifically in relation to the buildings designed by *Le Corbusier* (Charles-Édouard Jeanneret) at Marseilles (1947) and Chandigarh (1951) in which roughly worked forms were used deliberately to emphasise the heavy concrete structure. The term *brutalism* was coined in England in 1954 to describe concrete structures in this style by other architects, and it has since been applied also to steel structures which over-emphasise the structural surfaces, and sometimes the service pipes and ducts.

BÉTON TRANSLUCIDE GLASS–CONCRETE CONSTRUCTION with clear glass inserts.

BETTI'S THEOREM *See* RECIPROCAL THEOREM.

BEVEL A junction between two surfaces which is either greater or smaller than a right angle. *See also* CHAMFER and SPLAY.

BILL OF QUANTITIES A list of numbered items, which describe the work to be done on a building contract. Each item shows the quantity of work involved. When the contract is sent out to tender with a bill of quantities, the contractor is expected to submit a priced bill. His payment is based on these prices and the measured work actually done, which usually varies at least in part from the work originally envisaged. The preparation of the bill of quantities and the final measurement of the work is the work of a *quantity surveyor*. The system is widely used in Great Britain and South Africa, and to some extent in British Commonwealth countries (excepting Canada) and Germany. In the USA and Canada the tender documents consist only of the drawings and specifications.

BILLET OF STEEL An intermediate product in hot-working of steel. It has been rolled or forged down from the ingot, and will be further worked into sections or forgings. A large billet is known as a *bloom*.

BILLION In Europe and Australia, 1×10^{12}, *i.e.* a million million. In the USA, 1×10^9, *i.e.* a thousand million. Similarly, the European trillion is 10^{18}, while the US trillion is 10^{12}. The ISO recommends the use of the prefix *giga* (G), derived from the Greek word for giant, for 10^9, and the prefix *tera* (T), derived from the Greek word for monster, for 10^{12}. It has made no recommendations for abbreviations above 10^{12}.

BI-METALLIC STRIP A strip fused together from two metals with widely differing coefficients of thermal expansion. It consequently deflects with a change in temperature. It is used as a control element in THERMOSTATS.

BINARY ALLOY An alloy containing two principal elements.

BINARY ARITHMETIC An arithmetic system based on the two digits 0 and 1. It is generally used by electronic computers, because the digits correspond to the two possible conditions of an electrical circuit (go or no go). The binary digits, or BITS, are converted by the computer to decimal arithmetic, so that it can be operated by the conventional decimal system.

binary system	decimal system
0	0
1	1
10	2
11	3
100	4
101	5 etc.

See also BOOLEAN ALGEBRA.

BINARY CODED DECIMAL The representation of an ALPHANUMERIC CHARACTER in binary digits.

BINARY DIGIT A digit in BINARY ARITHMETIC, *i.e.* either 0 or 1. It is commonly abbreviated to *bit*.

BINOMIAL FUNCTION Any function containing two, and only two, parameters, of which one is generally variable.

BISCUIT Tiles, earthenware products, etc., in the intermediate stage of manufacture after the first firing, but before glazing.

BIT (*a*) An interchangeable cutting tool used in a CARPENTER'S BRACE, in a rock drill, etc. (*b*) The tip of a soldering iron, which is heated and tinned. (*c*) An abbreviation for BINARY digit. (*d*) The smallest unit of information recognised by a DIGITAL COMPUTER.

BITUMEN A black sticky mixture of hydrocarbons used for waterproofing basements and flat roofs, and for damp-proof courses. It is obtained from natural deposits (*asphalt*) and from the distillation of petroleum.

BITUMINOUS FELT *See* ROOFING FELT.

BLACK BODY The designation of a theoretical surface which absorbs all the radiation falling on it and does not transmit or reflect any radiation. Its nearest practical approximation is the inside of a hollow sphere with a matt black surface at a uniform temperature, viewed through a small hole. *Black-body radiation* is the quality and quantity of the radiation emitted by an ideal black body.

Black-body temperature is the temperature of a body whose emissivity is unity, *i.e.* an ideal black body. *See also* STEFAN-BOLTZMANN LAW.

BLACK BOLT A hot-formed bolt for making site connections in steel structures. Because of imperfections in the surface, the holes in the pieces to be connected must be made a little larger than the bolt diameter, and perfect contact between the bolt and the sides of the holes is not possible; consequently the permissible loads are lower than for HIGH-TENSILE BOLTS or BRIGHT BOLTS.

BLACK LEAD Graphite.

BLACK MORTAR Mortar with the addition of ash, either because a black colour is required for pointing, or to reduce cost.

BLANC FIXE A white pigment, consisting of barium sulphate.

BLANK WINDOW A walled-up window.

BLAST FURNACE A tall, cylindrical, refractory-lined furnace used for reducing iron ore, *i.e.* for the production of PIG iron.

BLAST FURNACE SLAG A by-product of steel manufacture which is sometimes used as a substitute for Portland cement. It consists essentially of the silicates and alumino-silicates of calcium, which are formed in the blast furnace in molten form simultaneously with the metallic iron. Blast-furnace slag is blended with Portland cement clinker to form PORTLAND BLAST-FURNACE SLAG CEMENT. It is also used for topping

27

on BUILT-UP ROOFING, as LIGHT-WEIGHT AGGREGATE, and for making MINERAL WOOL.

BLEEDING (*a*) Accumulation of water and cement on the top of concrete due to settlement of the heavier particles. Bleeding may be caused by too high a water content, by overworking of the concrete near the surface, by excessive traffic on the wet concrete, or by improper finishing of the surface. The result is the formation of LAITANCE, (*b*) Exudation of gum, resin or sap from the surface of timber.

BLIND NAILING *See* SECRET NAILING.

BLINDING GLARE GLARE so intense that for an appreciable time no object can be seen.

BLOATED CLAY Expanded clay, used as LIGHTWEIGHT AGGREGATE.

BLOCK A masonry unit, usually larger than a brick. It is generally hollow, and may be made of burnt clay, terra cotta, concrete etc.

BLOOM (*a*) *See* BILLET. (*b*) An efflorescence or coating (which can be removed by rubbing or brushing) on a masonry wall, a painted or varnished surface, etc.

BLUE METAL *See* METAL (*b*).

BLUEPRINT (*a*) A contact print on ferro-prussiate paper made from a drawing on transparent material; it can be printed in daylight, and developed in water. At one time this was the most common method for copying drawings, but it has been largely displaced by other processes (*e.g. dye line*) which produce dark lines on white paper. (*b*) Hence, a master plan for a project, irrespective of the form in which it is presented.

BLUING Increasing the apparent whiteness of a white pigment by adding a small amount of blue.

BM Bending Moment.

BOARD FOOT A measure for timber, 1 inch thick by 12 inches square. A *Petersburg* or *Petrograd standard* equals 1980 board feet, or 165 cubic feet.

BODY-CENTRED CUBIC LATTICE A crystal structure which has an atom at each corner of a cube, and one in the centre. It may be imitated by packing spheres in horizontal layers.

BOILED OIL Linseed oil used in quick drying paint. It is heated for a short period to about 500°F (260°C), *not* boiled, and a small quantity of drier (*e.g.* manganese dioxide or LITHARGE) is added.

BOLE The main stem of a tree.

BOLT *See* BLACK BOLT, BRIGHT BOLT and HIGH-TENSILE BOLT.

BOND (*a*) The system in which bricks, blocks and stones are laid in overlapping courses in a wall in such a way that vertical joints in any one course are not immediately above the vertical joints in an adjacent course (Fig. 5). The basic distinction is between *English* (or *Old English Bond*) which consists of alternate layers of HEADERS and STRETCHERS, and *Double Flemish Bond*, which consists of alternate headers and stretchers in each layer. There are numerous variations on these two standard types. *English Cross Bond* (also known as St Andrew's Cross Bond) introduces a single header placed into alternate stretcher courses; this produces diagonal lines which are more decorative than the vertical lines of Old English Bond. In *Common*

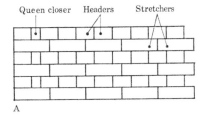

A

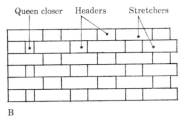

B

FIG. 5. Bond for bricks and blocks: (a) Old English Bond; (b) Double Flemish Bond.

Bond (also known as *American Bond* and as *Scotch Bond*) all courses are stretchers, except the fifth, sixth or seventh course, which is a header course. In *Single Flemish Bond* the Flemish bond is seen on one side only. The other side is usually English Bond. In *Diagonal Flemish Bond* a course of headers alternates with a course of headers and stretchers, and the bond shows a diagonal pattern, as in English Cross Bond. The terms are, however, by no means standard, and different names are in use. Most bonds require a QUEEN CLOSER to line up the joints at the corners. (b) The

adhesion or grip exercised by concrete or mortar on surfaces to which it is required to adhere. The most important bond is between concrete and reinforcing bars. This is largely produced by the shrinkage of the concrete, which creates a normal pressure between the concrete and the steel, and this produces a frictional force resisting pull-out of the steel. *See also* DEFORMED BAR. (c) Adherence of the plaster to the wall, and between the various coats of plaster. (d) Attraction between atoms which causes them to aggregate into larger units. The principal bonds are IONIC, COVALENT (or homopolar), metallic, and molecular (or van der Waals) bonding.

BOND LENGTH *See* TRANSFER LENGTH.

BOND STONE A long stone, used as a *header*, running through the thickness of the wall to give additional transverse bond.

BOND TIMBER Horizontal timbers used as a bond for a brick wall. The battens are sometimes secured to them. Bond timbers are liable to rot unless suitably protected.

BONDED TENDON A prestressing tendon which is bonded to the concrete. In pre-tensioned members this is achieved directly by casting the concrete around them. In post-tensioned members the annular spaces around the tendons are grouted after stressing.

BOO-BOO A mistake.

BOOLEAN ALGEBRA An algebra of logic, where a proposition may be either true or false, and therefore suited to BINARY ARITHMETIC.

BOOM *See* JIB.

BOOSTER FAN A fan used to step up the static pressure in an air distribution system in order to serve a remote area, used only intermittently (*e.g.* a conference room).

BORAX Sodium metaborate.

BORE The internal diameter (ID) of a hole, pipe, etc.

BORED PILE A pile formed by pouring concrete, usually containing some reinforcement, into a hole bored in the ground, as opposed to a *precast pile* driven into the ground with a pile driver.

BOREHOLE (OR CORE) SAMPLES Samples obtained by boring or drilling for the purpose of determining the nature of the foundation material. In the case of *clay*, it is necessary to obtain *undisturbed samples*, since the properties of clay are greatly affected by working. *See also* PENETRATION TEST.

BORROWED LIGHT A window in an internal wall.

BOSS An ornamental, projecting knob or block, *e.g.* the carved keystone at the intersection of the ribs of a Gothic vault.

BOUNDARY CONDITIONS FOR STRUCTURAL PROBLEMS The known conditions of displacement, slope, force or moment at the edges of a structural member, which may be utilised for determining the constants of integration resulting from the

solution of a differential equation. In architectural structures, the principal application is to shell structures. The stresses in the shell cannot normally be determined, unless the restraints at the *edges* of the shell are known. *See also* EDGE BEAM and MEMBRANE THEORY.

BOUNDARY LAYER The layer of fluid adjacent to its boundary with a solid. Inside this layer the velocity of the fluid falls to zero at the boundary. The flow in the boundary layer may be STREAMLINE or *turbulent*. Boundary-layer problems arise mainly with turbulent flow, because separation of the eddies produces more eddies, and this causes loss of energy. *See also* WINDTUNNEL.

BOURDON GAUGE A pressure gauge consisting of a tube bent into an arc, which tends to straighten out under internal pressure. It actuates a pointer which moves over a scale.

BOUSSINECQ PRESSURE BULB A bulb formed by the ISOSTATIC LINES in a semi-infinite elastic solid carrying a single concentrated load. The analysis, primarily used for determining the stresses in the soil beneath a heavy foundation, was published by the French mathematician J. Boussinecq in 1885.

BOW GIRDER A girder curved horizontally in plan, *i.e.* an arch turned through a right angle. It serves as a spandrel on a curved facade, to support *balconies* etc. (Fig. 6). A bow girder is subject to combined bending and torsion.

BOW WINDOW A curved projecting window, usually on the ground floor. *See also* ORIEL WINDOW.

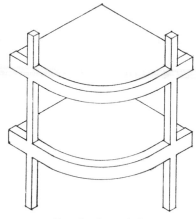

FIG. 6. Bow girder.

BOW'S NOTATION *See* RECIPRO-CAL DIAGRAM and Fig. 78.

BOWSTRING GIRDER (OR TRUSS) A tied arch, which can be used like a girder, since the horizontal reactions are internally absorbed by a tie (the bowstring). The curved top chord is stiffened by light diagonal members. A *Belfast truss* is a wooden bowstring girder for spans up to 50 ft (15 m).

BOX BEAM *See* BOX GIRDER.

BOX FRAME A rigid frame formed by load bearing walls and floor slabs. It is suitable for buildings which are permanently divided into small repetitive units.

BOX GIRDER A hollow beam whose cross-section is shaped like a box. It uses material where it is most highly stressed both by bending moments and by twisting moments. Consequently it is used for large spans and for locations where eccentric loading etc. causes torsion.

BOX GUTTER A gutter of rectangular cross section, built behind a parapet or in a roof valley.

BOXED HEART A piece of square-sawn timber, cut so that the pith, or central part, is cut out, *e.g.* by *back sawing*. This is done in most Australian and some other hardwoods in which the heart is unsound.

BOYLE'S LAW 'The volume V occupied by a given mass of any gas at constant temperature varies, within moderate ranges of pressure, inversely as the pressure P to which it is subjected, *i.e.* $PV =$ constant'. It was proposed by the seventeenth century English natural philosopher, the Hon. Robert Boyle. *See also* CHARLES' LAW.

BRAB Building Research Advisory Board, Washington.

BRACE *See* CARPENTER'S BRACE.

BRACING The ties and struts used for supporting and strengthening a frame, *e.g.* to resist horizontal loads.

BRASS A copper–zinc alloy. *See* ALPHA BRASS and BETA BRASS.

BRAZILIAN TEST *See* SPLITTING TENSILE TEST.

BRAZING A process for joining two pieces of metal by means of *brazing solder*. Copper–zinc (brass), copper–zinc–silver, and nickel–silver alloys are used as brazing solders, and their melting point is generally above 500°C, but well below the melting point of the metal to be brazed. *See also* SOLDERING and WELDING.

31

BREAKDOWN A term used for the separation of an EMULSION into its constituents.

BREASTSUMMER Originally a long and heavy timber beam (or *summer*) carrying the frontage of a building (or *breast*). It is a very large lintel supporting a masonry or brick wall. *Also called bressumer*. The term is still used for steel or concrete girders which provide an opening in a load-bearing wall, *e.g.* over a shop window.

BRESSUMER *See* BREASTSUMMER.

BRI Building Research Institute, Washington.

BRICK A building block, generally small enough to be lifted comfortably with one hand, and usable in a bonded wall. The biblical brick was made of ADOBE reinforced with straw; but burnt bricks have been found which appear to be even older. The modern CLAY brick is hard-burnt, and sometimes glazed. Bricks are also made of *concrete* and of CALCIUM SILICATE. The size of bricks is standardised, but standards vary from place to place. The length with due allowance for mortar joints, is usually twice the width so that HEADERS and STRETCHERS can be combined into BOND (Figs. 5 and 94). FACE BRICKS are specially made to have an agreeable colour or finish, and they are often weaker than common bricks. (*See, for example,* SAND-FACED BRICK.) *See also* SUCTION RATE.

BRICK-AND-STUD WORK *See* BRICK NOGGING.

BRICK CONSTRUCTION Construction in load-bearing brick, as opposed to BRICK VENEER or BRICK NOGGING.

BRICK EARTH A sandy clay suitable for making bricks.

BRICK FACING *See* BRICK VENEER.

BRICK NOGGING Brickwork infilling between the studs of wooden frame or a framed partition. *Also* called *brick-and-stud work*.

BRICK ON EDGE A header laid on its edge, or thinnest dimension (which is normally its height). This makes more economical use of bricks, but produces a thinner and weaker wall. Also called *rowlock*.

BRICK VENEER A veneer of bricks (stretchers) built outside a timber frame, which supports the load. A brick veneer house looks like a brick building, but is essentially a timber-framed structure. Also called *brick facing*.

BRIGHT BOLT A steel bolt which has been turned to fit exactly into the holes of the steel pieces to be joined, as opposed to a BLACK BOLT.

BRIGHTNESS *See* LUMINOSITY.

BRILLIANCE A bright, clear, ringing sound, rich in harmonics. It comes from the relative prominence of the treble and the slowness of its decay.

BRINELL HARDNESS TEST A test for hardness, named after its originator, the Swedish nineteenth century engineer J. A. Brinell. It consists of indenting the surface with a hard 10 mm diameter ball, subjected to a load of 10 kg for 30 seconds. The ball is then removed, and the indentation measured. The *Brinell hardness number* is the ratio of the load to the surface area of the indentation. Smaller loads and different sized balls may be used for softer materials.

BRISE SOLEIL A sun break, or SUN-SHADING device. It is used in tropical and sub-tropical latitudes to protect from the heat of the sun a wall facing south (north in the southern hemisphere, north and south near the equator), or west. *Brise soleil* generally implies screens, horizontal or vertical projections which serve a decorative purpose, even if under a functional disguise. The term is rarely used for projecting balconies or unobtrusive louvres.

BRITISH SYSTEM OF MEASUREMENT *See* FPS UNITS.

BRITISH THERMAL UNIT *See* BThU.

BRITTLE COATING A technique used in experimental stress analysis. A rough form is the observation of the cracking of mill scale on hot-rolled steel during a test to destruction to observe the direction of the principal stresses and the area where yield has occurred. More sensitive results in the elastic range can be obtained by coating a model of the structure with a *brittle lacquer*. Proprietary lacquers are available which crack at a definite strain, and can thus be used to determine not merely the direction, but also the magnitude of the principal tensile stresses (*Stresscoat*). Brittle lacquers can also be used to analyse residual stresses due to heat-treatment etc. A hole about $\frac{1}{8}$ in (3 mm) diameter and about $\frac{1}{16}$ in ($1\frac{1}{2}$ mm) deep is drilled at each check point. The stress relaxation around the hole produces a crack pattern which indicates the pattern of the locked-in stresses. Brittle lacquers are not as sensitive as strain gauges, but they give a clear visual image.

BRITTLE–DUCTILE RANGE The range of temperature over which a material (particularly steel) may change from brittle to ductile, or vice versa. Above that range it is entirely ductile, and below it is entirely brittle.

BRITTLE LACQUER *See* BRITTLE COATING.

BRITTLENESS Lacking ductility. A brittle material ruptures with little or no PLASTIC DEFORMATION. Brittle failure occurs by the rupture of inter-atomic bonds, and this occurs more readily in tension than in shear (or diagonal shear resulting from compression). Hence brittle materials have a much lower strength in tension than in compression. However, since plastic failure does not occur, their compressive strength is often high. Brittle failure is typical of concrete, brick and other ceramics; it also occurs in metals, particularly in high-carbon steels and in cast iron. Low temperature increases the tendency to brittle failure, as does rapid application of the load (shock). *See also* FRACTURE and GRIFFITH CRACK.

BROKEN JOINTS Joints arranged, as in a BOND, so that they do not fall in a straight line and weaken the structure.

BROKEN WHITE Off-white, generally with a touch of cream.

BRONZE An alloy of copper and tin, in varying proportions. Small quantities of zinc, nickel, phosphorus, aluminium and lead are sometimes added.

BRS Building Research Station, Garston, England.

BRUNELLESCHI, F. *See* RENAISSANCE.

BRUTALISM *See* BÉTON BRUT.

BSI British Standards Institution, London.

BThU OR BTU British Thermal Unit. The amount of heat required to raise the temperature of 1 lb of water through 1°F. *See also* CALORIE and JOULE.

BTU *See* BThU.

BUBBLE MODEL A demonstration model devised in the 1930s by the English physicist W. L. Bragg to illustrate crystal structure. Uniform soap bubbles, about $1\frac{1}{2}$ mm in diameter, are blown on to a water surface. These are then disturbed to form grain boundaries and dislocations.

BUCKINGHAM'S THEOREM *See* PI-THEOREM.

BUCKLE To load a structural member, notably a column or strut, until it bends suddenly sideways (Fig. 7). The material is not necessarily damaged by buckling, but it loses its *elastic stability*. *See* EULER FORMULA, LATERAL BUCKLING, LOCAL BUCKLING and TORSION BUCKLING.

FIG. 7. Buckling.

BUILDING BLOCK *See* BLOCK.

BUILDING BOARD Board used for ceilings and for the interior lining of walls, generally less than $\frac{1}{2}$ in (13 mm) thick and more than 2 ft (0·6 m) wide. The term is not precisely defined, and includes boards made from a wide range of materials.

BUILDING PAPER Heavy paper, sometimes reinforced with fibres and waterproofed with bitumen. *See also* KRAFT PAPER.

BUILT-IN BEAM OR SLAB A condition of support which prevents the ends from rotating in the plane of bending. It does not imply longitudinal constraint. Built-in beams and slabs are HYPERSTATIC, and the geometry of the end-restraints supplies the solution. At each built-in support the SLOPE is nil, and there are thus as many equations as there are known restraints. *See* SLOPE DEFLECTION.

BUILT-UP ROOFING Flat roof built with multiple layers of ROOFING

FELT, as distinct from a roof built with a single layer. Also called *composition roofing* or *roll roofing*.

BULB ANGLE A steel or aluminium angle section enlarged at one end, *i.e.* a section intermediate between an angle and a channel (Fig. 8).

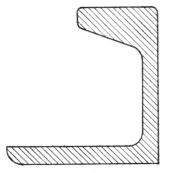

FIG. 8. Bulb angle.

BULB OF PRESSURE *See* BOUSSINECQ PRESSURE BULB.

BULB TEE A strengthened steel or aluminium T-section, used particularly as a sub-purlin.

BULK DENSITY The weight of a porous material per unit volume, including the voids. *See also* SPECIFIC GRAVITY and APPARENT VOLUME.

BULK MODULUS OF ELASTICITY The ratio of the triaxial (tensile or compressive) stress, equal in all directions, to the corresponding change in volume. The most common example is that of hydrostatic pressure and the corresponding volumetric strain. The bulk modulus K is related to the direct modulus of elasticity, E, and to Poisson's ratio, σ, by

$$E = 3K(1 - 2\sigma)$$

BULK VOLUME OF A POROUS SUBSTANCE The total volume, including closed and open pores. *See also* APPARENT VOLUME and TRUE VOLUME.

BULKING Increase in the volume of sand when it is in a damp condition, as compared with its volume when dry. It must be allowed for when measuring sand by volume, instead of by weight. Bulking increases with increasing moisture content, and then declines again; completely inundated sand occupies practically the same volume as dry sand.

BULL HEADER A brick made with one long corner, or arris, rounded. It is used for rounded sills and corners.

BULLNOSE The rounding of an ARRIS.

BULL'S EYE (*a*) A small circular or oval opening, or window. (*b*) The centre of a disc of CROWN GLASS.

BUNGALOW A one-storey house, lightly built with a sloping roof and a verandah, originally in Bengal, India.

BUNSEN BURNER A type of burner, widely used in laboratories, in which the amount of air to be mixed with gas can be adjusted before burning. The idea is attributed to R. W. Bunsen, a German nineteenth century chemistry professor.

BUOYANCY The reduction in the weight of a body immersed in a liquid, particularly in water, due to the upward pressure exerted by the liquid. If the body floats in the liquid, its weight is equal to the weight of the liquid displaced. This is known as the *principle of Archimedes*, after the Greek philosopher who discovered it in Sicily in the third century BC.

BUOYANT FOUNDATION A reinforced concrete raft foundation so designed that the weight of the load carried by it (generally its own weight and that of the building) equals the weight of the soil and water displaced. It is particularly useful in fine-grained soils whose WATER TABLE is near the surface.

BURETTE A cylindrical graduated glass tube fitted with a ground glass stop cock, used for the measurement and delivery of small volumes of liquid in a laboratory.

BURGERS VECTOR *See* DISLOCATION IN A CRYSTAL.

BURL A FIGURE in wood caused by an adjacent knot, enlarged rootstock or other large excrescence. It is decorative on veneers, but may be a source of weakness in boards. *See also* CROTCHWOOD.

BURLAP A coarse fabric of hemp or jute. It is frequently used to cover concrete during curing to reduce evaporation. Also called HESSIAN.

BURLINGTON, LORD *See* PALLADIANISM.

BURNHAM, DAVID H. *See* CHICAGO SCHOOL.

BURR The rough or sharp edge left on metal by a drill, saw, or other cutting tool.

BUS BAR A bare, *i.e.* uninsulated, electrical conductor, from which circuits can be tapped.

BUSH-HAMMER FINISH A concrete finish obtained by roughening the hardened surface with a pneumatically operated hammer, which has a serrated face.

BUSH'S ANALOGY An analogy between the mathematical equations for SLOPE DEFLECTION and between the relations of current, resistance and voltage in certain electrical circuits, which is utilised in several analogue computers for the solution of rigid frames.

BUTT JOINT A joint between two pieces of material which are in line, with or without cover plates (Fig. 9). By contrast, the two pieces to be joined are not in line, but overlap, in a *lap joint. See also* END JOINT.

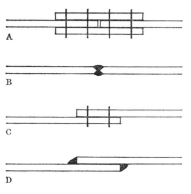

FIG. 9. Butt joint and lap joint: (A) bolted butt joint with cover plates; (B) welded butt joint; (C) bolted lap joint; (D) welded lap joint.

BUTT STRAP A cover plate used in a BUTT JOINT.

BUTTERFLY ROOF A roof consisting of two sloping surfaces, connected at the *lower* edges.

BUTTRESS A projecting structure built against a wall to resist a horizontal reaction (Fig. 21). A *flying buttress* is suspended in the air for the same purpose. *See also* COUNTERFORT.

BYZANTINE DOME *See* SQUARE DOME.

C

C Chemical symbol for *carbon*.

°C *Degree Celsius or Centigrade* The temperature scale fixed by the boiling point of water (100°) and its freezing point (0°), suggested by the Swedish physicist Celsius in 1740. To convert to the Fahrenheit scale (°F), multiply by 1·8 and add 32°.

Ca Chemical symbol for *calcium*.

CABLE, ELECTRIC The conductor through which an electric appliance or lamp receives its power.

CABLE LAY *See* LAY.

CABLE, PRESTRESSING *See* TENDON.

CABLE STRUCTURE *See* SUSPENSION ROOF.

CABLE, SUSPENSION *See* SUSPENSION CABLE and Fig. 97.

CADMIUM PLATING Plating with metallic cadmium, applied to steel bolts used in conjunction with aluminium to prevent ELECTROCHEMICAL CORROSION.

CADMIUM YELLOW Cadmium sulphide, a permanent pigment ranging in colour from pale yellow to orange.

CAISSON A watertight chamber used for construction in waterlogged ground, or below water.

CAISSON PILE A cast-in-place concrete pile, made by driving a tube, excavating it, and filling the hole with concrete.

CALCAREOUS Containing calcium, or more commonly calcium carbonate ($CaCO_3$).

CALCIMINE *See* KALSOMINE.

CALCINE An old-fashioned term, dating from the days of alchemy and early chemistry, for altering the composition of a substance by heating it below the temperature of fusion.

CALCITE The crystalline form of calcium carbonate, $CaCO_3$. It is a common constituent of limestone, marble, and some igneous rocks.

CALCIUM A silvery-white metal. Its chemical symbol is Ca, its atomic number is 20, its atomic weight is 40·08, its specific gravity is 1·55, its valency is 2, and its melting point is 851°C. Its oxide is QUICKLIME.

CALCIUM-ALUMINATE CEMENT *See* HIGH ALUMINA CEMENT.

CALCIUM CARBONATE One of the abundant minerals in the earth's crust ($CaCO_3$). It is the material of which limestone, chalk and marble are composed, and a principal raw material for cement and mortar.

CALCIUM CHLORIDE An ACCELERATOR used for concrete.

CALCIUM HYDROXIDE Slaked lime ($Ca(OH)_2$).

CALCIUM OXIDE Quicklime (CaO).

CALCIUM-SILICATE BRICK A light-coloured yellow brick made principally from sand and lime. It is usually hardened by AUTOCLAVING. Also called a *sand-lime brick*.

CALCIUM SULPHATE Anhydrite ($CaSO_4$).

CALCULUS A method of calculation, from the simplest to the most complex. The term is often taken synonymously with the differential and integral calculus, which is concerned with the evaluation of DIFFERENTIAL COEFFICIENTS and INTEGRALS.

CALDARIUM The hot room in an Ancient Roman bath.

CALIBRE Originally the diameter of a cannon-ball or bullet. Hence the bore of the gun, or of any pipe.

CALIFORNIA BEARING RATIO A standard test for determining the bearing capacity of a foundation. It is defined as the ratio of the force per unit area required to penetrate the foundation soil with a circular piston (area 3 in^2 = 1935 mm^2) at a rate of 0·05 in (1·27 mm) per minute, to the force required for penetration of a standard material (usually crushed rock). The measurement is commonly made after 2 minutes penetration.

CALK *See* CAULK.

CALLIPERS A pair of steel legs, joined by a pivot. They may be *external* (curved convex for measuring the outside diameter) or *internal* (curved concave for measuring the inside diameter of a tube).

CALORIE (*cal.*) The quantity of heat required to raise 1 gram through 1°C. The term is, however, commonly used without prefix for the large or kilocalorie, which equals 1000 cal. In the SI SYSTEM the calorie has been replaced by the JOULE. For conversion to FPS units, 1 kilocalorie = 3·968 BThU.

CALORIFIC VALUE The amount of heat liberated by the complete burning of a unit weight of a fuel.

CALORIFIER A closed tank in which water is heated by submerged hot pipes.

CALORIMETER (*a*) An instrument for measuring the heat exchange during a chemical reaction. It is particularly used for measuring the heat produced by the combustion of a material. (*b*) A vessel containing the liquid used in calorimetry.

CALORIMETRY The measurement of thermal constants, such as specific heat, latent heat, or calorific value. This is generally done by observing the rise of temperature which a heat exchange causes in a liquid (commonly water) contained in a calorimeter.

38

CAM A mechanism for converting circular into regular or irregular linear motion. It consists of a wheel, of a carefully designed non-circular form, attached to a shaft. For example, cams are used to open and close the valves in an internal-combustion engine.

CAMBER A slight upward curvature of a structure to compensate for its anticipated deflection.

CAME *See* LEADED LIGHT.

CAMEO A striated precious stone (such as ONYX) or a shell carved in relief to exhibit the various colours in the layers. Hence, any modelled relief exhibiting different colours.

CAMPANILE A bell tower, particularly one which is tall and detached from the building.

CAMPBELL, COLEN *See* PALLADIANISM.

CANADA BALSAM A yellowish liquid of pine-like odour, soluble in ether, chloroform and benzene. It is used for lacquers and varnishes, and also as an adhesive for lenses, because its refractive index is almost identical with that of most optical glasses.

CANDELA (*cd*) Unit of luminance; also called the *new candle*. It is a fundamental unit defined in terms of the luminance of a standard radiator, at the temperature of the solidification of platinum, with which other radiators may be compared.

CANDELA, FELIX *See* HYPERBOLIC PARABOLOID SHELL.

CANDLE POWER An obsolete unit of illumination whose place has been taken by the CANDELA.

CANEPHORAE *See* CARYATIDS.

CANOPY A roof-like covering, usually projecting, over an entrance or window, or along the side of a wall.

CANTILEVER A projecting beam, truss or slab supported only at one end.

CANTILEVER BRIDGE A bridge continuous over several spans which is made isostatic by the insertion of hinges, using the GERBER BEAM principle. It is particularly useful where bridges have to be supported on poor foundations which may settle, since the bridge can take up small foundation movements without hyperstatic stresses.

CANTILEVER, PROPPED *See* PROPPED CANTILEVER.

CANTILEVER RETAINING WALL A wall retaining soil by cantilever action, as distinct from a retaining wall which spans between COUNTERFORTS or buttresses as a continuous slab. The wall normally consists of three cantilevers (Fig. 10): the wall which resists the horizontal EARTH PRESSURE, the *heel* and the *toe*, both of which resist vertical earth pressures.

CAPACITANCE *See* FARAD.

CAPACITANCE STRAIN GAUGE A strain gauge which utilises the principle that the capacitance of a parallel plate condenser is changed by varying the separation of the plates.

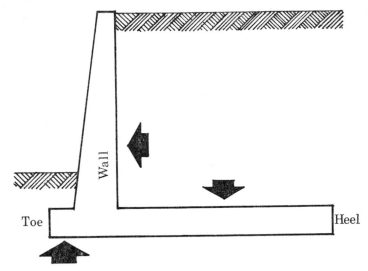

Fɪɢ. 10. Cantilever retaining wall.

CAPILLARY ACTION The action of a CAPILLARY TUBE, when dipped into a bucket of water, to cause the level in the tube to rise above that of the bucket. It is caused by SURFACE TENSION.

CAPILLARY TUBING Tubing with a very fine bore.

CAPILLARY WATER Water held by CAPILLARY ACTION in the soil above the water table.

CAPITAL *See* COLUMN CAPITAL.

CAPPING CABLES Short TENDONS introduced into *statically indeterminate prestressed structures* in the zone of negative bending moment (Fig. 11).

CARAT (*a*) For weighing precious stones, 1 carat = 0·2 gram. (*b*) For measuring the gold content of an alloy, pure gold = 24 carat, so that

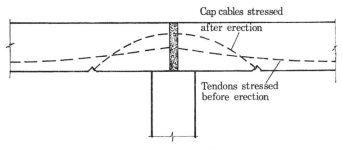

Fɪɢ. 11. Capping cables.

22 carat gold contains 2 parts of alloying metal.

CARBON A non-metallic element. Its chemical symbol is C, its atomic number is 6, its valency is 4, its atomic weight is 12·01, and its melting point is above 3500°C. It occurs in several allotropic forms: in crystalline form as diamond and as graphite, and in amorphous form as charcoal, coke, etc. Its specific gravity depends on its form.

CARBON BLACK Finely divided amorphous carbon, used as a mineral pigment in plastics, concrete, paint, etc. Unlike most organic pigments, it does not fade. It is produced by burning petroleum or natural gas in a supply of air insufficient for complete combustion.

CARBON STEEL *See* ALLOY STEEL and Fig. 71.

CARBONATION Chemical reaction between calcium compounds and carbon dioxide; calcium carbonate is produced. The reaction occurs slowly when carbon dioxide is absorbed from the atmosphere.

CARBORUNDUM Silicon carbide (SiC), an abrasive, not to be confused with CORUNDUM.

CARBURISING The introduction of carbon into the surface of steel by holding it at a suitable temperature in contact with carbon and nitrogen. *See also* CASE-HARDENING.

CARCASS In building construction the load-bearing part of the building without the finishes. Also spelled *carcase*.

CARD PUNCH A device, usually operated in conjunction with a typewriter, to punch holes into the standard cards used by digital computers to record information. Normally cards contain the information both in the form of BITS, represented by the holes, and typed on in ALPHA-NUMERIC characters. A *tape punch* operates in the same manner, except that the information is continuous on a single paper tape, and it cannot be read from the tape in alphanumeric characters.

CARD READER *See* TAPE READER.

CARPENTER'S BRACE A cranked hand tool used for turning the drilling BIT, to make holes in timber. It has been largely superseded by electrical drills.

CARPORT A shelter for a car attached to or near a house, which contains a roof, but no door. It usually does not have walls on all sides.

CARRELL A small enclosure in the stack of a library, designed for individual study.

CARRY-OVER MOMENT 'If joint A in a continuous structure (Fig. 12) is rotated by a moment M_A, while the member is firmly restrained at B, then a carry-over moment is produced at B due to the rigid restraint. Similarly, if the joint C is moved sideways (given a sideways TRANSLATION or a SIDE-SWAY) while the joint D is firmly restrained, then a (larger) carry-over moment is produced at D due to the restraint.' Carry-over moments are used in the MOMENT DISTRIBUTION METHOD.

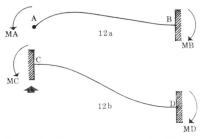

FIG. 12. Carry-over moments: (a) due to rotation; (b) due to translation (sidesway).

CARTESIAN CO-ORDINATES
The co-ordinates conventionally used for plotting a curve, measured perpendicularly from two axes at right angles to one another (Fig. 13). They are named after the seventeenth century mathematician René Descartes. The horizontal axis is called the *abscissa* and the vertical axis the *ordinate*. In three dimensions, Cartesian co-ordinates are usually called the x-, y- and z-axes.

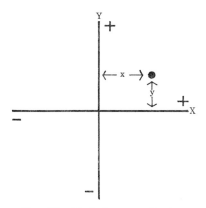

FIG. 13. Cartesian co-ordinates.

CARYATIDS Figures of women used as supports in classical architecture, and taking the place of columns. Correspondingly used male figures are called *atlantes* or *telamones*. *Canephorae* are female figures carrying baskets on their heads, used as columns.

CASE HARDENING CARBURISING the exterior of a low-carbon steel, so as to increase its surface hardness without impairing its overall DUCTILITY. For example, the carbon may be absorbed from a molten bath of sodium cyanide, and the steel is then QUENCHED to produce a hard case.

CASEIN GLUE Adhesive manufactured from milk powder.

CASEMENT WINDOW A window contained in cased frames which is hinged and opens outwards; as opposed to a SASH WINDOW. The more common type of casement window is hinged vertically, or *sidehung*. The *top-hung* casement is hinged horizontally, and must be held open with a *casement stay*.

CASINO Originally a small country house or summer house. Later a public room or building for social occasions, and more recently for gambling.

CAST IN PLACE Cast liquid in its permanent location, where it hardens as part of the building, as opposed to PRECAST. Monolithic concrete must be cast in place.

CAST IN SITU Cast in place, as opposed to precast.

CAST IRON Iron with a total carbon content between 1·8 and 4·5 per cent (Fig. 71). It is one of the two traditional forms of iron, the other being *wrought iron*. Steel is intermediate between the two in carbon content, but prior to the invention of the *Bessemer* process it could only be produced at great expense. Cast iron was used extensively in the nineteenth century for structural members and for railings. It is hard, brittle, and easy to cast into moulds. *Grey cast iron* contains some free carbon in the form of GRAPHITE; in *white cast iron* all the carbon is present as iron carbide (Fe_3C).

CAST STONE Precast concrete, particularly when it is cast in blocks to resemble natural stone. *Also called reconstituted stone.*

CASTELLATED Decorated with battlements.

CASTELLATED BEAM A steel beam formed by cutting a rolled steel joist along the web in a zig-zag shape. The two halves are then welded together at the crests of the cuts. The resulting beam is deeper, but it has a series of holes in the web. The section modulus can be doubled by this technique.

CASTIGIANO'S METHOD See STRAIN ENERGY METHOD.

CASTING RESIN A THERMOSETTING RESIN or a COLD-SETTING RESIN.

CATALYST A substance which causes or ACCELERATES a chemical reaction without being itself transformed in the process.

CATENARY The curve assumed by a freely hanging cable of uniform section, due to its own weight. The stresses in it are purely tensile. Its mathematical equation is $y = a \cosh (x/a)$, where a is a constant.

CATENARY ARCH An arch shaped like an inverted catenary, so that the stresses due to its own weight are purely compressive. *See also* PARABOLIC ARCH.

CATHEDRA Greek word for chair.

CATHEDRAL A church which contains the chair of a bishop. *See also* ABBEY.

CATHODE *See* ANODE.

CATHODE RAY TUBE A device in which a narrow beam of electrons, emitted from an ELECTRON GUN, impinges on a fluorescent screen. The beam is subjected to transverse magnetic or electrostatic fields, whose intensities control the position of the luminous spot. It can thus be used for the graphic display of measurements, or to translate an image into numerical data with a LIGHT PEN.

CATHODIC CORROSION *See* ELECTROCHEMICAL CORROSION.

CATION *See* ANION.

CATWALK An elevated narrow walkway.

CAULKING The process of making a joint watertight. The term originally implied stopping up the joints with *oakum* (loose fibre got by picking old rope) and melted pitch. It is now also applied to stopping with lead, mastics, rubber and other flexible fillers. Also spelt *calking*.

CAUSTIC POTASH Potassium hydroxide (KOH).

CAUSTIC SODA Sodium hydroxide (NaOH).

CAVITY SHELL *See* DOUBLE-WALLED SHELL.

CAVITY WALL A wall built of an inner and outer leaf, or *wythe*. In Britain and Australia it usually consists of two wythes of $4\frac{1}{2}$ in (114 mm) brickwork with a 2 in (50 mm) cavity, making an 11 in (280 mm) wall. The cavity drains any water penetrating one $4\frac{1}{2}$ in brick wythe, and the air space provides good thermal and acoustic insulation. The two wythes are tied by metal wall ties at intervals. The inner wythe carries the floor joists and the ceiling joists. Cavity walls are not used in very cold climates, such as the North East of America, because the cavity would fill with ice.

CBR California bearing ratio.

cc (*a*) cubic centimetre. (*b*) centre to centre.

CEBS Commonwealth Experimental Building Station, Sydney.

CEBTP Centre Expérimental de Recherches et d'Études du Bâtiment et des Travaux Publics, Paris.

CEI Council of Engineering Institutions, London.

CEILING DIFFUSER An air outlet from an air conditioning duct, which diffuses the air over a larger area to produce an even distribution and avoid draughts. It is often combined with a *luminaire*.

CEILING FAN A slowly rotating overhead fan with a wide sweep. It moves large volumes of air at a low speed, and it is widely used in hot-humid climates to improve thermal comfort. *See also* PUNKAH.

CEILING JOIST A joist which carries the ceiling above it, but not the floor or roof over the ceiling.

CELL, ELECTRICAL A combination of two electrodes in an electrolyte.

CELLA The main body of a classical temple; it excludes the portico.

CELLULAR CONCRETE Lightweight concrete containing a substantial proportion of air or gas bubbles. It is produced either by adding a foaming agent (such as detergent) or a gas-forming agent (such as aluminium powder) to the mix. Cellular concrete, which contains sand, is lighter than LIGHTWEIGHT-AGGREGATE CONCRETE, but heavier than water. Cellular concrete which floats on water can be made from cement, water and a foaming or gas-forming agent; this material has insufficient strength for structural purposes, and is used mainly as a thermal insulator and as a stiffener for a light-gauge steel or aluminium structure. *Also called aerated concrete, or foamed concrete.*

CELLULAR PLASTIC *See* EXPANDED PLASTIC.

CELSIUS SCALE *See* °C.

CEMENT *See* COLOURED CEMENT, EXPANSIVE CEMENT, HIGH-ALUMINA CEMENT, HYDRAULIC CEMENT, LOW-HEAT CEMENT, NATURAL CEMENT, PORTLAND CEMENT, PORTLAND BLAST-FURNACE SLAG CEMENT, PORTLAND-POZZOLAN CEMENT, and WHITE CEMENT.

CEMENT CLINKER The product of burning the raw cement mix. Cement is made by finely grinding the clinker. PORTLAND CEMENT mix is normally burnt at a temperature of approximately 2600°F (1400°C) in a rotary KILN.

CEMENT GROUT *See* GROUT.

CEMENT GUN A machine for placing mortar or concrete through a nozzle under pressure. The mixture of cement and small aggregate is forced by compressed air through a hose to the nozzle, where water is added from a separate pipe. The resulting material is known as *gunite*, *pneumatically applied mortar*, or *shotcrete*.

CEMENT MORTAR *See* MORTAR.

CEMENT PAINT A paint which can be used over cement. It is either a mixture of cement and pigment, or a paint based on alkali resistant vehicles such as *casein* or *tung oil*.

CEMENT SLURRY A liquid mixture of water and cement.

CEMENTATION Injecting cement grout under pressure, *e.g.* into fissured rock.

CEMENTITE The iron carbide (Fe_3C) constituent of cast iron and steel. It is crystalline, hard and brittle.

CENOTAPH An empty tomb; hence a memorial to a person or persons buried elsewhere.

CENTI The Latin prefix for hundred, used now for one hundredth, *e.g.* 1 centimetre = 0·01 metres.

CENTIGRADE SCALE *See* °C.

CENTIMETRE *See* CGS UNITS.

CENTRAL PROCESSOR UNIT The heart of a computer, which contains the computing unit and the internal store.

CENTRE OF GRAVITY *See* CENTROID.

CENTRE OF MASS *See* CENTROID.

CENTRE OF PRESSURE The point of action of the resultant force acting on an area subjected to liquid pressure. Since liquid pressure increases with depth, its location is below the centroid.

CENTRIFUGAL FAN A fan with an impeller of *paddle-wheel* form, in which the air enters axially at the centre and is discharged radially by centrifugal force. By contrast, an *axial-flow fan* is a propeller fan, with the blades mounted on the axis. *See also* FAN NOISE.

CENTRING Curved temporary support for an arch or dome.

CENTROID The point of any plane figure through which all CENTROIDAL AXES pass. It is often referred to as the *centre of gravity* or *centre of mass* of the figure, since a piece of cardboard or sheet metal of this shape balances if hung freely from the centroid.

CENTROIDAL AXIS An axis of any plane figure about which the moment of the area is nil.

CERAMIC MOSAIC Ceramic tiles, arranged in . patterns on a paper backing, and sold in sheet units ready for placing.

CERAMIC TILE *See* CLAY TILE.

CERAMIC VENEER Large units of thin TERRA COTTA, generally moulded by EXTRUSION.

CERAMICS (*a*) In building, any component made from burned clay, such as brick, terra cotta, ceramic tile (glazed or unglazed), stoneware pipe, and other pottery. (*b*) In solid-state physics, compounds of metallic and non-metallic elements; these include clay, cement, and natural stone.

CG Centre of gravity.

CGS UNITS The units of the traditional metric system, based on centimetre, gram and second. Another term is *MKS units*, based on the metre, kilogram and second. *See also* SI UNITS and FPS UNITS.

CHAIR, BAR *See* BAR CHAIR.

CHALET Originally a herdsman's hut in the Swiss mountains; hence a house built in the style of an Alpine cottage.

CHALK A soft LIMESTONE.

CHALK LINE A length of string which has been thoroughly coated with chalk dust. It is pulled tight across a piece of timber or other material, and plucked, thus producing a straight chalky line which serves as a guide during cutting.

CHALKING Disintegration of paint and other coatings, which produces loose powder at, or just beneath, the surface.

CHAMFER A right-angled corner cut symmetrically, *i.e.* at 45°. *See also* BEVEL and SPLAY.

CHAMFER STRIP An insert placed into an inside corner of concrete formwork to produce a chamfer.

CHANCEL The eastern part of a church, reserved for the clergy and choir, and separated from its main body by a screen or rail.

CHANCEL ARCH An arch at the west end of a chancel.

CHANNEL SECTION A metal section shaped [.

CHARACTER *See* ALPHANUMERIC CHARACTER.

CHARLES' LAW 'The volume of a given mass of gas, kept at a constant pressure, increases by 1/273 of its volume at 0°C for each degree rise of temperature'. *Also known as Gay-Lussac's law. See also* BOYLE'S LAW.

CHARPY TEST An impact test carried out on a notched specimen fixed at both ends. The energy absorbed during fracture is measured by a pendulum.

CHARTERED ENGINEER A title restricted in the United Kingdom and in Australia by law to corporate members of certain engineering institutions. It is roughly equivalent to State registration in North America.

CHASE A groove cut into a wall or floor to receive a small pipe, conduit, cable or flashing. A very large chase is a *duct*.

CHECK *See* SEASONING CHECK.

CHERT A very fine-grained, hard siliceous rock, which sometimes includes the remains of siliceous organisms. It tends to splinter when it is fractured. *See also* FLINT.

CHEVRON (*a*) The meeting of two rafters at an angle at the ridge of a roof. (*b*) Hence a decoration consisting of two lines meeting at an angle, or a zigzag pattern of lines.

CHIAROSCURO The disposition of light and shade in the pictorial composition of a painting. The term is also used in sculpture and photography.

CHICAGO SCHOOL A group of architects who established in Chicago in the 1880s and 1890s a functional style of commercial architecture. It introduced the SKYSCRAPER, the principle of steel framing, and a façade treatment of verticals and horizontals without classical or Gothic ornament. Its principal exponents were Jenney, Burnham and Root, Hollabird and Roche, and Adler and Sullivan.

CHIMNEY COWL A revolving metal ventilator over a chimney.

CHINA CLAY A pure white form of hydrated aluminium silicate, resulting from the decomposition of *felspars* contained in igneous rock. It is the raw material for the best quality pottery (*porcelain*). Also called *kaolin*.

CHINA WOOD OIL *See* TUNG OIL.

CHINESE WHITE Zinc oxide (ZnO). A permanent, non-poisonous white pigment.

CHINOISERIE European imitations or evocations of Chinese art, which were popular in the late seventeenth, eighteenth and early nineteenth century.

CHIP BOARD *See* PARTICLE BOARD.

CHIPS Small broken fragments. The term is commonly employed for small-size aggregates used for decorative concrete surface finishes (*e.g.* marble chips).

CHLORINATED RUBBER A white powder formed when natural rubber is treated with chlorine under heat and pressure. It is soluble in coal-tar solvents, and produces paint films of exceptionally good chemical resistance.

CHLORINATION Disinfectant treatment of drinking water, and sometimes sewage, with a source of chlorine, such as bleaching powder.

CHLORINE A highly reactive gas, and a powerful oxidising agent frequently used in the form of bleaching powder. Its chemical symbol is Cl, its atomic number is 17, its atomic weight is 35·5, and its boiling point is $-34\cdot6°C$.

CHORD A horizontal member in a truss.

CHROMA *See* COLOUR.

CHROMIUM A bright, silvery metallic element, used in stainless steel and other alloys. Its chemical symbol is Cr, its atomic number is 24, its atomic weight is 52·0, its specific gravity is 7·14, and its melting point is 1830°C.

CHROMIUM PLATING Electroplated surface of chromium applied as a protective finish, which is extremely hard. Chromium plating for steel is usually done over a coating of nickel; this in turn is electrodeposited on a coating of copper, which is the first coat on the steel.

CHU *Centigrade Heat Unit.* The heat required to raise 1 lb of water through 1°C.

CIAM *Congrès Internationaux d'Architecture Moderne.* The first Congress was organised in 1928 by a group led by *Le Corbusier* and the art historian *Siegfried Giedion.* Altogether ten congresses were held, and the organisation was formally dissolved after disagreement over its objectives in 1959. In its most vital periods, 1930–34 and 1950–55, the congresses provided a meeting ground for the leading proponents of the modern style of architecture; however, their concern, in spite of their opposition to the *Beaux-Arts* system, was essentially with aesthetic matters. *See also* INTERNATIONAL STYLE.

CIB Conseil International du Bâtiment pour la Recherche l'Étude et la Documentation, Rotterdam; the international co-ordinating body for building research.

CIDB Le Centre d'Information et de Documentation du Bâtiment, Paris.

CIE Commission Internationale de l'Éclairage. The international co-ordinating body for lighting research.

CIE STANDARD OVERCAST SKY *See* MOON-SPENCER SKY.

CINQUECENTO *See* QUATTROCENTO.

CINQUEFOIL *See* FOIL.

CINVA Centro Interamericano de Vivienda y Planeamiento, Bogotá, Colombia.

CIRCLE The curve generated by a point equidistant from another point. It has a constant radius and constant curvature. It can consequently be drawn with a compass or even a piece of string, a factor which evidently commended it to Gothic designers, who set out POINTED ARCHES approximating catenary arches with circular arcs. The Roman use of semi-circular arches is less logical, and probably explained by philosophical considerations; since the circle was regarded as the perfect curve, its use for structures may have been considered as self-evident.

CIRCLE, GREAT OR SMALL *See* GEODETIC LINE.

CIRCUIT BREAKER A device for opening an electric circuit automatically in case of an overload. It is a more elaborate device than a FUSE, and is required for larger currents. *See also* SOLENOID.

CIRCULAR ARCH OR SHELL An arch or shell with a constant radius of curvature, *i.e.* forming part of the circumference of a circle, as opposed to an elliptical, parabolic or CATENARY arch. A *semi-circular* arch is one forming a complete semi-circle. (Fig. 73).

CIRCULAR FUNCTIONS The functions obtained from the radius of a circle and its horizontal and vertical projections (*see* Fig. 14 *and Appendix D*). *Also called trigonometric functions.*

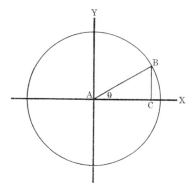

Fig. 14. Circular functions: $\sin \theta =$ BC/AB; $\cos \theta =$ AC/AB; $\tan \theta =$ BC/AC; $\cot \theta =$ AC/BC; $\sec \theta =$ AB/AC; $\operatorname{cosec} \theta =$ AB/BC.

CIRCULAR MEASURE *See* RADIAN.

CIRCULAR SAW A saw with teeth spaced around the edge of a circular disc, as opposed to a BAND SAW.

CIRCUMFERENCE The curve that forms an encompassing boundary, especially of anything rounded.

CIRCUS (*a*) In Ancient Rome, a large, *oblong* arena surrounded by rising tiers of seats, for the performance of public spectacles. (*b*) In modern architecture, a circular arena surrounded by tiers of seats and covered by a tent or permanent roof, particularly for the performance of acrobatic or equestrian acts. (*c*) A circle placed at the junction of two or more streets.

CITC Canadian Institute of Timber Construction, Ottawa.

Cl Chemical symbol for *chlorine*.

CLADDING (*a*) A synonym for curtain wall. (*b*) Covering a structural material with a protective surfacing material, *e.g. alclad* which is aluminium alloy (for strength) covered with pure aluminium (for corrosion resistance).

CLAPBOARD A long thin board, graduating in thickness from one end to the other, used for SIDING. The thick end overlaps the thin portion of the board. Called *weatherboard* in England and in Australia.

CLAPEYRON'S THEOREM *See* THEOREM OF THREE MOMENTS.

CLARITY *See* ACOUSTIC DEFINITION.

CLASP *Consortium of Local Authorities Special Programme*. A group of Local Education authorities in England, who combined in 1957, under the leadership of *Sir Donald Gibson*, to design and use a system of *prefabricated schools*.

CLASSIFICATION OF SOILS *See* PARTICLE SIZE ANALYSIS *and* ATTERBERG LIMITS.

CLAW HAMMER A hammer with a split, claw-shaped peen, used by carpenters for drawing nails. *See also* BALL-PEEN HAMMER.

CLAY A fine-grained COHESIVE soil produced either by the decomposition of rock, or as a sedimentary deposit. It generally consists of hydrated silicates of aluminium with various impurities, and is in part COLLOIDAL. When clay is mixed with coarser grained soils, the clay fraction is usually considered the part which is finer than 2 μm (0·002 mm) diameter. For engineering purposes, clays are classified by their PLASTICITY INDEX, since it is not practicable to analyse them either by sieving or by sedimentation (*see* PARTICLE SIZE ANALYSIS). Their unconfined compressive strength depends greatly on the water content, ranging from less than 5 psi (35 kN/m^2) for a very soft clay to more than 40 psi (275 kN/m^2) for a very stiff clay. It may be seriously reduced by *remoulding*. Soils containing clay may cause *settlement* of the foundation. Clay is the principal raw material for the manufacture of *brick*. *See also* CHINA CLAY.

CLAY PUDDLE *See* PUDDLE.

CLAY TILE (*a*) Roof tile made from clay. (*b*) QUARRY TILE for flooring or wall surfacing. (*c*) Glazed clay tile for flooring and wall surfacing.

CLEAR TIMBER Timber practically free from defects.

CLEARSTOREY The portion of a high room extending above the single-storey height of an adjacent portion of the building, and containing *highlight windows*. Banister Fletcher states that the word is probably derived from the French *clair* (light). Clearstorey windows have been used extensively above the side aisles for lighting the naves of churches, but

the term is also applicable to highlight windows for secular buildings. Also spelt *clearstory*, *clerestorey* and *clerestory*.

CLEAT *See* ANGLE CLEAT.

CLEAVAGE FRACTURE A fracture along the cleavage planes, characteristic of a BRITTLE fracture, and showing little plastic deformation. It usually occurs abruptly, without warning. The failure of glass is a typical example.

CLEPSYDRA Water clock used by the ancient Greeks, which measures time by the discharge of water.

CLERESTORY *or* **CLERESTOREY** *See* CLEARSTOREY.

CLIMBING FORMWORK Formwork which is raised or pulled, either in stages or in a continuous operation, to speed the placement of concrete. Also called *sliding formwork* or *slipform*.

CLINKER, CEMENT *See* CEMENT CLINKER.

CLINOMETER A hand-held instrument for measuring vertical angles on a sloping site, also called an *Abney level*. *See also* DUMPY LEVEL.

CLIP *See* U-BOLT.

CLOISTER Originally an enclosed space. Hence (*a*) a monastery, and (*b*) a quadrangle surrounded by a covered walk.

CLOSE GRAIN Wood with narrow growth or annual rings. The opposite is *coarse grain*.

CLOSE-PACKED HEXAGONAL LATTICE An arrangement of atoms in crystals which may be imitated by close-packing spheres of identical diameter on a hexagonal grid. It is based on a prismatic unit which has a rhombus (included angles 60° and 120°) as its base and, ideally, a height 1·63 times the length of an edge of the base (Fig. 15). Several metals crystallise on this form. They have an atom at each corner of the prismatic unit, and other atoms located at half-height directly above the centre of one of the two equilateral triangles into which the base may be divided.

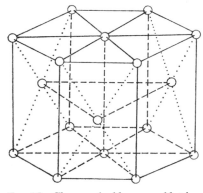

Fig. 15. Close-packed hexagonal lattice.

CLOSED SYSTEM *See* INDUSTRIALISED BUILDING.

CLOSER *See* QUEEN CLOSER.

CLOUD *See* ACOUSTICAL CLOUD.

CLOUT NAIL A nail with a large flat head, used for fastening sheet metal.

CLUSTERED PIER A masonry pier, particularly in the *Gothic* style, consisting of several shafts bundled together.

CNIT EXHIBITION HALL A double-walled shell structure, erected in Paris in 1958 with a span of 720 ft (237 m). The structural design, by N. Esquillan, set a new record.

COARSE AGGREGATE The larger size of aggregate used for mixing concrete, as opposed to *fine aggregate*. *See also* CONCRETE AGGREGATE.

COARSE GRAIN *See* CLOSE GRAIN.

COARSE-GRAINED SOIL A soil in which sand and gravel predominate.

COARSED ASHLAR *See* ASHLAR.

COB WALLING A term synonymous in some places with ADOBE and in others with PISÉ.

COEFFICIENT OF EXPANSION *See* THERMAL EXPANSION

COEFFICIENT OF FRICTION *See* FRICTION and ANGLE OF FRICTION.

COEFFICIENT OF VARIATION (CV) The ratio of the STANDARD DEVIATION of a series of results to their *mean*. In the manufacture of materials it may be taken as a measure of quality control. *See also* FREQUENCY DISTRIBUTION CURVE.

COFFER A recessed panel in a ceiling.

COFFERDAM A watertight enclosure built of piles or clay, for the purpose of providing dry ground for excavating foundations.

COHESIONLESS SOIL A granular soil which consists of clean sand and/or gravel. From COULOMB'S EQUATION, its shear strength depends entirely on the normal pressure, and it is zero on a free surface.

COHESIVE SOIL A sticky soil which contains an appreciable proportion of fine-grained particles (CLAY). Its strength, determined from COULOMB'S EQUATION, depends on the foundation pressure; however, even when there is no normal pressure, a cohesive soil has a shear strength equal to its internal cohesion. This cohesion is very sensitive to the water content of the soil. *See also* CONSOLIDATION.

COIN *See* QUOIN.

COKER, E. G. *See* PHOTOELASTICITY.

COLD *See* THERMAL TRANSMITTANCE.

COLD CATHODE LAMP *See* FLUORESCENT TUBE.

COLD CHISEL A chisel made sufficiently hard for cutting cold metal.

COLD-DRAWN WIRE Wire made from rods which have been hot-rolled from steel billets, and then cold-drawn through a die. This increases the strength, but also lowers the ductility of the steel. Cold-drawn wire is extensively used for reinforced and for prestressed concrete; for the latter diameters range generally from 2 mm (0·080 in) to 7 mm (0·276 in).

COLD RIVETING Closing the head of a RIVET by pressure without heating it. It is much simpler than hot-riveting but in building practice it is restricted to aluminium rivets.

COLD-SETTING RESIN A resin which becomes rigid due to chemical reaction with a *hardener* at room temperature. Cold-setting resins usually set more quickly when heated. *See also* THERMO-SETTING *and* THERMO-PLASTIC.

COLD-WORKED STEEL REINFORCEMENT Steel bars, wires or sections which, subsequent to hot-rolling have been subjected to rolling, twisting or drawing at room temperature.

COLD WORKING The shaping of a metal while at room (or at a slightly elevated) temperature which is below the temperature of recrystallisation, as opposed to *hot working*. It includes cold-forging, cold-rolling, and wire-drawing. Cold working may be used to produce a desired shape with a better surface finish than can be obtained by hot-working. Thus light-gauge steel and aluminium are usually cold-rolled, while normal structural steel sections can only be produced by hot-rolling. Cold-working always increases the strength, and it may be used for *work-hardening*, particularly of high-carbon steel wires by wire-drawing. The latter is a normal method for producing cables for suspension structures and tendons for prestressed concrete.

COLLAPSE A flattening of the cells of timber during drying, which is manifested by excessive or uneven shrinkage. It is liable to occur in certain Australian hardwoods, such as brush box, mountain ash and messmate stringybark. In quarter-sawn timbers it produces a wash-board effect, and in back-sawn timbers it produces an unusually high degree of shrinkage. *Reconditioning* is a steam treatment, carried out in a sealed chamber for about six hours at about 200°F (93°C) which restores the timber to its normal condition.

COLLAPSE LOAD *See* LIMIT DESIGN.

COLLAR BEAM The horizontal member in a timber roof, connecting the two opposite rafters at points which are much higher than the wall-plate (Fig. 16). A *collar-beam roof* thus gives more headroom than one with KING-POST or other conventional trusses, but less than a rigid frame. Although commonly called a collar beam or collar tie, the horizontal member of a collar beam roof is actually in compression.

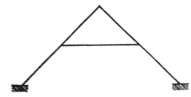

FIG. 16. Collar beam.

COLLIMATION LINE The line of sight, or optical axis, of a telescope, *e.g.* in a survey instrument. When properly adjusted, it passes through the *cross hair* or *reticule*.

COLLOID A substance consisting of very fine material, 10^{-9} to 10^{-7} m (1–100 nm) in diameter. When mixed with water, the particles are too fine to settle. If undisturbed, they remain in suspension to form a GEL. CLAYS are partly colloidal.

COLLOIDAL GROUT A grout which has an artificially induced ability to retain the dispersed solid particles in suspension.

COLONIAL ARCHITECTURE A style derived from British Georgian architecture. It is applied to American and Australian architecture of the eighteenth and early nineteenth century which follows the classical tradition, and to modern revivals.

COLONNADE A series of columns placed at regular intervals, usually supporting an architrave.

COLOPHONY *See* ROSIN.

COLOR *See* COLOUR.

COLOUR Classified according to the *Munsell Book of Color*; this was first published by A. H. Munsell in 1905, and is now produced by the Munsell Color Co., Baltimore (9th edition, 1941). Munsell arranges colour according to hue, value and chroma. *Hue* is designated by the basic colours: red, yellow, green, blue and purple, and by combinations identified by combinations of the letters R, Y, G, B and P. *Value* correlates with the lightness of the colour perceived to belong to it, and it ranges from 0/ for ideal black to 10/ for ideal white. The *chroma* scale correlates the saturation of the colour, and it ranges from /1 in arbitrary steps to express departure from the equivalent grey. The *power* scale is the composite of Munsell value and chroma; thus 4/14 means a colour slightly darker than the middle value between black and white, and 14 arbitrary steps from the equivalent grey. The colours are then matched against the standard colours in the *Book of Color* for identification.

COLOUR CONTRAST *See* CON-TRAST.

COLOUR RENDITION The effect which the spectral characteristics of a light have on the appearance of coloured objects illuminated by it.

COLOURED CEMENT Portland cement blended with a pigment, which does not chemically react with any of the components of concrete. Certain pigments, which are suitable for internal use, fade if used externally. Ordinary (grey) cement is satisfactory for the darker colours. WHITE CEMENT is required with some lighter pigments.

COLUMN *See also* PILLAR.

COLUMN ANALOGY An analogy between the equations for SLOPE DEFLECTION and those for load and moment in short eccentrically loaded columns, published by Hardy Cross (who also devised the MOMENT DISTRIBUTION METHOD) in 1930. It is used for the design of rigid frames.

COLUMN CAPITAL (*a*) The head or crowning feature of a column. In the classical ORDERS, and also in Egyptian and Gothic architecture, capitals were elaborately decorated. (*b*) In modern concrete construction, the enlargement at the head of a column, built as an integral unit with the column and the FLAT SLAB. It is designed to increase the shearing resistance of the flat slab (Fig. 31).

COLUMN, COMBINATION *See* COMBINATION COLUMN.

COLUMN, COMPOSITE *See* COMPOSITE COLUMN.

COLUMN, LONG A column whose load bearing capacity must be reduced because of its SLENDERNESS RATIO.

COLUMN, SHORT A column whose load bearing capacity need not be reduced because of its SLENDERNESS RATIO.

COLUMN STRIP The portion of a FLAT SLAB or FLAT PLATE over the column. Most building codes define the column strip as consisting of the two adjacent quarter panels on each side of the column centre line (Fig. 31).

COMBINATION COLUMN A pipe column filled with concrete, or a structural steel column encased in concrete reinforced with mesh. *See also* COMPOSITE COLUMN.

COMBINED FOOTING A foundation supporting more than one column.

COMBINED WATER Water in mineral matter which is chemically combined, and driven off only at temperatures above 110°C.

COMBUSTIBLE A material which burns. If placed in a hot furnace (usually at 750°C) it raises the temperature of the furnace. A combustible material may or may not be *flammable*, *i.e.* burn with a flame. A material which does not support combustion is *non-combustible*.

COMFORT ZONE The range of temperature, humidity, etc. at which people may rest or work comfortably, particularly in a hot climate, shown as a loop on a comfort chart. Concepts of thermal comfort vary considerably, depending on acclimatisation, whether comfort at rest or work is considered, whether work is manual or mental, and whether the climate is hot or cold, humid or arid. *Comfort charts* usually have the dry bulb temperature as an abscissa, and the wet-bulb temperature as an ordinate. The additional variable, plotted on a diagonal scale, may be either the speed of the air movement, or the amount of radiant heat. *See* EFFECTIVE TEMPERATURE, EQUIVALENT TEMPERATURE, KATA-THERMOMETER, DEGREE-DAY, EQUATORIAL COMFORT INDEX and WET-BULB GLOBE THERMO-METER INDEX.

COMMON BOND   *See* BOND (*a*).

COMMON RAFTER   *See* RAFTER.

COMMON WALL   A wall forming part of two properties, and equally owned or leased by both parties. Also called a *party wall*.

COMPACTING FACTOR TEST A method for determining the WORKABILITY of freshly mixed concrete. The concrete is placed in a container of standard size, and allowed to fall under standard conditions into another container. Fully compacted concrete has a factor of one. The test is more precise, but also more time-consuming, than the SLUMP TEST.

COMPENSATED BALANCE   A spring balance in series with a turnbuckle, used to measure a force in a structural *model*. As the spring balance extends or contracts, the turnbuckle is adjusted to keep the geometry of the structure correct.

COMPENSATING ERROR   An error due to an *accidental* cause, which may be either positive or negative, and is therefore likely to be self-compensating if sufficient data are taken. By contrast an error due to a *systematic* cause is always in the same direction, and therefore *cumulative*.

COMPILING ROUTINE   A routine whereby a digital computer can construct a PROGRAM from its existing SOFTWARE. Also called a *compiler*.

COMPLEMENTARY ANGLE   An angle which equals the difference between a given angle and a right angle (90°). For example, 60° is the complementary angle to 30°. Similarly, a *supplementary angle* is the difference between a given angle and 180° (two right angles).

COMPLEX NUMBER   *See* IMAGI-NARY NUMBER.

COMPONENT OF A FORCE   *See* RESOLUTION OF A FORCE.

COMPOSITE COLUMN   A concrete column, reinforced with a structural steel section, surrounded by longitudinal bars and spiral reinforcement (Fig. 17). The structural steel section provides a large area of steel without offering the same obstruction to concreting as a great number of bars, and the column therefore carries a greater load, for the same size, than one reinforced with bars only. *See also* COMBINATION COLUMN and COMPOSITE ORDER.

COMPOSITE CONSTRUCTION (*a*) A type of construction made up of

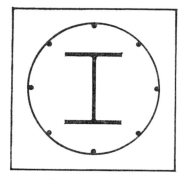

FIG. 17. Composite column.

different materials. (*b*) Specifically, structural steelwork and reinforced concrete designed as a single structural system.

COMPOSITE GIRDER (*a*) A PLATE GIRDER. (*b*) A girder of COMPOSITE CONSTRUCTION (*b*).

COMPOSITE ORDER The last of the five Roman ORDERS. The proportions are borrowed from the Corinthian, with elements from the Tuscan, Roman Doric and Ionic orders.

COMPOSITION ROOFING *See* BUILT-UP ROOFING.

COMPOUND CURVE Normally defined as a curve consisting of two (or more) circular arcs, which have different radii, and a common tangent at their point of junction.

COMPRESSED FIBRE BOARD *See* HARDBOARD.

COMPRESSION A direct push in line with the axis of a body, and therefore the opposite of TENSION.

COMPRESSION CYCLE A refrigeration cycle. It utilises two phenomena: (*a*) The evaporation of a liquid refrigerant absorbs heat to lower the temperature of its surroundings; and (*b*) the condensation of the refrigerant vapour rejects heat to raise the temperature of its surroundings. The refrigerants in use are mostly compounds of carbon, chlorine and fluorene (CCl_3F, CCl_2F_2, $CHClF_2$, $C_2Cl_3F_3$, $C_2Cl_2F_4$).

COMPRESSION FAILURE Failure under the action of a compressive force, either due to the material reaching its LIMITING STRENGTH, or due to BUCKLING, or due to a combination of both.

COMPRESSION REINFORCEMENT Reinforcement used near the compression face of the concrete; it requires ties to prevent buckling.

COMPRESSION WOOD A region of excessively dense wood; it is very brittle, and shows abnormal longitudinal shrinkage.

COMPRESSIVE STRENGTH OF CONCRETE *See* CUBE STRENGTH and CYLINDER STRENGTH.

COMPRESSIVE STRENGTH OF SOIL *See* UNCONFINED COMPRESSION TEST and TRIAXIAL COMPRESSION TEST.

COMPUTER *See* DIGITAL COMPUTER and ANALOGUE COMPUTER.

COMPUTER CARD, TAPE *See* CARD, TAPE.

COMPUTER GRAPHICS *See* DIGITAL PLOTTER and LIGHT PEN.

COMPUTER HARDWARE, SOFT-WARE *See* HARDWARE and SOFT-WARE.

COMPUTER STORE *See* ADDRESS, BACKING STORE, IMMEDIATE ACCESS STORE, and MEMORY.

COMPUTER SYSTEM The units which comprise a working DIGITAL COMPUTER, *i.e.* a CENTRAL PROCESSOR UNIT, together with one or more INPUT, OUTPUT, and BACKING STORE units, depending on the size of the system.

CONCAVE A curve which bends inwards, like the inside of a circle, ellipse, etc. It is opposite of *convex*, which denotes a curve bending outwards.

CONCEALED GUTTER A BOX GUTTER.

CONCEALED LIGHTING An artificial light source, recessed into a ceiling or wall, or concealed behind a decorative facing or a *pelmet*.

CONCEALED NAILING *See* SECRET NAILING.

CONCENTRIC COLUMN LOAD A load which compresses a column without bending, as opposed to an ECCENTRIC column load. Also called *axial load*.

CONCORDANT TENDONS Tendons in statically indeterminate prestressed concrete structures which do not produce secondary moments. They must be coincident with the line of pressure produced by the tendons.

CONCRETE *See also* CAST-IN-PLACE, CELLULAR CONCRETE, GRANO-LITHIC CONCRETE, GREEN CONCRETE, LEAN CONCRETE, NO-FINES CONCRETE, PLAIN CONCRETE, PRESTRESSED CONCRETE, READY-MIXED CONCRETE, REINFORCED CONCRETE, TRANSLUCENT CONCRETE, and VACUUM CONCRETE.

CONCRETE ADMIXTURE *See* ACCELERATOR, AIR-ENTRAINING AGENT, and RETARDER.

CONCRETE AGGREGATE The inert component of concrete. Heavyweight, or normal, aggregate consists of sand (FINE AGGREGATE) and gravel, crushed gravel or crushed stone (COARSE AGGREGATE). *See also* LIGHTWEIGHT AGGREGATE.

CONCRETE BLOCK, BRICK A *block* or *brick* moulded in sand and cement, often with the addition of a mineral pigment.

CONCRETE JOIST CONSTRUCTION *See* RIBBED SLAB.

CONCRETE MIXER *See* BATCH MIXER.

CONCRETE PILE *See* PILE.

CONCRETE PUMP A pump which pushes concrete through a pipeline. It is used in conjunction with a CEMENT GUN, and it may also be economical for transporting concrete, particularly over difficult ground.

CONCRETE QUALITY CONTROL *See* QUALITY CONTROL.

CONCRETE TERRAZZO Concrete made with marble aggregate, and frequently with WHITE CEMENT, and subsequently ground smooth for decorative floor or wall surfaces. It may be precast or cast-in-place.

CONDENSATE Liquid formed by the condensation of vapour.

CONDENSATION The formation of water on a surface, due to the air temperature falling below its dew point. The water content of SATURATED AIR falls with falling temperature, so that the dew point may be reached even though the moisture content remains constant. Condensation is particularly likely in cool weather when the temperature drops at night, since the RELATIVE HUMIDITY then tends to be high, even in relatively dry climates.

CONDENSATION GROOVE A groove to collect the condensation on the inside of windows, from which the moisture escapes to the outside by means of WEEP-HOLES.

CONDENSER (*a*) An apparatus for condensing vapours, *e.g.* in a steam engine, or in the refrigeration plant of an air conditioning unit. (*b*) A lens or mirror used in an optical system to collect light and direct it on to a projecting lens. (*c*) A capacitor in an electrical circuit.

CONDENSER MICROPHONE A microphone which consists of an extremely thin foil, stretched close to a fixed backing plate. The sound pressure fluctuations cause displacement of the diaphragm, and hence change the capacitance between the plates.

CONDOMINIUM American term for an APARTMENT which is sold and not rented.

CONDUCTANCE, CONDUCTION, CONDUCTIVITY, THERMAL *See* THERMAL CONDUCTANCE, *etc.*

CONDUIT (*a*) A natural or artificial channel for conveying liquids. (*b*) A tube, usually of plastic or metal, which encloses electrical wires or cables. Conduits are partly used for protection, and partly to allow cables to be pulled through after the concrete or other building material has been placed in position.

CONDUIT BOX A junction box serving as an outlet and as a place from which to pull wires through the conduits.

CONE A figure whose base is a circle, with sides tapering uniformly towards a point. A *truncated cone* is one which is cut off before reaching the point. *See also* CONIC SECTIONS.

CONFINED COMPRESSION TEST *See* TRIAXIAL COMPRESSION TEST.

CONGLOMERATE A rock composed of (usually rounded) pieces of pre-existing rock cemented together.

CONIC SECTIONS The CIRCLE, the ELLIPSE, the PARABOLA, and the HYPERBOLA.

CONIFER A tree belonging to the botanical group *gymnospermae* which bears cones. It includes all the softwoods used in building, particularly the pines and firs.

CONNECTING ROD *See* RECIPROCATING ENGINE.

CONOID *See* SURFACE OF TRANSLATION, RULED SURFACE, and NORTH-LIGHT SHELL.

CONSISTENCY INDEX A ratio for comparing the stiffness of cohesive soils. It is defined as

$$\frac{\text{liquid limit} - \text{water content of sample}}{\text{liquid limit} - \text{plastic limit}}$$

CONSISTENCY LIMITS The LIQUID and PLASTIC LIMITS of COHESIVE SOILS, which describe its range of workability. Also called *Atterberg limits*.

CONSISTENCY OF CONCRETE *See* WORKABILITY.

CONSOLIDATION The gradual settlement of a COHESIVE SOIL under the weight of the structure which it carries. It results from the squeezing of water from the pores of the soil, and is a problem only in clays and other soils of low water permeability. *See also* OEDOMETER and DIFFERENTIAL SETTLEMENT.

CONSOLIDOMETER *See* OEDOMETER.

CONSTANTAN An alloy of about 55 per cent copper and 45 per cent nickel. Junctions of copper and constantan are widely used in THERMOCOUPLES for temperature measurement.

CONSTITUTIONAL DIAGRAM *See* PHASE DIAGRAM.

CONSTRUCTIVISM The name chosen by a group of artists formed in Moscow in 1920, and culminating in the proclamation of the *Constructivist International* by *El Lissitzky* in 1922. In spite of its emphasis on efficient structure, it was essentially concerned with sculpture.

CONTACT PRINT *See* BLUEPRINT.

CONTINUITY *See* CONTINUOUS BEAM and RIGID JOINT.

CONTINUOUS BEAM A beam which is continuous over intermediate supports, and thus HYPERSTATIC as opposed to a SIMPLE BEAM (Fig. 18). Continuity over a support implies that the slopes on both sides of the support are equal and opposite. *See* THEOREM OF THREE MOMENTS.

CONTINUOUS MIXER *See* BATCH MIXER.

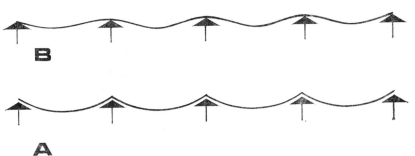

FIG. 18.   Continuous beam. (A) Four simply supported beams; (B) beam continuous over four spans.

CONTRACTION JOINT A joint, usually vertical, which allows shrinkage of concrete or brick walls to take place in a predetermined location, and so avoid objectionable shrinkage cracks elsewhere. *See also* EXPANSION JOINT.

CONTRAFLEXURE A change in the direction of bending of a beam (Fig. 19). The slope has a maximum value at this point, and therefore the differential coefficient of the slope $d\theta/dx = 0$. From NAVIER'S THEOREM and the geometry of the figure,

$$\frac{d\theta}{dx} = \frac{M}{EI}$$

so that the bending moment is zero at the point of contraflexure. Consequently it is possible to insert a hypothetical 'hinge' at this point and remove a redundancy. This device is used to simplify the design of some HYPERSTATIC structures.

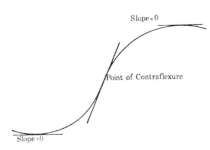

Slope = 0

Point of Contraflexure

Slope = 0

FIG. 19. Point of contraflexure.

CONTRAST In illuminating engineering, the subjective assessment of the difference in appearance of two parts of a field of view seen simultaneously or successively.

CONURBATION A term coined by Patrick Geddes in 1910 to denote a group of towns linked together by their geographic proximity.

CONVECTIVE HEAT TRANSFER The transmission of heat by natural or forced motion of a liquid or gas, *i.e.* by movement of the particles, as opposed to thermal conduction and radiation.

CONVERGENT SERIES *See* INFINITE SERIES.

CONVERSION FACTOR A number which converts the units of one system into that of another.

CONVERSION OF TIMBER The process of sawing timber from the log.

CONVERTER *See* BESSEMER PROCESS.

CONVEX *See* CONCAVE.

COORDINATES *See* CARTESIAN CO-ORDINATES.

COPAL A hard natural resin, derived from the gum of tropical trees or from recently fossilised gum. It is used for varnishes and paints.

COPING A capping of stone, brick, or concrete for the top of a wall. It frequently projects beyond either or both faces of the wall, partly for protection from the weather, and partly for decoration.

COPING SAW A bow saw with a narrow blade, which can be used for cutting sharp curves in timber.

COPOLYMERISATION Addition POLYMERISATION involving more than one type of MER.

COPPER A metallic element in the first group of the periodic system. Its chemical symbol is Cu, its atomic number is 29, its valency is 1 or 2, it has an atomic weight of 63·54 and a specific gravity of 8·96, and it melts at 1083°C. Copper is used in many alloys, notably brass, bronze and aluminium bronze. Because of its high electrical conductivity and good corrosion resistance it is used extensively as an electrical conductor.

CORBEL A projection of masonry, brick or concrete from a wall face, which serves as support for a lintel, beam or truss (Fig. 20). It is, in effect, a short cantilever.

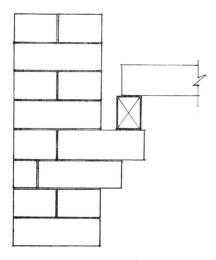

Fɪɢ. 20. Corbel.

CORBELLING Masonry or brickwork consisting of a series of *corbels*, each projecting a little more than the one below. It is used for supporting ORIEL WINDOWS and forming chimney stacks. At one time it was used for forming arches. The *corbelled*

arch differs from the true *arch*, in that the joints are horizontal, instead of being normal to the line of thrust.

CORBUSIER, LE *See* BÉTON BRUT, CIAM, FUNCTIONALISM, MODULOR, and PILOTIS.

CORD A measure for timber. 1 cord = 128 ft³ (4 ft × 4 ft × 8 ft).

CORE SAMPLES *See* BOREHOLE SAMPLES.

CORINTHIAN ORDER One of the three Greek and one of the five Roman ORDERS. It is characterised by a bell-shaped column capital, decorated with stylised ACANTHUS leaves.

CORK The bark of the cork-oak. It is used for stopping bottles, in granulated form as an insulating material, and as a floor-surfacing material which has good heat and sound insulation.

CORK TILES Tiles made from compressed cork. They form a cheap floor-covering with good insulating properties.

CORK WOOD Balsa wood.

CORNICE (*a*) In classical architecture, the projecting section of an entablature. (*b*) An overhanging moulding at the top of an outside wall which throws water clear off the wall. (*c*) A ceiling moulding at its junction with the walls.

CORPS DE LOGIS The main body of a large building, as opposed to its wings and outhouses.

CORROSION Destruction of material, particularly metal, by chemical means. *See also* DECAY and PRIMER.

CORRUGATED SHEET Roof sheet which has been corrugated, partly for flexural stiffness and partly to provide drainage channels for rainwater. The most common type is 'corrugated iron', which is normally galvanised mild steel. Corrugated sheets are also made from aluminium, asbestos cement, PVC, and glass-fibre reinforced translucent resin.

COR-TEN STEEL A high-strength, low-alloy steel which does not need to be protected by paint because, like aluminium, it forms a protective oxide coating through weathering. The colour changes from light brown (after about 1 month) to dark brown or purple (after 1 to 2 years). However, the oxidation process may produce brown streaks on concrete or other materials at a lower level.

CORTILE An enclosed courtyard without a roof in an Italian-style building.

CORUNDUM Aluminium oxide (Al_2O_3) used as abrasive. Not to be confused with *carborundum*. *See also* EMERY.

cos The cosine of an angle. *See* CIRCULAR FUNCTIONS.

cosec Cosecant of an angle, cosec $\theta = 1/\sin \theta$.

cosh Hyperbolic cosine. *See* HYPERBOLIC FUNCTIONS.

COSINE WAVE *See* SINE WAVE.

COT The cotangent of an angle. *See* CIRCULAR FUNCTIONS.

COTTAGE ORNÉ A deliberately picturesque rustic building, usually with roughly hewn timber and a thatched roof, erected for a wealthy or aristocratic client in the late eighteenth or early nineteenth century.

COULOMB (C) Unit of electrical charge, named after the 18th century French scientist and military engineer. It is the quantity of electricity transported by 1 ampere in 1 second.

COULOMB'S EQUATION A relation between the shear strength of a cohesive soil, v, and the normal (foundation) pressure, f, devised by the French engineer C. A. Coulomb in 1776.

$$v = c + f \tan \theta$$

where c is the cohesion of the soil, and θ is its *angle of shearing resistance*. This is a straight limiting line, inclined at an angle θ, for a series of MOHR CIRCLES. The equation can also be applied to the compression failure of concrete; the cohesion is contributed mainly by the cement paste, and the shearing resistance mainly by the aggregate. *See also* MOHR FAILURE CRITERION and COWAN CRITERION.

COUNTERFORT A pier at right angles to a RETAINING WALL, on the side of the retained material and therefore not visible. It can be used only to strengthen a structure capable of tensile resistance, such as reinforced concrete, since the earth pushes the wall away from the counterfort. A BUTTRESS can also be used to strengthen a plain-concrete or masonry retaining wall (Fig. 21).

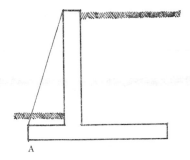

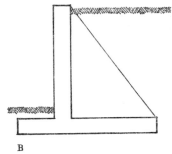

B

FIG. 21. Retaining wall (A) with buttress and (B) with counterfort.

COUNTERSINK To make a depression just sufficient to receive the head of a screw, rivet, or some other part of a joint which normally projects. It can be done in timber, metal or any other material, using a conically shaped cutting tool, or *bit*.

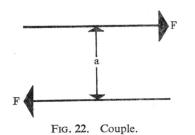

FIG. 22. Couple.

COUPLE A pair of equal and parallel forces, not in line and oppositely directed. The moment of a couple equals the magnitude of one of the forces F times the perpendicular distance between them, a (Fig. 22).

COVALENT BOND Atomic BONDing by sharing electrons.

COVE A concave quadrant moulding, *i.e.* a hollow CORNICE.

COVER In reinforced concrete, the thickness of concrete overlying the steel bars nearest to the surface (Fig. 33). An adequate layer is needed to protect the reinforcement from rusting.

COVER FILLET OR STRIP A moulding used to cover a joint.

COVERING POWER *See* HIDING POWER.

COWAN FAILURE CRITERION A FAILURE CRITERION, proposed by H. J. Cowan in 1952, for the transition of the failure of concrete from a slow CRUSHING FAILURE to an abrupt CLEAVAGE FRACTURE under the action of combined stresses. Failure occurs when a MOHR CIRCLE touches the limiting lines in Fig. 23. The lines X–X conform to COULOMB'S EQUATION; circle A which represents a simple compression test, and all circles touching lines X–X indicate a crushing failure. Line Y–Y conforms to RANKINE'S CRITERION; circle B which represents a simple tension test, circle C which represents a simple torsion test, and all circles touching line Y–Y, indicate a cleavage fracture. Circle D shows the transition from cleavage to crushing failure.

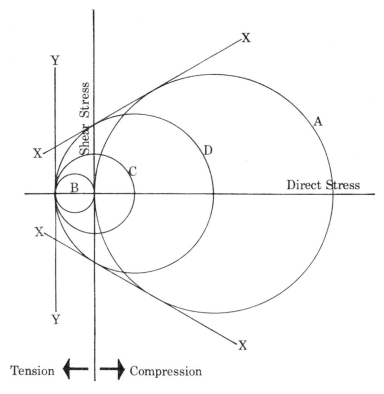

FIG. 23.   Cowan failure criterion.

COWL   A metal cover, often capable of rotating, and fitted with louvres. It is fixed on a roof ventilator or chimney to improve the natural ventilation or draught.

CPM   Critical path method.

cps   Cycles per second. (The SI unit is the HERTZ.)

Cr   Chemical symbol for *chromium*.

CRACK   *See* CRAZING, GRIFFITH CRACK, and MICRO-CRACK.

CRACKED SECTION   In concrete design, a section which is designed on the assumption that the concrete has no resistance to tensile stresses, and that the tension is taken by the reinforcement.

CRANK   A bar with a right-angled bend, which gives it leverage, *e.g.* a crank handle for starting an internal combustion engine, or a crank shaft in a RECIPROCATING ENGINE.

CRASHED TIME   The fastest time in which an activity in a building project can be performed by making more labour and materials available than would normally be done. If the activity lies on the CRITICAL PATH, it may be more economical to perform the activity in crashed time than in normal time. The critical path must be recalculated if any operation is speeded up, since it may no longer lie on it; however, this would not necessarily imply that the use of crashed time is uneconomical.

CRAZING The development of fine, random cracks caused by shrinkage on the surface of plaster, cement paste, mortar, or concrete. They are particularly noticeable if the material is finished to a smooth surface with a steel trowel. Crazing is more likely to occur in concrete if the cement is brought to the surface by excessive trowelling.

CREEP Time-dependent deformation due to load. *See* RHEOLOGICAL MODEL. (*a*) In structural metals, creep occurs only at elevated temperatures (in steel above 300°C). It is caused by the increased mobility of atomic particles at higher temperatures. *See also* PLASTIC FLOW. (*b*) In concrete, a sustained load squeezes water from the cement GEL at ordinary temperatures. Creep deformation may be two or three times as great as the *elastic* deformation, and it causes a substantial redistribution of stress, often transferring load from the concrete to the steel reinforcement. Because of creep, the EFFECTIVE MODULUS OF ELASTICITY of concrete is reduced, and the MODULAR RATIO increased. Creep is a major cause of LOSS OF PRESTRESS in both POST-TENSIONED and in PRE-TENSIONED concrete.

CREOSOTE OIL A liquid dis-

tilled from coal tar between 240 and 270°C, which is used as a timber preservative.

CRESCENT A concave row of houses.

CRIPPLING LOAD OF A COLUMN The load at which a slender column BUCKLES.

CRITICAL ACTIVITY *See* ACTIVITY.

CRITICAL PATH The route through the NETWORK, from starting point to terminal point, which is critical for the project (Fig. 67). The *critical time path* is the longest route through the network. The *critical resources path* may also have to be determined, if either labour or materials are limited.

CRITICAL PATH METHOD (CPM) A method of scheduling complex building contracts to determine the critical operations which would, if neglected, lead to delays, and to determine the cost of finishing the work more quickly, or the saving in cost if it were finished more slowly. As in a progress chart, the normal time required for all operations is worked out, and the inter-relation of the various operations is plotted as a NETWORK. The CRITICAL PATH through the network is then determined, which gives the completion period if all operations are performed on time. The operations which lie on this path are critical, and it may be worth while to perform them in CRASHED TIME, by making more material or labour available; this may, however, alter the critical path. For a large contract the method requires the use of a DIGITAL COMPUTER. *See also* PERT.

CRITICAL VELOCITY *See* REYNOLDS' CRITICAL VELOCITY.

CROSS-CUT SAW A saw whose teeth have been set and sharpened to cut *across* the grain of the wood.

CROSS HAIR *See* COLLIMATION LINE.

CROSS-LINKING OF POLYMERS The tying together of adjacent chains.

CROSS METHOD *See* MOMENT DISTRIBUTION METHOD.

CROSS-SECTIONAL AREA The area of a cross section, *i.e.* of a section cut transversely to the longitudinal axis of a member.

CROSS VAULT OR GROIN VAULT A vault resulting from the intersection at right angles of two BARREL VAULTS of identical shape (Fig. 24).

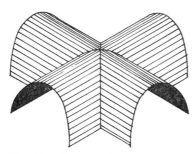

FIG. 24. Cross vault or groin vault.

CROSSING The part of the church at the intersection of the NAVE, CHANCEL, and TRANSEPT. It is often surmounted by a tower.

CROTCHWOOD The portion of a tree where a large limb branches from the trunk. The fibres of the branch produce a curly grain, and a veneer cut from crotchwood is highly FIGURED. *See also* BURL.

CROWN The highest point, or *vertex*, of an arch (Fig. 25) or dome. The stone at the crown of a masonry arch is called the *keystone*, and the arch becomes self-supporting only after the keystone has been placed in position. Hence it was frequently given prominence in the design, or especially decorated. Since the keystone need only resist the horizontal thrust, it is less heavily loaded than any of the stones lower down the arch; the view held by many Renaissance architects that the keystone requires a higher-quality material is mistaken. *See also* SPRINGINGS.

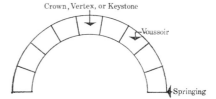

FIG. 25. Voussoir arch, showing location of crown and springings.

CROWN GLASS (*a*) Glass made by blowing a mass of molten material, which is then flattened into a disc, and spun into a circular sheet. It is limited in size to about 4 ft 6 in (1·4 m) diameter. Since the glass does not come into contact with any other material during manufacture it is free from the defects which occur in CYLINDER GLASS. The process was introduced into England in the seventeenth century, and the sash window with brilliantly clear, often slightly curved crown glass became one of the characteristics of Georgian architecture. The centre of the disc, or *bull's eye*, was originally used for inferior work. Later it was employed deliberately, often using cast imitations, for the windows of 'rustic' cottages. The crown glass process became obsolete in the nineteenth century. (*b*) GLASS of the alkali–lime–silica type, as opposed to FLINT GLASS.

CRT Cathode-ray tube.

CRUCKS Pairs of large curved timbers used for the principal framing of barns and primitive houses. They form a pointed arch, taking the place of both the posts and the rafters.

CRUSHING FAILURE The compression failure (due to diagonal shear) of many brittle materials which results in the production of crushed material or debris. Since this takes time, it is much slower than a CLEAVAGE FRACTURE. The compression failure of concrete is a typical example.

CRUSHING STRENGTH *See* CUBE STRENGTH and CYLINDER STRENGTH.

CRYSTAL A body whose atoms are arranged in a definite pattern. The regular arrangement gives rise to the characteristic crystal faces. *See* SPACE LATTICE (*b*).

CRYSTAL MICROPHONE A microphone which contains a PIEZO-ELECTRIC crystal, such as Rochelle salt or barium titanate. A diaphragm exposed to the acoustic pressure is mechanically linked to the crystal, which is thus deformed and a piezoelectric voltage produced.

CRYSTAL PALACE A large glass house, 1850 ft (560 m) long, designed by Joseph *Paxton* to house the 1851 International Exhibition in London. It was subsequently dismantled, re-erected in a London suburb, and used as a place of entertainment until destroyed by fire in 1936. Its outstanding success led to the building of Crystal Palaces in New York and other cities in the later nineteenth century. The Crystal Palace occupies an important place in the history of INDUSTRIALISED BUILDING, because this enormous building was constructed in a few months from factory-made parts.

CRYSTAL STRUCTURE *See* SPACE LATTICE (*b*).

CRYSTALLINE FRACTURE Fracture taking place between crystals, and exposing their faces. In a tension test particularly, this is evidence of *brittleness*.

c/s Cycles per second. (The SI unit is the hertz.)

CSA Canadian Standards Association, Ottawa.

CSIR Council for Scientific and Industrial Research, a term employed in India, Pakistan, South Africa, and formerly in Australia.

CSIRO Commonwealth Scientific and Industrial Research Organisation, Australia.

CTESIPHON ARCH A catenary arch. The name comes from a Persian palace, built in AD 550, from mud brick and now in ruins, which includes a CATENARY ARCH spanning 85 ft (26 m).

ctg Continental abbreviation for cotangent (cot).

CTSB Centre Scientifique et Technique du Bâtiment, Paris.

Cu Chemical symbol for *copper*.

CUBE *See* POLYHEDRON.

CUBE STRENGTH (OF CONCRETE) In most European countries the compressive strength of concrete is ascertained by crushing a cube; this is expressed as the ultimate load per unit cross-sectional area. The size of the cube varies from 4 in (100 mm) to 8 in (200 mm). Because of the greater probability of the presence of flaws (*see* GRIFFITH CRACK) in large pieces of a brittle material, the average cube strength decreases as the size of the cube increases. Consequently the strength of concrete must be stated in terms of the size of the cube. For the same reason, the (6 in × 12 in) CYLINDER STRENGTH of concrete is only about 85 per cent of the 6 in cube strength.

CUBIC LATTICE *See* BODY-CENTRED CUBIC LATTICE *and* FACE-CENTRED CUBIC LATTICE.

CULLS In forestry, trees which are removed individually, because they are deformed, too closely spaced, etc. They provide one source of raw material for FIBRE BOARD.

CUMEC Cubic metres per second.

CUMULATIVE ERROR *See* COMPENSATING ERROR.

CUP-AND-CONE FRACTURE The typical fracture of a DUCTILE material, *e.g.* structural steel, in

tension (Fig. 26). The bar first elongates plastically, and the consequent *necking* reduces the cross-sectional area until the ultimate tensile stress of the material is reached, when a brittle fracture across the bar occurs. After failure one of the two parts of the bar is cup-shaped and the other cone-shaped.

FIG. 26. Cup-and-cone fracture.

CUPOLA A small dome, rising above a roof.

CURING (*a*) The maintenance of an appropriate humidity and temperature in freshly placed concrete to ensure the satisfactory hydration of the cement, and proper hardening of the concrete. It may be necessary to provide a source of heat in very cold weather, or some cooling in very hot conditions. Evaporation of water is reduced by placing covers over the concrete, applying a *curing membrane* (a liquid sealing compound) to the exposed surface, or sprinkling the concrete periodically with water. (*b*) The chemical change, resulting in additional linkages between the molecules, which occurs when an ACCELERATOR is added to a COLD-SETTING RESIN, or when a THERMO-SETTING plastic is heated above the critical temperature. The material gains the required strength and hardness through curing. *See* POLYMERISATION.

CURL A fine, curved figure in the grain of wood, frequently obtained by the conversion of CROTCHWOOD.

CURTAIN WALL A thin wall supported by the structural frame of the building, and not dependent on the load-bearing quality of the wall below it.

CURVATURE *See* SLOPE.

CURVILINEAR Consisting of, or bounded by curved lines.

CUSEC Cubic feet per second.

CUSP (*a*) A point where two branches of a curve meet, have a common tangent, and terminate. (*b*) An ornament of roughly this shape, used in Gothic tracery.

CV Coefficient of variation.

CYANIDE HARDENING CASE HARDENING of steel by immersion in a bath of molten sodium cyanide (NaCN).

CYBERNETICS The science of automatic control.

CYCLOID A curve generated by a point on a circle rolling over a straight line. It can be rotated to form a dome which, like a semi-elliptical dome, has vertical springings. *See also* EPICYCLOID.

CYCLOPEAN CONCRETE *See* PLUM.

CYCLOPEAN MASONRY (*a*) Masonry composed of very large irregular blocks, particularly in prehistoric architecture. (*b*) RUSTICATED masonry in squared blocks, popularised by the Renaissance in the fifteenth century, which either have the rough hewn texture resulting from quarrying, or an artful imitation thereof.

CYLINDER A solid bounded by a curved surface terminating in two parallel and equal plane figures (which are generally circles).

CYLINDER GLASS Glass made by blowing a mass of molten material into a hollow cylinder, about 15 in (380 mm) diameter and 7 ft (2·1 m) long. This cylinder, while soft, is cut lengthwise, laid out on a pre-heated iron table, and placed in an annealing furnace to flatten. The glass so produced is cheaper than CROWN GLASS or PLATE GLASS, but less perfect. *See also* SHEET GLASS.

CYLINDER STRENGTH (OF CONCRETE) In most parts of America and Australia the compressive strength of concrete is ascertained by crushing a 6 in (150 mm) diameter cylinder, 12 in (300 mm) long, and expressed as the ultimate load per unit cross-sectional area. *See also* CUBE STRENGTH.

CYLINDRICAL SHELL *See* BARREL VAULT, BEAM THEORY OF SHELLS, DEVELOPABLE SURFACE, EDGE BEAM, NORTHLIGHT SHELL, RULED SURFACE, SURFACE OF TRANSLATION, and Figs. 32 and 68.

CYMA *See* OGEE MOULDING.

D

D1S   Timber dressed one side only.

DADO   In classical architecture, the portion of the pedestal between the base and the cornice. In modern buildings, any border or panelling over the lower half of the walls of a room, which is above the skirting.

DAIS   A raised platform at one end of a large room.

DAMAGE RISK CRITERION (DRC)   An empirical curve, relating the frequency to the upper limit of sound pressure (in DECIBELS), above which loss of auditory acuity or deafness is likely to result.

DAMASCENE STEEL   Sword steel made in India and Persia, which found its way to Europe through Syria during the Middle Ages. The process consisted of heating together wrought iron and charcoal, and cooling extremely slowly. The resulting steel had about $1\frac{1}{2}$ per cent of carbon, and it was shaped into a sword by repeated hammering and reheating. The blade had an excellent cutting edge, great toughness, and a characteristic 'watered' pattern. The Damascene process was later introduced into Toledo, Spain. In the nineteenth century the term was applied to the Spanish art of inlaying steel with gold. The pattern was scratched into the surface, and pure gold wire was hammered into it. The steel was then heated to a blue colour.

DAMP COURSE   *See* DAMP-PROOF COURSE.

DAMP-PROOF COURSE (DPC)   An impervious layer inserted into a pervious wall (such as a brick wall) to exclude water. A DPC is required a few inches above ground level (below the level of a timber floor), below a roof parapet (above the roof level) and sometimes above the openings for doors and windows. Traditional materials for a DPC are slate, vitrified brick and lead. BITUMEN with a core of lead, copper, zinc, aluminium or plastic is now more common. A DPC is sometimes, misleadingly, called a *damp course.*

DAMPED OSCILLATION   A mechanical (*e.g.* acoustical) or electrical oscillation in which there is an appreciable diminution of amplitude during successive cycles. Reducing vibrations by damping is analogous to reducing movement by friction. *See also* DECAY FACTOR.

DANTER'S THERMAL ANALOGUE   An analogy between the equations for heat transfer through a slab of material, and that of certain electrical circuits, which is utilised in ANALOGUE COMPUTERS for determining the change in room temperature due to the varying solar heat outside.

DAR   Dressed all round (all four sides of a piece of timber dressed).

DARCY'S LAW   'The velocity of percolation of water in saturated soil = hydraulic gradient × coefficient of PERMEABILITY.'

DATA MEDIUM   A medium used to contain data. The most common types used in conjunction with digital computers are CARDS and paper TAPE, into which the data are punched in binary notation.

DATUM LINE, or LEVEL A reference *line* or *level*, used to locate other lines or levels in a survey.

DAYLIGHT *See* ARTIFICIAL SKY, DAYLIGHT FACTOR, GLOBOSCOPE, LIGHTING, MOON-SPENCER SKY, and SOLARSCOPE.

DAYLIGHT FACTOR A factor describing the efficiency of a window, used in the design of rooms for natural lighting. It is defined as the ratio of the illumination at any point in a room to the illumination at the same instant on a horizontal plane exposed to the unobstructed sky. The *externally received component* (*ERC*) of the daylight factor is that due to external reflecting surfaces, illuminated directly or indirectly. It is therefore the 'equivalent SKY COMPONENT' of the external obstructions at the reference point, and is a function of the luminance of these obstructions relative to that of the sky which is obscured. The *internally received component* (*IRC*) of the daylight factor is that due to the internal reflecting surfaces.

DAYLIGHT PROTRACTOR A transparent mask which is used in the graphical determination of the daylight factor. The most common pattern is the one devised by the BRS.

db Dry bulb.

dB Decibel.

dBA DECIBEL readings on the A scale.

DBR Division of Building Research, Melbourne or Ottawa.

DBT Dry-bulb temperature.

DC Direct current.

DE STIJL A group of artists and architects, formed in 1917, of whom *Piet Mondrian* is the best known. It published a magazine of the same name, which publicised many avant-garde movements of the 1920s, *e.g.* the manifesto of the CONSTRUCTIVIST International. Some of its architect members later became prominent proponents of the INTERNATIONAL STYLE, notably *J. J. Oud* who also contributed to the 1927 exhibition of the DEUTSCHER WERKBUND.

DEAD-END ANCHORAGE The anchorage opposite to the jacking end of a tendon when POST-TENSIONING is carried out from one end only.

DEAD LIGHT A window which does not open, as opposed to an *open light*.

DEAD LOAD A load which is permanently applied to a structure, and acting at all times, as opposed to a LIVE LOAD.

DEAD SHORE *See* SHORE.

DEBUGGING The process of eliminating errors from a digital computer program.

DECA *See* DEKA.

DECAGON *See* POLYGON.

DECARBURISATION Removal of carbon from the surface of steel.

DECAY (*a*) The decomposition of timber, particularly by fungi. *See also* CORROSION and PRESERVATIVE. (*b*) The damping of an oscillation, particularly a sound wave.

DECAY FACTOR The factor expressing the rate of decay of oscillations in a DAMPED oscillatory system, defined as the natural logarithm of the ratio of two successive amplitude maxima, divided by the time interval between them. It is a measure of the damping, for example, of acoustical RESONANCE.

DECI Latin prefix for 1/10, *e.g.* 1 decimetre = 0·1 m.

DECIBEL The practical unit of sound measurement, equal to 0·1 BEL. It has a (decimal) logarithmic scale; *i.e.* if the sound level increases from X to Y, then the gain is $10 \log_{10} X/Y$ decibels. The decibel as such is not an absolute unit of sound measurement. The scale based on the *threshold of audibility* is the PHON. *Sound level meters* are calibrated in one of three weighted networks. The A-network (*dBA scale*) corresponds to the 40-phon contour of loudness at the various frequencies; this scale has shown the best correlation between pure tones and bands of noise. The B and C networks (dBB and dBC scales) are now rarely used.

DECIDUOUS TREE A tree which loses its leaves in winter. In Northern Europe the term is largely synonymous with hardwood, since all native hardwoods lose their leaves and none of the commercially used softwoods lose their needles in winter. However, most hardwoods native to the tropics and sub-tropics are evergreen.

DECIMAL ARITHMETIC Arithmetic based on the ten digits 0, 1, 2, 3, 4, 5, 6, 7, 8, 9. It originated from counting on the ten digits of one's fingers. Since $10 = 2 \times 5$, this is not the most natural arithmetic system, and DUODECIMAL ARITHMETIC has evident advantages. Electronic digital computers use BINARY ARITHMETIC.

DECIMAL SYSTEM *See* SI UNITS.

DECLINATION *See* AZIMUTH.

DECORATED STYLE A phase of the Gothic style in England between the EARLY ENGLISH and the PERPENDICULAR. It began in the late thirteenth and continued into the second half of the fourteenth century, Wells Cathedral being a typical example. The use of ogee curves, and a maximum of decoration covering the surfaces are characteristic of the DECORATED style, which was popular in the Neo-Gothic Revival of the nineteenth century.

DEFIBRATOR Machine for disintegrating wood into fibres.

DEFINITION In acoustics, the degree to which individual sounds in a musical performance stand apart, one from another. Also called *clarity*.

DEFLECTED TENDON A tendon whose eccentricity, with reference to the CENTROID of the section, varies along the length of the beam. The object of deflecting, or *draping*, the tendon is to set up a bending moment which opposes the moment due to the imposed load, and thus reduces the stresses in prestressed concrete under load. *See also* LOAD BALANCING.

DEFLECTION The flexural deformation of a structural member. Although *elastic* deflection is recoverable, it may damage brittle finishes, such as plaster, if it is excessive. Limits are frequently imposed, $L/325$ and $L/250$ being commonly specified figures (where L is the span). The deflection of concrete and timber increases over a period of time due to CREEP.

DEFLECTOMETER An instrument for measuring deflection. The most common device is a DIAL GAUGE.

DEFORMATION Change of shape. It may be *elastic* (instantly recoverable), *plastic* (permanent), or *viscous* (recoverable after a time-interval).

DEFORMED BAR A reinforcing bar with surface deformations which provide an anchorage with the surrounding concrete, and thus increase the BOND with the concrete. The deformed surface is produced during the hot-rolling of the bars from indentations in the rolls of the steel mill. The deformations must provide sufficient anchorage without setting up excessive stress concentrations in the concrete. Most building regulations permit the doubling of the bond stress, as compared with a plain bar, if the deformations conform to the required standard. *See also* INDENTED WIRE.

DEFORMETER *See* BEGGS' DEFORMETER.

DEGRADATION *See* DEPOLYMERISATION.

DEGREE-DAY A unit employed in estimating the fuel consumption and specifying the heating load for a building in winter. For any one day, when the temperature is below a specified value (usually 59°F (15°C) in Europe and 65°F (18½°C) in America), there exist as many degree-days as the mean temperature for the day is below the specified value. The total for the winter is then compiled.

DEGREE OF COMPACTION The degree of density of a soil sample, given by the ratio $(V_l - V_s)/(V_l - V_d)$, where V_s = voids ratio of sample, V_l = voids ratio of soil in its loosest state, and V_d = voids ratio of soil in its densest state.

DEGREE OF POLYMERISATION Mers per average molecular weight.

DEGREE OF SATURATION A measure of the VOIDS in a soil which are filled with water (the remainder being filled with air). It is the ratio of the volume of water-filled voids to the total volume of voids.

DEGREES OF FREEDOM The number of variables, defining the state of a system, which may be fixed at will.

DEHUMIDIFICATION OF AIR Removal of moisture from the air. In AIR CONDITIONING plants it is accomplished by cooling the air below the DEW POINT, normally with a spray of chilled water, and draining off the condensate. In order to reduce the humidity to the desired level, it is usually necessary to cool the air below the ultimately required temperature and then reheat it. This is one reason why air conditioning is more expensive than heating. For small quantities of air, for example in instruments sensitive to moisture, the water vapour may be removed by an adsorbent, such as SILICA GEL. *See also* REFRIGERATION CYCLE.

DEHUMIDIFIER *See* HUMIDIFIER.

DEKA (*da*) The Greek word for ten. Prefix for 10 times, *e.g.* 1 dekagram (a unit popular in Austria) = 10 grams. Also spelt deca (not to be confused with deci).

DELAMINATION Separation of layers in a laminated assembly, either through failure in the adhesive or through failure at the interface of the adhesive and the lamination.

DELIQUESCENCE The liquefying of certain salts due to their absorption of water from the air. Bricks or plaster containing chlorides may show an appearance of dampness due to this property.

DEMAND FACTOR A factor by which the actual wattage of an electrical installation may be multiplied in designing the wiring system, to allow for the fact that the larger the house, the less likely is it that all lights, etc., will be switched on at the same time. Demand factors are generally specified in electrical codes.

DEMEC STRAIN GAUGE A demountable strain gauge developed by the Cement and Concrete Association in England. It consists of a long bar with one fixed and one movable point which operates a DIAL GAUGE through a lever. The points engage plugs with holes glued to the structure under test. To compensate for temperature, the Demec gauge is compared at intervals with a standard bar made of an alloy with a very low coefficient of thermal expansion. The *Tensotast* made by Dr Huggenberger in Switzerland, the Portuguese LNEC deformeter, and the American Berry gauge are based on the same principle.

DENDRITE A tree-shaped crystal, caused by the tendency of some metals to grow by branches developing from a nucleus. The secondary branches growing from the primary branches produce the dendritic structure.

DENSE CONCRETE Concrete made in the conventional way, as opposed to LIGHTWEIGHT CONCRETE. It generally weighs 120–160 lb/ft^3 (1900–2600 kg/m^3).

DENSITY Weight per unit volume. *See also* SPECIFIC GRAVITY.

DEPOLYMERISATION Reduction of polymers to smaller molecules, also called *degradation*.

DERRICK Originally the gallows at Tyburn in London, named after a sixteenth century hangman. Now a crane consisting of a jib set up obliquely, with its head steadied by guy ropes. There are many different types ranging from small, hand-operated derricks, to large power-operated derrick cranes. *See also* TOWER CRANE and GANTRY CRANE.

DESICCATION Drying, *e.g.* of timber in a kiln.

DESIGN LOAD The working load for which the building is designed.

DESK CALCULATOR A digital computer small enough to be placed on a desk. It may be a hand or electrically operated machine using decimal or binary arithmetic.

DESTRUCTIVE DISTILLATION DISTILLATION of a solid substance, accompanied by its decomposition.

DETECTOR *See* FIRE DETECTOR.

DETERMINANT A square array of quantities, and the sum of the products formed by evaluating it in accordance with certain rules.

DETRUSION Shear strain.

DEUTSCHER WERKBUND A society founded in 1907 by the German architect *Hermann Muthesius*, disbanded with the advent of National Socialism, and revived after the Second World War. Muthesius had come to admire the ARTS AND CRAFTS movement during a stay in England from 1896 to 1903, but unlike *William Morris* he encouraged co-operation with industry, and accepted the economic importance of mass production. As superintendent of the Prussian schools of arts and crafts, he exercised considerable influence over architectural education (*see* BAUHAUS). The Werkbund organised an exhibition of industrial art in Cologne in 1914, for which pioneer modern architects, including *Walter Gropius*, *Adolf Meyer*, and *Bruno Taut* designed some of the earliest glass-walled buildings. The 1927 Werkbund exhibition included a housing exhibition in *Weissenhof*, a suburb of Stuttgart, superintended by *Mies van der Rohe*, to which most well-known exponents of the modern style contributed.

DEVELOPABLE SURFACE A curved surface which can be flattened into a plane surface without shrinking, stretching or tearing. CYLINDRICAL or SINGLY-CURVED surfaces are developable. DOUBLY CURVED surfaces are non-developable. *See also* GAUSSIAN CURVATURE.

DEVIATION The amount by which one observation of a set of observed values differs from their mean. *See also* STANDARD DEVIATION.

DEW POINT The temperature at which condensation of water vapour in the air takes place, *i.e.* the temperature at which the air is fully saturated.

DEW-POINT HYGROMETER An instrument which measures the relative humidity of the atmosphere by determining the dew point. A refrigerating effect is produced on a silvered bulb by evaporating ether inside it, until dew appears on the silvered surface, where it can easily be recognised. The best-known type is the *Regnault hygrometer*.

DEXTRIN A water-soluble gum made from starch, used as an adhesive.

DF Degree of freedom.

DIAGONAL (*a*) A straight line connecting two non-adjacent angles of a quadrilateral, polygon, or polyhedron. (*b*) A strut or tie running at an angle to both the horizontal and the vertical in a lattice girder.

DIAGONAL COMPRESSIVE STRESS One of the principal stresses resulting from the combination of horizontal and vertical SHEAR STRESSES in a beam. Beams with slender webs, such as steel plate girders, must be provided with STIFFENERS to prevent web BUCKLING due to the diagonal compression.

DIAGONAL CRACK An inclined crack caused by the DIAGONAL TENSILE STRESSES in a brittle material, such as concrete. It starts on the tension face and gradually disappears as it passes into the compression zone of the beam. (Fig. 27). Small diagonal cracks are acceptable in concrete; but when the permissible shear stress is exceeded, SHEAR REINFORCEMENT must be provided across the cracks.

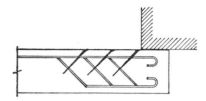

FIG. 27. Diagonal tension cracks and shear reinforcement at the end of a reinforced concrete beam.

DIAGONAL TENSILE STRESS One of the principal stresses resulting from the combination of horizontal and vertical SHEAR STRESSES in a beam or slab. In brittle materials, such as concrete, it causes DIAGONAL CRACKS (Figs. 27 and 76).

DIAL GAUGE An instrument for measuring deflection. It consists of a plunger whose movement is enlarged by a train of gears, operating a rotating pointer. It can readily measure a deflection of 1×10^{-4} in or 1×10^{-3} mm. A dial gauge may be used as a *deflectometer*, as a measuring device in a *proving ring*, etc.

DIAMAGNETIC Pertaining to bodies which are repelled by a magnet.

DIAMETRAL COMPRESSION TEST *See* SPLITTING TENSILE TEST.

DIAMOND *See* CARBON and MOHS' SCALE.

DIAMOND MESH *See* EXPANDED METAL.

DIAMOND PYRAMID HARDNESS An indentation hardness test. A square diamond pyramid is forced by a standard load into the surface of the specimen to be tested, and the diagonal of the indentation produced is measured.

DIAMOND SAW A saw, usually a circular saw about 6 ft (2 m) in diameter, which has industrial diamonds on its cutting edge. It can be used for cutting FREESTONE.

DIAPHRAGM A relatively thin, usually rectangular element of a structural member which is capable of withstanding shear in its plane. It serves to stiffen the structural member.

DIAPHRAGM PUMP A pump in which a flexible partition (or diaphragm) of rubber, leather or canvas is operated by a rod. The diaphragm takes the place of the piston in the usual *reciprocating* pump. The pump is very robust, and it can be used for pumping water containing mud, sand and even small stones.

DIASTYLE *See* INTERCOLUMNIATION.

DIATOMACEOUS EARTH A whitish powder consisting mainly of the frustules of diatoms (which are microscopic plants). It consists of nearly pure hydrous amorphous silica, and is resistant to heat and chemical action. It is used in insulating materials and fireproof cements, in filters, and also as an absorbent in the manufacture of explosives. Nobel's discovery of the 'safe' explosive dynamite consisted of the observation that nitroglycerin is absorbed by diatomaceous earth. Also called *kieselguhr*.

DICALCIUM SILICATE One of the principal components of PORT-LAND CEMENT. Its chemical composition is $2CaO.SiO_2$, or C_2S in the notation used by cement chemists. It is the component named *Belite* by Tornebohm in 1897, before the chemical composition of cement had been properly established.

DIELECTRIC HEATING The process of generating heat at high frequencies in non-conducting materials by placing them in a strong alternating electric field.

DIELECTRIC STRENGTH Electric breakdown potential of an electrical insulator, per unit thickness.

DIFFERENTIAL COEFFICIENT The derivative of a function, and the reverse of its INTEGRAL. Geometrically the differential coefficient is the SLOPE of the curve at the point under consideration.

DIFFERENTIAL EQUATION An equation involving derivatives, *i.e. differential coefficients*. It is of the *first order* if it contains only the first derivatives (*i.e.* dy/dx), of the *second order* if it contains also second, but no higher, derivatives (d^2y/dx^2) etc. The differential equation is *linear* if the derivatives of y are only multiplied by constants or functions of x, but not by functions of y. A *partial* differential equation is one containing more than two variables, *e.g.* x, y and z.

DIFFERENTIAL PULLEY BLOCK A lifting tackle consisting of two pulleys of slightly different diameters and an endless chain. The closer the diameter of the two pulleys, the greater the lifting power of the tackle, and the slower its speed of operation.

DIFFERENTIAL SETTLEMENT Uneven sinking of different parts of a building. It endangers the safety of a building if it causes it to tilt (the Leaning Tower of Pisa is a classical example), or if it places severe strain on some of the structural members. Differential settlement may be due to a clay foundation of variable consolidation characteristics, or to grossly uneven load distribution (*e.g.* due to a tower block) on a compressible clay. Settlement occurs whenever tall buildings are constructed on compressible soils (*e.g.* in Chicago or in Mexico City); however, it does not endanger the structure if the entire building settles uniformly. The lowering of the entrances becomes a nuisance (but not a danger) if even settlement reaches several feet.

DIFFRACTION Deviation of an X-RAY beam by regularly spaced atoms.

DIFFUSER A device used to alter the spatial distribution of the luminous flux from a source by diffusion. A perfectly *matt* surface is one in which the whole of the incident light is redistributed uniformly in all possible directions in such a way that the luminance is the same in all directions.

DIFFUSION (*a*) The movement of atoms or molecules of one material in another without chemical combination. (*b*) Alteration of the spatial distribution of a beam of light which, after reflection at a surface or passage through a medium, travels on in numerous directions.

DIGIT Originally one of the terminal divisions of the hand or foot (*i.e.* a finger or toe); hence one of the first ten numbers (represented by the numerals 0 to 9) which can be counted on the fingers of one's hands.

DIGITAL COMPUTER A computer which works by the addition of digits, as opposed to an ANALOGUE COMPUTER which operates by a continuously varying process. The *abacus* developed by the Chinese 5000 years ago, is a digital computer, as is an ordinary cash register, or a small hand-operated *adding machine*. However, the term is now commonly used for the *electronic digital computer*, a high-speed COMPUTER SYSTEM which uses BITS, or digits converted into binary notation.

DIGITAL PLOTTER A form of output device for a digital computer which presents the result in graphical form.

DIHEDRAL ANGLE The angle of inclination of two meeting or intersecting planes.

DIMENSION (*a*) The measured distance between two points. (*b*) A measure shown on a drawing which is intended to become a precise distance between two points in a building.

DIMENSIONAL ANALYSIS Analysis of structures and other problems by means of dimensional similarity (*see* PI-THEOREM). Some problems can be solved theoretically, others must be solved by MODEL ANALYSIS.

DIMENSIONAL CO-ORDINATION Design of building components to conform to a dimensional standard, usually a MODULE.

DIMENSIONAL STABILITY The term usually implies that a material has little moisture movement and creep, since thermal and elastic deformation are unavoidable.

DIMETRIC PROJECTION *See* PROJECTION.

DIMMER *See* THEATRE DIMMER.

DIN Deutsche Industrie Normung (German Standard).

DIORITE An IGNEOUS rock formed by plutonic intrusion; it contains no quartz and is more basic than GRANITE.

DIRECT CURRENT *See* ALTERNATING CURRENT.

DIRECT GLARE GLARE due to a luminous object situated in the same or nearly the same direction as the object viewed. *Indirect glare* is due to a luminous object in some other direction.

DIRECT MODEL ANALYSIS Analysis of structural problems and, to a lesser extent, studies of lighting, ventilation or acoustics, by means of models which are accurately scaled. They may be made from any convenient material, such as plexiglas. In the case of structures, the loads and the deformation measurements must be scaled precisely in accordance with the theory of DIMENSIONAL ANALYSIS (*see Models in Architecture*, Elsevier 1968). Loads are accordingly reduced to a magnitude which can be easily handled in a laboratory. Measurements of deformations are usually made with DIAL GAUGES and with ELECTRIC RESISTANCE STRAIN GAUGES. The rules of dimensional analysis also apply to non-structural models, *e.g.* as an acoustical model is reduced in size, the frequency of the sound must be increased proportionately.

DIRECT MODULUS OF ELASTICITY The MODULUS OF ELASTICITY, as opposed to the MODULUS OF RIGIDITY.

DIRECT STRAIN *See* STRAIN.

DIRECT STRESS *See* STRESS.

DISABILITY GLARE GLARE which impairs the vision without necessarily causing DISCOMFORT.

DISCHINGER, F. *See* JENA PLANETARIUM.

DISCOMFORT GLARE GLARE which causes discomfort without necessarily impairing the vision of objects.

DISCONTINUOUS CONSTRUCTION Construction which avoids continuous sound paths, particularly

through a steel frame, to provide the greatest amount of impact sound insulation with the least amount of insulating material. This sometimes conflicts with the structural requirements of the building, and it may be easier to substitute a more massive concrete structure which at the same time provides local insulation against airborne sounds. *See also* FLOATING FLOOR *and* SOUND INSULATION.

DISLOCATION IN A CRYSTAL An imperfection caused by a failure of the crystal planes in two portions to match. Dislocations are closely related to slipping action. The displacement vector around a dislocation is called the *Burgers vector*. In an *edge dislocation* there is a linear displacement (Fig. 28) accompanied by zones of tension and compression; the Burgers vector is at right angles to the edge dislocation line. In a *screw dislocation* there is a distortion, with associated zones of shear; the Burgers vector is parallel to the dislocation. *See also* BUBBLE MODEL and SPACE LATTICE.

DISPERSION The even distribution of finely divided drops in an EMULSION, *e.g.* an emulsion paint.

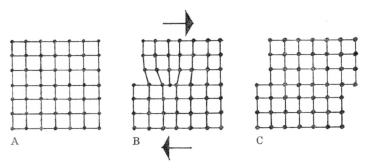

A   B   C

FIG. 28. (A) Perfect crystal; (B) dislocated crystal; (C) sheared, dislocation-free crystal.

DISPLACEMENT *See* BUOYANCY.

DISTEMPER A cheap paint with a binder of casein or some other glue; it usually is heavily pigmented, and thinned with water. *Washable distemper* contains some drying oil. *See also* TEMPERA PAINTING.

DISTILLATION Process of converting a liquid into vapour, condensing the vapour, and collecting the condensed liquid or *distillate*. It is used for separating a mixture of liquids with different boiling points. *See also* DESTRUCTIVE DISTILLATION.

DISTRIBUTION BOX A box, now usually of plastic, which contains a junction between several conduits, and gives access to them.

DISTRIBUTION REINFORCE-MENT Small-diameter bars placed in concrete slabs at right angles to the main reinforcement to spread a concentrated load. It also serves to control *temperature* and *shrinkage* cracking.

DIURNAL Belonging to each day, *e.g.* the *diurnal motion* of the sun, or the *diurnal rhythm* of the human body.

DODECAGON *See* POLYGON.

DODECAHEDRON *See* POLYHE-DRON.

DOLERITE An igneous rock formed by MINOR INTRUSIONS or thick lava flows. It is a fine-grained rock of basic composition.

DOLLY A block of hardwood used to cushion the blow from a PILE HAMMER.

DOLOMITE The double carbonate of calcium and magnesium, $CaMgCO_3$. It is used for REFRACTORY MATERIALS and as a lime for jointless flooring. Dolomitic limestone is also known as *magnesian limestone*.

DOME A vault of double curvature, both curves being convex upwards. Most domes are portions of a sphere; however, it is possible to have a dome of non-spherical curvature on a circular plan, or to have a dome on a non-circular plan, such as an ellipse, an oval, or a rectangle. In classical architecture domes were normally constructed of masonry (*see also* PANTHEON). A spherical dome is subject to hoop tension when the angle subtended at the centre of curvature exceeds 104°; consequently hemispherical domes, which subtend 180°, are subject to HOOP TENSION for the lowest 38°, and this presents problems in masonry construction (Fig. 29). Shallow domes also have tensile stresses in the masonry, since the horizontal component of the arch action has to be absorbed by a *tension ring*, whose elastic deformation is incompatible with that of the dome. In reinforced

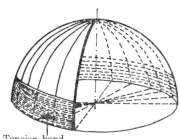

Tension band

FIG. 29. Hoop tension in hemi-spherical dome. (In a membrane shell the forces along the *meridians* are entirely compressive.)

concrete construction the tension is resisted by steel reinforcement, but the construction of spherical form-work from linear timber or steel sections is wasteful, so that the construction of curved roofs has shifted away from domes. *See* CROWN, LANTERN, GEODESIC and SQUARE DOME. *See also* SHELL CONSTRUCTION.

DORIC ORDER The oldest of the three Greek ORDERS, distinguished by a tapered shaft and the absence of a base. Roman Doric, one of the five Roman Orders, has a base, and also generally a more slender shaft.

DORMER WINDOW A vertical window inserted into a sloping roof. It usually has its own gable projecting through the main roof slope.

DORMITORY A large room or building containing a number of beds.

DOUBLE-ACTING ENGINE A RECIPROCATING ENGINE in which the working fluid (*e.g.* steam) acts on each side of the piston alternately, so that each stroke is a working stroke. In *single-acting* engines (which includes most internal combustion engines) only 1 in 2 or 1 in 4 strokes are working strokes.

DOUBLE-ACTING HINGE A hinge which allows a door to swing both ways, through 180°. It usually has a spring to make the door also self-closing.

DOUBLE-ACTING PUMP A reciprocating pump in which both sides of the piston act alternatively, giving two delivery strokes per cycle.

DOUBLE FLEMISH BOND *See* BOND (*a*).

DOUBLE GLAZING Glazing in which two layers of glass are separated by an air space for thermal and acoustic insulation. It has been used traditionally in the very cold climates of Northern and Eastern Europe, and of North-East America, with the windows contained in separate wooden frames. Double glazing is much more common in air-conditioned buildings, where it saves operating costs at the expense of a higher prime cost. The two layers of glass are then placed in one frame, and the space between is sealed. Venetian blinds, if used, may be placed in this space.

DOUBLE-HUNG WINDOW A SASH WINDOW with two vertically sliding sashes, each balanced by a set of sash weights.

DOUBLE-WALLED SHELL A structure with two parallel MEM-BRANES, joined by diaphragms at regular intervals. It can be used for long spans, since the two membranes, separated by the space between, provide bending resistance. Doubling the shell also improves the water-proofing, since any water penetrating the first membrane can be drained off before entering the interior of the building. *See* CNIT EXHIBITION HALL.

DOUBLE WINDOW *See* DOUBLE GLAZING.

DOUBLY CURVED SURFACE A surface curved in both directions, as opposed to a singly curved surface. Doubly curved surfaces are divided into domes, which have positive GAUSSIAN CURVATURE, and saddles which have negative Gaussian curvature.

DOUBLY PRESTRESSED CONCRETE Concrete prestressed in two mutually perpendicular directions. By this means the diagonal tensile stresses due to shear and torsion can be completely eliminated.

DOUBLY REINFORCED CONCRETE Concrete with both tension and COMPRESSION REINFORCEMENT.

DOUBLY RULED *See* RULED SURFACE.

DOUGLAS FIR A softwood widely used in the USA and Canada, and exported in large quantities to Europe and Australia, because it is obtainable in large sizes which are straight-grained and free from knots. It is light (about 33 lb/ft^3 or 530 kg/m^3) and comparatively strong. Also called *British Columbian pine, Oregon pine or just Oregon.*

DOVETAIL An interlocking joint between two pieces of timber, in which the interlocking pins are fan-shaped, like the tails of certain pigeons. They are thicker at the ends than at the root, and are therefore not easily pulled out. Dovetail joints

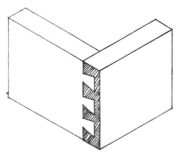

FIG. 30. Lap dovetail joint.

are used for joinery, drawers and boxes (Fig. 30).

DOWEL A pin of wood or metal, used in timber, masonry or concrete structures, usually to resist shear.

DPC Damp-proof course.

DRAFT *See* DRAUGHT.

DRAGON'S BLOOD A red resin, obtained from certain trees, particularly certain palms, which is insoluble in water, but soluble in alcohol and ether. It is used for tinting varnish.

DRAPED TENDONS *See* DEFLECTED TENDONS.

DRAUGHT A pressure difference in a room, and more particularly in a chimney or ventilator, which induces natural air movement. Also spelt *draft.*

DRC Damage risk criterion.

DRESSED LUMBER *See* DRESSED TIMBER.

DRESSED STONE Stone which has been squared all round and smoothed or rusticated on the face.

DRESSED TIMBER Timber which has been machined and surfaced at a timber mill. It is customary to state the nominal size, which is the size of the timber before dressing. The dressed timber is $\frac{3}{16}$ to $\frac{1}{2}$ in (5 to 13 mm) smaller. Also called *dressed lumber.*

DRIER Any compound which encourages oxidation of the drying oil in a paint or varnish.

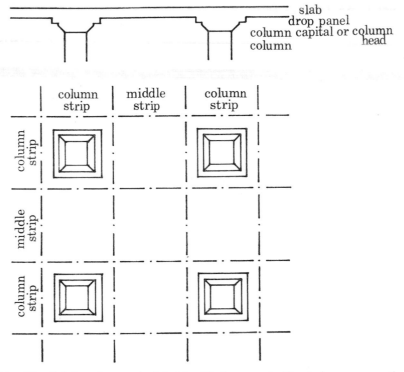

FIG. 31.   Reinforced concrete flat slab with drop panels. For design purposes, the slab is divided equally into column strips and middle strips.

DRIVEN PILE   *See* BORED PILE.

DROP PANEL   The portion of a FLAT SLAB which is thickened throughout the area surrounding the column or column capital, to reduce the magnitude of the shear stress, and thus obviate the need for shear reinforcement (Fig. 31).

DRUM   A vertical wall supporting a dome or cupola.

DRY-BULB THERMOMETER   A normal thermometer, which does not have its bulb wrapped in a damp wick, as opposed to a *wet-bulb thermometer*.

DRY CONSTRUCTION   Building without the use of wet plaster, wet concrete, or mortar on the site. However, the concrete may be precast, or the brickwork prefabricated; indeed a large proportion of dry construction consists of precast concrete units joined with dry fasteners and sealed with mastic. Dry construction avoids damage to finishes through concrete droppings, etc., and it is unnecessary to wait for the wet materials to dry. However, the cost is often higher.

DRY DENSITY The weight of soil or aggregate per unit volume, after it has been dried at 105°C.

DRY ROT Timber decay caused by a fungus, which flourishes only if the timber is *damp*. It is usually caused by inadequate ventilation.

DRY WALL Stones or blocks laid without mortar. Dry stone walls were used in prehistoric times before the invention of mortar, often by wedging small stones into the joints; they are still in use in some country districts, particularly for walls around fields. In recent years, a number of concrete BLOCK types have been designed for precision fitting without the use of mortar.

DRY WEIGHT *See* DRY DENSITY.

DRYING OIL An animal or vegetable oil (the most common being linseed oil) which forms a tough film by oxidation when exposed to air in a thin layer. The process can be speeded up by the use of a DRIER.

DRYING SHRINKAGE Contraction (of timber or concrete) caused by moisture loss.

DRYNESS As an acoustic quality, *see* LIVENESS.

DUCT (*a*) A pipe, conduit, or runway for electric or telephone wires. (*b*) A hole formed in a concrete member to accommodate post-tensioning tendons.

DUCT LINING An absorbent lining placed in an air conditioning duct to attenuate the noise caused by the equipment. Excessive absorption may reduce the BACKGROUND NOISE below the level desirable for speech privacy.

DUCTILITY The property of certain metals which enables them, when cold, to undergo large permanent deformations without rupture, as opposed to brittleness. *See also* PLASTIC MATERIAL.

DUMMY ACTIVITY An arrow in a NETWORK which is used as a logical connector, but does not represent actual work items. It is usually shown as a dotted line (Fig. 67).

DUMMY JOINT A groove cut into concrete where a shrinkage or temperature crack may be expected. The concrete cracks at the neatly cut groove, instead of forming an unsightly random crack.

DUMPY LEVEL A SURVEYOR'S LEVEL in which the telescope and the level tube are attached rigidly to the vertical spindle. Levelling is performed with three (in old instruments sometimes four) levelling screws, and the instrument is then level in any direction. This used to be the most commonly employed instrument. However, the *quickset* level is now becoming more common. In this, the vertical spindle is adjusted approximately, and the instrument is levelled for each sight. If only two readings are required at one station, this is the quicker method.

DUODECIMAL SYSTEM A system based on 12 units, *e.g.* 12 inches = 1 foot. It was used by the Babylonians, and it is arithmetically versatile, since 12 is divisible by 2, 3, 4 and 6. From a mathematical point of view it may be superior to the DECIMAL system (10 being divisible only by 2 and 5), based on counting on ten human fingers, which became firmly established during the Roman Empire. However, since the 'digital' computer (which is really a BINARY calculator) has a keyboard based on the decimal system, duodecimal units are unlikely to survive, except for time and angle measurement.

DUOMO Italian for Dome, particularly that of Florence Cathedral. *See* RENAISSANCE.

DUPLEX A building with two apartments for two separate families.

DURABILITY The ability of a material to resist abrasion, chemical attack, weathering action and other conditions of normal service. *See also* WEATHEROMETER.

DURALUMIN An aluminium alloy containing 3·5 to 5·5 per cent copper, 0·5 to 0·8 per cent of magnesium, 0·5 to 0·7 per cent of manganese, and up to 0·7 per cent of silicon. It is capable of AGE HARDENING at room temperature, and it can be cast, forged and rolled hot or cold.

DWV PIPING Drain, waste and vent piping.

DYE A colouring material which, unlike a PIGMENT, colours materials by penetration.

DYELINE *See* BLUEPRINT.

DYNAMIC LOAD A load which is not *static*, *e.g.* a *live load* due to moving machinery, earthquake, or wind. Dynamic loads are frequently converted into equivalent static loads by means of an IMPACT FACTOR; however, where they exercise a controlling influence on the design of the structure, a dynamic analysis is required. *See* EARTHQUAKE LOADING.

DYNAMIC MODULUS OF ELASTICITY The MODULUS OF ELASTICITY determined from the vibration of a specimen or structure, or from the velocity of a (*e.g.* ULTRASONIC) pulse. The test is NON-DESTRUCTIVE, since the specimen or structure is not destroyed by the test.

DYNAMIC PENETRATION TEST *See* STATIC PENETRATION TEST.

DYNAMIC RANGE The spread of sound levels over which music can be heard in a hall. At one end, normal audience noise should be unobtrusive, and at the other, the loudest levels produced by the orchestra should not cause uncomfortable sensations.

DYNAMICS The branch of the science of mechanics concerned with the action of forces and the motions they produce; as opposed to STATICS, which deals with the case when there is no motion, and the forces are in equilibrium.

DYNAMO A direct-current generator. *See also* ALTERNATOR.

DYNE The force required to produce in one second an acceleration of one centimetre per second in a mass of one gram. In the SI SYSTEM the dyne is replaced by the newton (1 dyne = 10^{-5} N).

E

e The base of the NATURAL LOGA-
RITHM; $e = 2.718\,28\ldots.$

E The common symbol for the
modulus of elasticity.

EARLY ENGLISH The first of
three phases into which English
GOTHIC architecture is conventionally
divided. It retains some of the features
of the NORMAN style.

EARLY WOOD The lighter wood
with thinner cell walls formed during
the earlier stages of the growth of
each GROWTH RING (*also called spring
wood*). The denser wood with thicker
walls, formed during the later stages,
is called *late wood*, or *summer wood*.

EARPHONE An electro-acoustic
TRANSDUCER for audition by one
person only.

EARTH PRESSURE The horizontal pressure exerted by soil on a
retaining wall, or vice versa. The
active, or minimum earth pressure is
that exerted by a retained soil on the
wall retaining it. The *passive*, or
maximum, earth pressure is exerted
by soil in front of a wall sunk in the
ground; it is the resistance offered
by that portion of the soil to the
movement of the wall (which is
pushed by the active pressure due to
the retained soil). *See also* RANKINE
THEORY.

EARTHQUAKE LOADING The
forces exerted on a structure by
earthquakes. Earthquakes occur only
in regions where there are suitable

GEOLOGICAL FAULTS, and many of
these do not pass through centres of
population; however, most of the
American West Coast, Japan, New
Zealand and several Mediterranean
countries are liable to severe earth
tremors. Earthquakes consist of
ground vibrations. The resulting
vertical forces are usually within the
loadbearing capacity of the structure,
but it must be designed to resist the
horizontal earth motion. When the
ground is moved beneath a structure,
the building tends to remain in the
original position because of its
inertia. Earthquake loading may
therefore, for small and medium-sized
buildings, be treated as a static
horizontal force, such as that specified by the Los Angeles City Building
Code of 1953:

$$F = (W_D + kW_L)\frac{0.60}{N + 4.5}$$

where W_D and W_L are the dead loads
and live loads acting on the portion
of the structure under consideration.
N is the number of storeys above the
storey under consideration, and k is
a constant (1.0 for tanks, 0.5 for
warehouses, zero for other buildings).

EARTHQUAKE SCALE *See* MER-
CALLI SCALE and RICHTER SCALE.

EAVES The part of a roof which
projects over a side wall (Fig. 45).

ECCENTRIC LOAD A load which
does not act through the CENTROID.
The *eccentricity* is the distance of the
line of action of the load from the
centroid. *See also* CONCENTRIC COL-
UMN LOAD.

ECHO Reflected sound arriving
more than 70 milliseconds after the

direct sound, which is heard by the human ear as a separate sound. Reflections received after a shorter interval contribute to the apparent loudness of the original sound source, and are not distinguished by the human ear as individual reflections. Echoes are particularly common in rooms with concave *sound mirrors*, such as domes and fan-shaped halls.

ÉCOLE DES BEAUX-ARTS A *School of Fine Art* in Paris, which contains France's principal school of architecture. Established in 1797, it became the most influential school of architecture in the nineteenth century, and its teaching strongly influenced the leading British and American schools. Its traditions have been severely criticised by some modern architects and educators, particularly since 1945, and the term *Beaux-Arts* is now generally intended to imply an out-dated approach to architectural education and design. *See also* CIAM.

ECONOMIC PERCENTAGE OF STEEL A term frequently used for the BALANCED PERCENTAGE OF REINFORCEMENT in a concrete beam. It is a misnomer, since the balanced percentage is not necessarily the economic one. Indeed, a smaller percentage is usually cheaper.

ECPD Engineers' Council for Professional Development, New York.

EDDY *See* STREAMLINE FLOW.

EDDY CURRENTS Currents introduced in surrounding masses of conducting material by circuits carrying alternating currents. They can result in a considerable loss of energy.

EDGE BEAM A beam at the edge of a shell or plate structure. The stiffness which it provides may greatly increase the load-bearing capacity of the structure. However, it may also complicate the BOUNDARY CONDITIONS. While some edge beams merely add to the stiffness and strength of the structure (often at the cost of making it appear unduly heavy), others are essential for stability. For example, the *transverse frames* of a cylindrical shell cannot be omitted, lest the shell collapse (Fig. 32).

FIG. 32. Cylindrical shells require transverse frames along the curved edges for stability; beams along the straight edges stiffen the shell, but are not essential.

EDGE DISLOCATION *See* DISLOCATION IN A CRYSTAL.

EDGE RESTRAINTS *See* BOUNDARY CONDITIONS.

EFFECTIVE DEPTH OF REINFORCED CONCRETE BEAM OR SLAB The distance of the *centroid* of the reinforcing steel from the compression face of the concrete (Fig. 33).

FIG. 33.   Effective depth of a reinforced concrete slab.

EFFECTIVE FLANGE WIDTH (*a*) The width of the concrete slab adjoining the rib, which is assumed to function as the flange element of a T-BEAM or L-BEAM section. (*b*) The width of plate adjoining the web, which is assumed to function as the flange element of a light-gauge metal section. *See also* STRESSED-SKIN CONSTRUCTION.

EFFECTIVE LENGTH (OR HEIGHT) OF COLUMN The distance between the points of inflection of a column when it buckles. It is the length used in the SLENDERNESS RATIO.

EFFECTIVE MODULUS OF ELASTICITY OF CONCRETE The deformation of concrete under load is partly due to (instantaneous) *elastic* deformation, and partly due to (time-dependent) *creep*. To simplify calculations, an effective modulus of elasticity is introduced which is the secant modulus in Fig. 83.

EFFECTIVE PRESTRESS The stress remaining in the concrete due to PRESTRESSING after all LOSSES have occurred. It normally includes the effect of the self-weight of the prestressed member, but excludes the effect of the superimposed dead loads.

EFFECTIVE SOUND PRESSURE The root mean square value of the pressure of a sound wave.

EFFECTIVE SPAN The span used in computing the bending moment in a beam. A common rule is to take the lesser of (1) the distance between the centres of the supports; and (2) the clear distance between the supports, plus the depth of the beam or slab.

EFFECTIVE TEMPERATURE The most commonly used criterion for determining the COMFORT ZONE, evolved by Houghton and Yaglou in 1924, and adopted with slight modifications by ASHRAE. It takes account of temperature, humidity and air movement, but ignores radiation. The effective temperature of an environment connotes that temperature of still air, saturated with water vapour, in which an equivalent sensation of warmth was experienced by the subjects of a long series of experiments, carried out in the laboratory of the ASHVE in Pittsburgh. *See also* EQUIVALENT TEMPERATURE.

EFFECTIVE WIDTH OF FLANGE OR SLAB *See* EFFECTIVE FLANGE WIDTH.

EFFICIENCY *See* MECHANICAL ADVANTAGE.

EFFLORESCENCE A deposit, usually white, formed on the surface of brick, block or concrete wall. It consists of salts leached from the surface of the wall. Although unsightly, it is harmless, and it can normally be removed by brushing. However, it is liable to lift any paint which has been applied to the area of wall where it occurs.

EFT Earliest finish time (of an activity in a NETWORK).

EGGSHELL GLOSS *See* GLOSS.

EIFFEL TOWER A 322 m (1056 ft) high tower, built for the Paris Exhibition of 1889 by Gustave Eiffel, a French engineer who also designed the ironwork of the Bon Marché store in Paris, of the Statue of Liberty in New York, and of many large bridges. The Eiffel Tower held the height record for 40 years, until the construction of the Chrysler Building in New York.

ELASTIC CONSTANTS The MODULUS OF ELASTICITY and POISSON'S RATIO.

ELASTIC DESIGN Design based on the assumption that structural materials behave elastically, and that the stresses therein should be as close as possible to, but not greater than, the MAXIMUM PERMISSIBLE STRESSES under the action of the WORKING LOADS. Also known as *working load design*. The main alternative approach is called *plastic design*, *ultimate strength design*, or LIMIT DESIGN.

ELASTIC LIMIT The limit of stress beyond which the strain is not wholly recoverable. In structural steel, the elastic limit is slightly lower than the YIELD POINT. (Fig. 93). In concrete, it is difficult to determine its precise location, because some CREEP occurs even under rapid loading.

ELASTIC LOSS OF PRESTRESS The LOSS OF PRESTRESS which is due to the elastic shortening of the concrete. It need be considered only for PRE-TENSIONED units. When members are POST-TENSIONED the elastic shortening occurs during the operation, and is automatically compensated.

ELASTIC MODULUS *See* MODULUS OF ELASTICITY.

ELASTIC SHORTENING *See* ELASTIC LOSS.

ELASTIC STABILITY *See* BUCKLE.

ELASTIC STRAIN *See* STRAIN.

ELASTICITY The ability of a material to deform instantly under load, and to recover its original shape instantly when the load is removed. *See also* HOOKE'S LAW, CREEP and PLASTIC DEFORMATION.

ELASTOMER A synthetic material with rubber-like qualities, which result from its coiled molecular structure.

ELBOW A sharp corner in a pipe or conduit, as opposed to a *bend*, which has a larger radius of curvature.

ELECTRIC ARC WELDING *See* ARC WELDING.

ELECTRIC EYE A *photoelectric* cell used as a detector.

ELECTRIC RESISTANCE STRAIN GAUGE A STRAIN GAUGE based on OHM'S LAW. It consists of a zig-zag wire (Fig. 34) or a zig-zag thin FOIL, attached to a paper or lacquer backing. This is glued to the part of the structure to be examined. Tensile strain causes the wire to elongate, become thinner, and consequently pass less current. Compressive strain causes an increase in current. The current is measured with a WHEATSTONE BRIDGE (Fig. 102), or some other suitable circuit. With careful

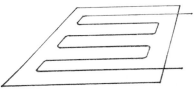

FIG. 34.   Electric resistance wire strain gauge.

choice of wire and adhesive, strains as low as 1×10^{-6} can be determined. *See also* STRAIN ROSETTE.

ELECTRICAL CURRENT *See* AMPERE.

ELECTRICAL INSULATION *See* INSULATION, OHM'S LAW and SEMI-CONDUCTOR.

ELECTRICAL RESISTANCE *See* OHM'S LAW.

ELECTROCHEMICAL CORRO-SION Corrosion caused by the gradual solution of the anode, when two electrochemically dissimilar metals are in contact in the presence of moisture. These then form a galvanic cell, which generates a direct current. Aluminium sheet used externally is subject to such damage if it is fixed with copper nails, or to a lesser extent with steel nails. Steel nails or bolts when used with aluminium should be plated with *cadmium* or *zinc*.

ELECTROCHEMICAL SERIES The sequence of elements in order of the electrode potential developed when immersed in a solution of normal ionic concentration. For commonly used metals the series is: (*cathodic end*) gold, platinum, silver, mercury, copper, hydrogen (*zero*), lead, tin, nickel, cadmium, iron, chromium, zinc, aluminium, magnesium, sodium, calcium, potassium, lithium (*anodic end*). The use of dissimilar metals in contact in the presence of water may lead to corrosion. *See also* NOBLE METALS.

ELECTRODE (*a*) The terminal through which a current is led into and out of an ELECTROLYTE. (*b*) A rod of metal, either bare or covered, used as a terminal in ARC WELDING.

ELECTRODE POTENTIAL *See* ELECTROCHEMICAL SERIES.

ELECTRODEPOSITION The deposition of a layer of metal or alloy on to another in an electrolyte consisting of a solution of salts of the metal to be deposited. *See* GALVANISING.

ELECTROLUMINESCENT PANEL An artificial light source consisting of a phosphor (usually zinc sulphide) contained in a sandwich panel, and excited by an alternating current. It can be used to illuminate large surfaces at a low, uniform brightness in circumstances where the spatial requirements of the lighting system must be kept to a minimum; however, it is more expensive than filament lamps or fluorescent tubes.

ELECTROLYTE A conducting medium (usually a solution) in which electric current flows by virtue of chemical changes. The process is known as *electrolysis*. *See also* ELECTROCHEMICAL SERIES.

ELECTROMAGNET *See* MAGNET STEEL.

ELECTROMOTIVE FORCE *See* VOLT.

ELECTRON (*a*) The Greek word for an *alloy* consisting of four parts of gold and one part of silver. This was widely used by the ancient Greeks, partly because they found it difficult to produce a purer gold (modern 'pure gold' is generally 22 carat, *i.e.* 11/12 pure). In nature gold commonly occurs alloyed with silver. (*b*) The Greek word for *amber*, which was thought to present an appearance similar to that of electron metal. In the sixteenth century Dr Gilbert, an English natural philosopher, coined the term *electricity* to denote attractions between certain bodies, including amber, after rubbing in a certain way. (*c*) Sub-atomic particle having a mass of 9.04×10^{-28} grams, *i.e.* 1/1840 that of the PROTON or NEUTRON, and bearing a negative electric charge. An electric current consists basically of the flow of electrons from the cathode to the anode.

ELECTRON GUN The assembly of electrodes in a CATHODE-RAY TUBE which produces the electron beam. It comprises a cathode from which electrons are emitted, an apertured anode, and one or more focusing electrodes. Those electrons which pass through the aperture form the beam.

ELECTRONIC DIGITAL COMPUTER *See* DIGITAL COMPUTER.

ELECTRO-OSMOSIS Flow of liquid through the pores of a membrane due to a difference of electric potential. The term is also used for WELL-POINT DEWATERING assisted by the passage of an electric current. When a direct current is passed through water-logged silt, water flows to the cathode. Cathodes are placed in the silt at intervals of about 30 ft (10 m), and water is pumped away from there; anodes are placed at intermediate points.

ELEMENT (*a*) A substance consisting entirely of atoms of the same atomic number. (*b*) The resistance wire constituting the heating unit of

an electric heater. (*c*) One of the four elements of Aristotle (earth, water, air, and fire), to which reference was commonly made in classical architectural and proto-scientific texts.

ELEVATION *See* PROJECTION.

ELEVATOR A cage or platform for the vertical transportation of persons or goods. Called a *lift* in Great Britain and in Australia. *See* SKY-LOBBY.

ELIZABETHAN ARCHITECTURE *See* TUDOR ARCHITECTURE.

ELLIPSE A closed loop obtained by cutting a right circular cone by a plane, if the whole of the section lies on one side of the vertex of the cone. It is also the locus of a point such that the sum of the distances of the point from two fixed points, or *foci*, is constant. Consequently an ellipse can be drawn by pinning the ends of a string at the foci, and running a pencil around with the string tightly stretched (Fig. 35). The greatest and

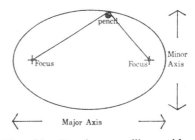

FIG. 35. Drawing an ellipse with a piece of string, pinned at the foci.

smallest diameters are the *major* and the *minor axes*. The equation of the ellipse is of the form $x^2/a + y^2/b = 1$, where a and b are constants. *See also* OVAL.

ELLIPSE OF STRESS *See* ELLIP-
SOID OF STRESS.

ELLIPSOID *See* SURFACE OF
REVOLUTION.

ELLIPSOID OF STRESS An ellipsoid representing the state of stress
at a given point in a body. Its semi-
axes are the vectors representing the
PRINCIPAL STRESSES at the point, and
any radius vector represents the
resultant stress on a particular plane
through the point. When one of the
principal stresses is nil (which includes
the *plane-stress* condition), the ellipsoid reduces to an *ellipse. See also*
MOHR CIRCLE.

ELLIPTICAL DOME *See* DOME.

ELLIPTICAL PARABOLOID *See*
SURFACE OF TRANSLATION.

ELONGATION *See* STRAIN.

ELUTRIATION A method of
separating grain sizes according to
the velocities at which they sink in a
liquid. *See* STOKES' LAW.

EMBOSSED Ornamental designs
or figures in RELIEF, *i.e.* raised above
the surface.

EMERY An impure form of
CORUNDUM. It consists mainly of
alumina and iron oxide.

EMF Electromotive force. *See* VOLT.

EMISSIVITY The ratio of the rate
of loss of heat per unit area of a
surface at a given temperature to the
rate of loss of heat per unit area of a
BLACK BODY at the same temperature
and with the same surroundings.

EMPIRE STATE BUILDING A
102 storey building, erected in New
York City in 1931 (structural design
by H. G. Balcom). Its height is 1250
ft (380 m), and it held the record as
the world's tallest building for 40
years.

EMULSIFIER A substance which
modifies the surface tension of
colloidal droplets, keeping them
suspended, and keeping them from
coalescing.

EMULSION A *colloidal* suspension
of one liquid in another. *See also*
BREAKDOWN.

EMULSION PAINT *See* LATEX.

ENAMEL *See* VITREOUS ENAMEL.

ENAMEL PAINT Hard-gloss paint
which contains a high proportion of
varnish, and consequently has less
pigment. For that reason it requires
one or two undercoats.

ENAMELLED BRICK A glazed
brick.

ENCASTRÉ Built in, and consequently restrained from rotation.
Also called *encastered.*

ENCAUSTIC TILES Tiles whose
coloured decoration has been fixed
by the application of heat. They were
much used in the Middle Ages, and
subsequently in Neo-Gothic churches.

END-BEARING PILE *See* BEARING
PILE.

END BLOCK A section at the end
of a POST-TENSIONED member, which
has been enlarged to reduce the
bearing stresses to permissible values.

END GRAIN The face of a piece of timber exposed when the fibres are cut transversely. It easily deteriorates when exposed unprotected to the weather.

END JOINT Joint formed between the ends of two pieces of material which are in line. The term is commonly used for timber in preference to BUTT JOINT.

END-LAP JOINT Joint formed between the ends of two pieces of timber, normally placed at right angles. Each piece is halved for a distance equal to the width of the other piece, so that the surfaces are flush in the assembled joint. Also called a *right-angled half-lap joint*.

ENDOTHERMIC REACTION A reaction in which heat is absorbed, as opposed to an *exothermic* reaction.

ENDURANCE LIMIT *See* FATIGUE STRENGTH.

ENERGY The capacity of a body for doing work. In classical physics, *mechanical energy* is divided into *kinetic energy*, due to its motion, and *potential energy*, due to its position. The latter consequently includes the gravitational HEAD of water and the STRAIN ENERGY stored in a stressed material. Mechanical energy can be converted into heat, sound, electrical energy, and chemical energy.

ENGAGED COLUMN A column partly embedded in a wall.

ENGLISH BOND *See* BOND (*a*).

ENSEMBLE The ability of the performers at a concert to hear one another so that they can play in unison. If the stage or orchestra pit is very wide and shallow, the two sides of an orchestra may not be able to hear each other.

ENTABLATURE In classical architecture, that part of the building which rests horizontally upon the columns. It is divided into the *cornice*, the *frieze* and the *architrave*.

ENTASIS An almost imperceptible swelling given to Greek and later classical columns to correct the optical illusion of concavity which would result if the sides were actually straight.

ENTHALPY The heat content per unit mass; *e.g.* the change in enthalpy of boiling water is the latent heat of the water.

ENTRAINED AIR *See* AIR-ENTRAINING AGENT.

ENTRAPPED AIR Air voids in concrete which are not purposely entrained. *See* AIR-ENTRAINING AGENT.

ENTROPY A name coined in 1865 by the German physicist R. Clausius for one of the quantitative elements of THERMODYNAMICS. He explained it as follows: 'A portion of matter at uniform temperature retains its entropy unchanged so long as no heat passes to or from it; but if it receives a quantity of heat without change of temperature, the entropy is increased by an amount of heat equal to the ratio of the mechanical equivalent of the quantity of the heat to the absolute measure of the temperature on the thermodynamic scale. The entropy of a system . . . is always increased by any transport of heat within the system; hence the entropy of the universe tends to a maximum.'

ENVELOPE The line formed by a series of common tangents to a family of related curves (*see* Fig. 36).

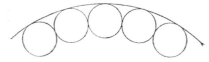

FIG. 36. Envelope.

EPHEMERIS A table giving the computed position of a heavenly body. *See* NAUTICAL ALMANAC.

EPICYCLOID A curve generated by a point on a circle rolling on another circle. It is particularly useful for the design of smoothly operating gears. *See also* CYCLOID.

EPOXY RESIN A group of thermosetting plastics based on the epoxide grouping

$$
\begin{array}{c}
O \\
/ \quad \backslash \\
-CH_2 - - - CH-
\end{array}
$$

The uncured resin consists of short-chain polymer molecules with an epoxide group at either end. The epoxide group is very reactive, and many substances can be used as HARDENERS. The uncured resin and the hardener are kept separately and mixed just before use. Epoxy resins are appreciably more expensive than the *polyesters*, but they are better suited for many applications. They adhere strongly to metals, glass, concrete, stone and rubber, and thus make good glues, and solvents for paints. They are resistant to abrasion, weather, acids and alkalis, and to heat up to 100°C (212°F). In particu-

lar, they have been found useful for repairing damaged concrete, and for joining new concrete to old; for this purpose they can be mixed with sand into an epoxy resin mortar.

EQ Symbol for = (equal) in many ALPHANUMERIC printouts.

EQUATOR *See* GEODETIC LINE.

EQUATORIAL COMFORT INDEX Criterion for determining the COMFORT ZONE, evolved by Webb in Singapore in 1952, also known as the *Singapore index*. It takes account of wet-bulb and dry-bulb temperature, and air movement, but ignores radiation. Its application is confined to the conditions which exist in the hot-humid equatorial zone, and it is intended for indoor conditions for people acclimatised to wet-bulb temperatures above 75°F (24°C).

EQUILATERAL Having all sides equal. *See also* ISOSCELES.

EQUILIBRANT A force required to keep an unbalanced system of forces in equilibrium.

EQUILIBRIUM *See* STATICS.

EQUILIBRIUM DIAGRAM *See* PHASE DIAGRAM.

EQUILIBRIUM MOISTURE CONTENT OF TIMBER The moisture content in the wood which balances that in the atmosphere, so that it neither gives off nor takes in any moisture from the surrounding air. It is expressed as a percentage of the oven-dry weight of the wood.

EQUILIBRIUM OF FLOATING BODIES *See* BUOYANCY.

EQUINOX The 21st of March and the 23rd of September, when the length of day equals the length of night, due to the fact that the sun crosses the celestial equator. *See also* SOLSTICE.

EQUIVALENT TEMPERATURE Criterion for determining the COMFORT ZONE, evolved by Dufton in 1929. It takes account of temperature, air movement and radiation, but ignores humidity. Dufton in 1932 devised the *eupatheoscope* for assessing the thermal environment in terms of equivalent temperature for research at the British Building Research Station. It consisted of a black, electrically heated cylinder, designed to have the same surface temperature as a clothed human under comfortable conditions; the heat input required for comfort was measured. *See also* EFFECTIVE TEMPERATURE.

ERC *See* DAYLIGHT FACTOR.

ERG The unit of energy in CGS UNITS. It is the work done when a force of one dyne moves one centimetre in the direction of the force. In SI UNITS the erg has been replaced by the joule, which equals 10^7 erg.

ERGONOMICS The study of the interaction between work and people: *e.g.* the design of chairs and tables to permit the performance of tasks with the least possible fatigue. *See also* ANTHROPOMETRY.

ERW Electric resistance welding.

ESCALATOR A moving staircase, consisting of an endless belt carrying a series of steps. It can be used for upward or downward movement; however, an escalator is frequently used for transporting people up, while a conventional stationary stair may be provided for downward movement.

ESQUILLAN, N. *See* CNIT EXHIBITION HALL.

ETCHING (*a*) Revealing the structure of metal by selective chemical attack. In PERLITE, for example, the ferrite lamellae are electro-positive to the cementite, so that they can be shown up by etching. (*b*) The process of biting lines into a metal plate by means of acid, and producing a picture from the plate. *Also* the picture produced by this process.

EUCALYPT An evergreen species of tree which sheds its bark. Many Australian hardwoods, with names such as white ash and red mahogony, are in fact eucalypts. Also called *gum tree*.

EULER FORMULA A formula for the BUCKLING of slender columns, published by the French mathematician Leonard Euler in 1757. The buckling load of a column (Fig. 77)

$$P = \frac{\pi^2 EI}{L^2}$$

where π = circular constant, 3·1416, E = modulus of elasticity, I = second moment of area, L = effective length.

Note that P is independent of the strength of material so that the buckling load is the same for low-strength and high-strength steel because the elastic modulus is the same. However, it is much lower for aluminium than for steel, because its elastic modulus is only about one

third. *See also* SHORT-COLUMN FOR-
MULA, RANKINE COLUMN FORMULA,
PERRY-ROBERTSON FORMULA, and
SECANT COLUMN FORMULA.

EUPATHEOSCOPE *See* EQUIVA-
LENT TEMPERATURE.

EUSTYLE *See* INTERCOLUMNIA-
TION.

EUTECTIC The alloy with the
lowest melting temperature in its
range of composition. It is readily
apparent on the PHASE DIAGRAM as a
point of intersection between two
descending LIQUIDUS curves in a
binary system, or three descending
liquidus curves in a ternary system.
The eutectic thus solidifies out of the
liquid as a mixture with a definite
composition.

EUTECTOID A mixture of two or
more constituents which forms on
cooling from a *solid solution*, and
transforms again on heating. PERLITE
is a typical eutectoid. The process is
similar to the formation of a EUTEC-
TIC, which solidifies from the liquid,
or melt.

EVAPORATION Conversion of a
liquid into a vapour, without necessarily reaching the boiling point.

EVAPORATIVE COOLING A
simple method of air conditioning
which can be employed in hot-arid
climates. The latent heat of the water
evaporated is absorbed from the hot
dry air, which is thus cooled. The
humidity is increased by the evaporation of the water.

EVENT A point in time representing the intersection of two ARROWS
in a network, *i.e.* the start or finish of

an activity (Fig. 67). An event has no
time duration.

EXACT NUMBER *See* APPROXI-
MATE NUMBER.

EXOTHERMIC REACTION One
which occurs with the evolution of
heat, as opposed to an *endothermic*
reaction.

EXPANDED CLAY *See* LIGHT-
WEIGHT AGGREGATE.

EXPANDED METAL Metal network formed from sheet metal by
cutting a pattern of slits, followed by
pulling the metal into a diamond
pattern. It is used as a metal *lath*, as
concrete reinforcement, and for the
making of screens. Also called
diamond mesh.

EXPANDED PLASTIC A very
light insulating material, obtainable
as a loose fill, in sheets or in blocks.
Expanded polystyrene, for example,
can be expanded 40 times by foaming.
The density of polystyrene or *rigid
polyurethane* foam is from 1 to 4 lb/
ft^3 (16 to 64 kg/m^3), its compressive
strength is between 15 and 60 psi (100
to 410 kN/m^2), and its thermal
conductivity is between 0·1 and 0·3
BThU-in/h/ft^2/°F (0·014–0·043 W/m/
°C). Also called CELLULAR PLASTIC,
FOAMED PLASTIC or RIGID FOAM.

EXPANDED CEMENT *See* EXPAN-
SIVE CEMENT.

EXPANSION BOLT An anchor
into masonry which consists of a bolt
operating inside a split cone. As the
bolt is turned, the cone expands and
wedges into the hole.

EXPANSION JOINT Separation between adjoining parts to allow for small relative movements, such as those caused by temperature change. *See also* CONTRACTION JOINT.

EXPANSION SLEEVE A tube covering a dowel bar, to allow its free longitudinal movement at a joint.

EXPANSIVE CEMENT A cement which, on setting, expands instead of shrinking. Depending on the amount of additive, the expansion may be merely sufficient to counter the shrinkage of PORTLAND CEMENT, or it may be sufficient to induce tensile stresses in the reinforcement, and thus POST-TENSION the concrete. The additives most commonly used to produce expansive cements are calcium sulphate ($CaSO_4$), free lime (CaO), and anhydrous alumino-sulphate ($4CaO.3Al_2O_3.SO_3$).

EXPERIMENTAL STRESS ANALYSIS *See* MODEL ANALYSIS, PHOTO-ELASTICITY, BRITTLE COATING, MOIRÉ FRINGES and ANALOGUE COMPUTER.

EXPLOSIVE RIVET A blind fastener (*i.e.* one usable in inaccessible locations) with a hollow shank which contains an explosive charge. The rivet shank is expanded by exploding the charge with a hammer blow after the rivet has been inserted. It is used particularly in aluminium structures.

EXPONENTIAL FUNCTION A function of the type $y = e^x$ where e is the base of the NATURAL LOGARITHM. It is thus the reverse of the natural logarithm. Exponential functions occur as a result of integration of terms containing $1/x$, and also in HYPERBOLIC FUNCTIONS. *See also* Appendix D.

EXPOSED AGGREGATE A decorative finish for concrete. The aggregate may be exposed by removing the outer skin of cement mortar from the surface before it has hardened, or the 'exposed' aggregate may be sprinkled on the wet concrete after placing.

EXPOSURE METER A device for estimating the exposure required for photography. It usually incorporates a PHOTOELECTRIC CELL.

EXTENSIBILITY The maximum tensile strain of which a material is capable. It is much lower in BRITTLE, as compared with DUCTILE, materials.

EXTENSOMETER *See* STRAIN GAUGE.

EXTERNALLY RECEIVED COMPONENT *See* DAYLIGHT FACTOR.

EXTRADOS The outer curve of an arch.

EXTRAPOLATE *See* INTERPOLATE.

EXTRUSION Producing a linear shape by pushing material through a die (Fig. 37). The process is used for the manufacture of sections in aluminium, as an alternative to hot-rolling. It is also used for plastics.

EYE BOLT A bolt with a ring forged on one end, or welded to one end.

EYE OF A DOME An opening at the top of a dome to admit light. It may be glazed or open to the sky. *See also* LANTERN.

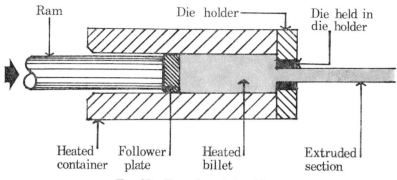

Ram   Die holder   Die held in die holder

Heated container   Follower plate   Heated billet   Extruded section

FIG. 37. Extrusion of aluminium.

F

φ   *See* PHI.

°F   *Degree Fahrenheit.* The scale was originally intended to run from 0° for the freezing temperature of a mixture of water and common salt to 100° for the blood temperature in the human body. However, it is now defined by the freezing point of water (+32°) and the boiling point of water (212°). To convert to the Celsius scale (°C) we subtract 32° and then divide by 1·8.

FABRIC   *See* MESH REINFORCEMENT.

FACE BRICK   *See* BRICK and SAND-FACED BRICK. In England frequently called a *facing brick*.

FACE-CENTRED CUBIC LAT-TICE   A rystal structure which has an atom at each corner of a cube and one at the centre of each face. It may be imitated by packing spheres to line up both vertically and horizontally.

FACIA   *See* FASCIA.

FACTOR OF SAFETY   Factor used in ELASTIC DESIGN to provide a margin of safety against collapse and serious structural damage. It includes an allowance for inaccurate assumptions in the loading conditions, for inadequate control over the quality of materials, for imperfections in workmanship, and for minor approximations made in the structural theory. It does not allow for arithmetical errors. The factor is usually laid down in the building code, and it depends on the control which can be exercised over design, materials and construction, and over the use to which the building will be put. In the design of buildings it currently ranges from 1·6 to 3·0. Much lower factors are used in the structural design of aircraft, where stricter control is possible and necessary. *See also* MAXIMUM PERMISSIBLE STRESS, *and* LOAD FACTOR.

FAHRENHEIT SCALE   *See* °F.

FAIENCE   Originally the French word for *porzellana di Faenza*, a fine painted and glazed earthenware made in Faenza in Italy. In buildings, prior to the twentieth century, the term generally denoted glazed TERRA

COTTA, which had been fired once without and once with the glaze. The term is now used for any decorative glazed tiles and even for glazed plastic floor mosaic.

FAILURE The condition when a structure or material ceases to fulfil its required purpose. The failure of a structural member may be caused by PLASTIC DEFORMATION, FRACTURE, or excessive DEFLECTION. The non-structural failure of a material may be due to *weathering, abrasion* or *chemical action.*

FAILURE CRITERIA Theories proposed for the prediction of the *ultimate strength* of structural materials under the action of combined stresses. The most important are the RANKINE CRITERION, COULOMB'S EQUATION, the MOHR CRITERION, the COWAN CRITERION, and the HENCKY-VON MISES CRITERION.

FALSE CEILING A ceiling *suspended* or *hung* from the floor above which hides its SOFFIT, and provides a space for cables and ducts.

FALSE HEADER A half-brick, which completes the visible bond, but is not a header.

FALSE SET The rapid development of rigidity in freshly mixed cement paste, mortar or concrete, with little evolution of heat. Plasticity can be regained by further mixing without addition of water. *See* SETTING and FLASH SET.

FALSEWORK A temporary structure erected to support work during construction, and subsequently removed.

FAN *See* CENTRIFUGAL FAN and BOOSTER FAN.

FAN NOISE The noise caused by fans used in ventilating and air conditioning equipment. Axial-flow fans have an approximately uniform sound spectrum, but the sound power level of CENTRIFUGAL FANS decreases at the higher frequencies. *See also* DUCT LINING.

FAN VAULT A vault covered with complicated decorative tracery. It was frequently used in the *Perpendicular Gothic* style.

FANLIGHT Originally a fan-shaped window, with sash bars radiating like the ribs of a fan, located over a door. It was much used in Georgian and Colonial architecture, and in its revivals. The term is now applied to any window located over a door.

FAO Finish all over.

FARAD (F) Unit of electrical capacitance, named after Michael Faraday, an English scientist who discovered the laws of electromagnetic induction, and first produced electromotive power in 1831. The capacitance is 1F if a capacitor is charged with 1 coulomb, and the difference of potential between the plates is 1 volt.

FASCIA A plain horizontal band on the surface of a building. In classical design, fascias were employed both in the CORNICE and in the ARCHITRAVE. The term is now used for the exposed EAVE of a building. Also spelt *facia.*

FASCIA BOARD In timber construction, a wide board fixed to the wall, the wall plate, or the ends of the RAFTERS. It usually carries the gutter.

FAT MORTAR A mortar which sticks to the trowel, as opposed to a *lean mortar*.

FATIGUE The tendency of materials to fracture under many repetitions of a stress considerably lower than the ultimate static strength. The term *fatigue* is reserved for a large number of repetitions, while *repeated loading* is commonly used for a small number of load repetitions.

FATIGUE STRENGTH The greatest stress which can be sustained without failure for a given number of stress cycles. Also called *endurance limit*.

FAULT *See* GEOLOGICAL FAULT.

Fe Chemical symbol for *iron* (ferrum).

FEATHER EDGE A fine taper.

FELSPAR A group of minerals consisting mainly of aluminium silicates. They are contained in granite and other rocks, and decompose into clay. Also spelt *feldspar*.

FELT *See* ROOFING FELT.

FELTING *See* AIR FELTING.

FENESTRATION The arrangement of the windows and other openings on the walls of a building, particularly with reference to the appearance of the façade.

FERRITE A solid solution in which ALPHA IRON, as distinct from GAMMA IRON, is the solvent. It may contain in solid solution up to 30 per cent of chromium and up to 15 per cent of silicon, but no more than 0·03 per cent of carbon. Ferrite is the principal constituent of low carbon steels and of many alloy steels.

FERROCEMENTO Concrete made with several layers of finely divided reinforcement, instead of the conventional larger bars. The material was invented and has been used almost exclusively by *P. L. Nervi*.

FERROCONCRETE An old-fashioned term for reinforced concrete.

FIBER *See* FIBRE.

FIBONACCI SERIES An infinite series, credited to the twelfth century Italian mathematician Leonardo Fibonacci of Pisa, in which successive pairs of numbers are added together to form the next number. Thus the simplest Fibonacci series is 1, 2, 3, 5, 8, 13, 21, 34, . . . The intervals increase rapidly, and it has therefore been suggested by Le Corbusier in the MODULOR and later by Ezra Ehrenkrantz as a basis for a system of preferred dimensions in MODULAR CO-ORDINATION. The Fibonacci series was used in the THEORY OF PROPORTIONS in conjunction with the GOLDEN SECTION.

FIBRE BOARD Building board made from felted wood or other fibres, and a suitable binder. It is a generic term which includes ACOUSTIC BOARD, FLAXBOARD, HARDBOARD, INORGANIC FIBRE BOARD, INSULATING BOARD, PARTICLE BOARD, PEG BOARD and SOFTBOARD. Some of these terms overlap. *See also* BAGASSE *and* CULLS.

FIBRE STRESS A term used to denote the longitudinal direct (tensile or compressive) stress in a beam. The *extreme fibre stress* is the stress in the fibre most remote from the NEUTRAL AXIS, and it is thus the maximum tensile or compressive stress. *See* NAVIER'S THEOREM.

FIBREGLASS Strictly the fibre made from glass. However, the term also denotes glassfibre-reinforced synthetic resin (commonly *polyester*). The fibres are finely distributed in the plastic as reinforcement, either as a woven fabric or as a random mat, and the material bears some resemblance to FERROCEMENTO. It has been used for corrugated roof sheets, (transparent or opaque), canopies, radomes, boat hulls, car bodies, etc., and also as permanent, re-usable formwork for concrete.

FIBRO (*a*) Australian term for asbestos cement sheet. (*b*) Hence, low cost houses built with a timber frame and an external lining of asbestos cement sheet, usually with an internal plasterboard or particle board lining.

FIBROUS PLASTER Gypsum plaster reinforced with *sisal* fibres or with canvas. In Australia large sheets capable of supporting their own weight are made from fibrous plaster by specialist craftsmen, called *fibrous plasterers*.

FIBTP Fédération Internationale du Bâtiment et des Travaux Publics, Paris.

FID Fédération Internationale de Documentation, The Hague.

FIELD MOISTURE EQUIVALENT The minimum moisture content at which a drop of water placed on a smoothed soil surface is not immediately absorbed, thus giving the soil a shiny appearance.

FIELD WELDING, FIELD BOLTING, FIELD RIVETING *See* SITE WELDING, etc.

FIGURE The natural grain of timber, particularly when it is cut as a veneer. Highly-figured timber, although it makes attractive veneer, is often structurally weak. *See also* BURL, CROTCHWOOD, and STRAIGHT GRAIN.

FILAMENT Electrical conducting material in the form of a fine wire.

FILAMENT LAMP An electric lamp in which a filament in a glass bulb, filled with an inert gas, is raised to incandescence by the passage through it of an electric current.

FILARETE'S TRATTATO D'ARCHITECTURA A book written between 1461 and 1464 by Antonio Alverlino (called Filarete). It played an important part in popularising the classical concepts of the early Renaissance (*arte antica*) which the author contrasted with the Gothic architecture of Lombardy (*arte moderna*). However, the designs proposed for the ideal city of *Sforzinda* were quite impracticable. *See also* ALBERTI and VITRUVIUS.

FILLET WELD A weld of approximately triangular cross section joining two surfaces approximately at right angles to one another.

FILM (*a*) A thin, not necessarily visible, layer of material. (*b*) A photo-sensitive emulsion, before or after processing, on a flexible base.

FINAL SET *See* INITIAL SET.

FINE AGGREGATE The smaller size of aggregate used for mixing concrete, as opposed to COARSE AGGREGATE. It usually consists of SAND, but small-size crushed stone is sometimes used. *See also* CONCRETE AGGREGATE.

FINE GRAINED SOIL Soil consisting predominantly of fine sand, silt or clay.

FINENESS MODULUS An empirical factor obtained by adding the total percentage of a granular sample retained on each of a specified series of sieves, and dividing the sum by 100. It is most commonly employed for concrete aggregates, but can also be used for paint and cement. *See also* PARTICLE SIZE ANALYSIS and SPECIFIC SURFACE.

FINGER JOINT An END JOINT made up of several meshing tongues or fingers of wood, made with a finger-jointing machine, and normally glued.

FINIAL A decorative detail at the uppermost extremity of a pinnacle or gable, or an ornamental cap for a spire.

FINK TRUSS A commonly used isostatic roof truss, suitable for spans of 40 to 50 ft (12 to 15 m) (Fig. 38). It has two trussed rafters, each divided into four parts by purlins. Also called a *Belgian* or *French truss*.

FIRE BARRIER Any element of a building, such as a wall, floor or ceiling, so constructed as to delay the passage of fire from one part of a building to another.

FIRE DETECTOR A device which detects a fire by some automatic control mechanism, usually a THERMO-STATICALLY controlled circuit which gives a fire alarm when the temperature rises above a pre-determined level.

FIRE DOOR A door made of fire-resisting material, generally metal plated, held open by a fusible link, which melts in a fire permitting the door to close, and thus delays or prevents the spread of fire by confining it to one compartment.

FIRE HYDRANT A connection to a water main for extinguishing a fire.

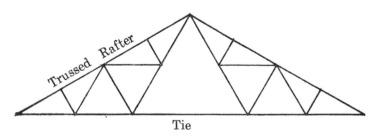

FIG. 38. Fink truss.

FIRE LOAD The amount of heat generated if the contents and combustible parts of a building were to be *completely* burnt. It is calculated in BThU per square foot (J/m^2) of floor area, assuming the burning material to be spread uniformly over the floor. The fire load depends on the type of occupancy.

FIRE PROTECTION OF STEEL STRUCTURES Although steel does not support combustion, it is more vulnerable to fire than a heavy timber section (which burns, but is protected by the charcoal formed). At about 750°F (400°C) it suffers significant loss of strength; at 1000°F (540°C) only 50 per cent, and at 1200°F (650°C) only 20 per cent of its cold strength remains. The most common protection is a layer of concrete, sometimes reinforced, the dimensions being specified in the building code. This greatly adds to the weight and the cost of the steel frame, and it is a principal reason for the use of reinforced concrete for multi-storey buildings. Lightweight fire-protection, such as sprayed VERMICULITE or SPRAYED ASBESTOS, may be used instead; however, it is sometimes more expensive, in spite of the saving in weight. Recent research has suggested that steel can be protected from fire without covering the frame. One method is to use only steel plates of great thickness, which do not heat up sufficiently quickly to permit failure of the frame before the fire burns out; another is to fabricate the steel into tubes through which water is circulated.

FIRE RESISTANCE GRADING *or* **RATING** The grading of building components according to the minutes or hours of resistance in a *standard*

fire test. The higher the FIRE LOAD, the higher the fire-resistance grading required.

FIRE-RESISTING Attribute of a material which does not burn. The term has replaced *fireproof*, since no material is completely proof against the effect of fire. *See also* SLOW-BURNING CONSTRUCTION.

FIRE RETARDANT PAINT A paint based on silicone, casein, borax, polyvinylchloride, urea formaldehyde or some other substance, a thin coating of which reduces the rate of flame spread of a combustible material, usually by intumescence under the effect of heat.

FIRE TEST *See* FIRE RESISTANCE GRADING.

FIREDAMP *See* METHANE.

FIREPROOF *See* FIRE-RESISTING.

FIRST-ANGLE PROJECTION *See* PROJECTION.

FIRST FLOOR, FIRST STOREY *See* FLOOR.

FIRST MOMENT OF AREA *See* MOMENT OF AN AREA.

FIXED-ENDED BEAM Built-in beam, or encastré beam.

FIXED-POINT ARITHMETIC A system of recording numbers in a digital computer so that the location of the decimal point is the same for all numbers. Small fractions which cannot be accommodated must be rounded off to the nearest lower digit, and this can be a serious source of error. In *floating point arithmetic,* the location of the decimal point can be varied to suit the number, but its location must be recorded for each number.

FLAMBOYANT Characterised by waving, flame-like forms. The flamboyant style is the name given to late French GOTHIC architecture.

FLAME CUTTING Cutting with an oxyacetylene torch.

FLAME-RETARDANT PAINT *See* FIRE-RETARDANT PAINT.

FLAME SPECTROPHOTO-METER An instrument used to determine the presence of chemical elements by the colour intensity of their unique flame spectra.

FLAME SPREAD The rate at which a flame spreads under intense radiant heat. The flame spread on wall and ceiling linings is classified by a standard test.

FLAMMABLE *See* COMBUSTIBLE.

FLASH POINT The lowest temperature at which a substance ignites when a flame is put to it.

FLASH SET The rapid development of rigidity in freshly mixed cement paste, mortar or concrete, usually with considerable evolution of heat. Plasticity cannot be regained by further mixing without the addition of water. *See* SETTING and FALSE SET.

FLASH WELDING A resistance welding process. After the current has been turned on, the two parts are brought together. This produces arcing which expels small particles

of metal (flashing), and thus protects the metal from oxidation.

FLASHING Sheet metal used to cover open joints in exterior construction, such as joints in parapets or roof valleys, to prevent ingress of water.

FLAT British and Australian term for a rented APARTMENT. The building is known as a *block of flats. See also* MAISONETTE.

FLAT (*of a painted surface*) *See* GLOSS.

FLAT ARCH An arch with a level *extrados* and *soffit* made from wedge-shaped stones or bricks; also called a JACK ARCH. It has been superseded by steel and reinforced concrete lintels.

FLAT JACK A hydraulic jack consisting of light gauge metal, bent and welded to a flat shape, which expands under internal pressure. It is particularly useful for prestressing concrete in confined locations, *e.g.* the springings or the crown of an arch.

FLAT PLATE A concrete slab reinforced in two directions, and supported directly on the columns without column capitals, drop panels, beams or girders. (Figs. 39 and 76). Because of its economy, it is particularly popular for lightly loaded (*e.g.* residential) buildings. By contrast, a FLAT SLAB has column capitals to reduce the shear around the columns, and it is used for heavier loads.

FLAT ROOF A roof with a slope less than 10°. *See also* BUILT-UP ROOFING and PONDING.

104

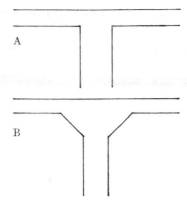

FIG. 39. (A) Flat plate and (B) flat slab.

FLAT SLAB A concrete slab reinforced in two or more directions, and supported directly on COLUMN CAPITALS without beams or girders (*Figs.* 31 and 39); also called *mushroom construction* because of the shape of the enlarged capitals. It is frequently used in conjunction with a *waffle slab*. In contrast, a FLAT PLATE has no column capitals.

FLAX A plant whose blue flowers are succeeded by seed pods from which *linseed oil* is made. Linen is made from flax fibres.

FLAXBOARD FIBRE BOARD manufactured from flax.

FLEMISH BOND *See* BOND (*a*).

FLEXIBILITY METHOD A matrix solution for HYPERSTATIC STRUCTURES, intended for evaluation by electronic digital computer, in which the equations are framed in terms of the redundant actions.

FLEXURAL RIGIDITY Measure of the stiffness of a member in resisting bending. It is usually taken as the product *EI*, where *E* is the MODULUS OF ELASTICITY, and *I* the SECOND MOMENT OF AREA.

FLEXURE Bending.

FLEXURE TEST *See* MODULUS OF RUPTURE.

FLIGHT A series of stairs unbroken by a landing.

FLINT Concretions of SILICA found in some limestone and dolomite beds. They are believed to be organic in origin, deriving from siliceous sponges. Flints were used extensively in stone-age tools (*see also* OBSIDIAN). In certain parts of Europe, particularly in Southern England, split flints (called *knapped flints*) were used in a characteristic vernacular architecture. *See also* CHERT.

FLINT GLASS GLASS containing lead, as opposed to CROWN GLASS. It was traditionally used for cut glass, because it was easier to engrave, and for this reason the best source of silica, *i.e.* FLINT, was used. The name is today often used for any high-quality glass, of whatever chemical composition.

FLIP-FLOP A device or circuit with two stable states. The circuit remains in one of these two states until a signal is applied that causes it to change.

FLITCH A large piece of converted timber, suitable for re-sawing into smaller sizes.

FLOAT The difference between earliest (*EFT*) and latest (*LFT*) finish times of an activity in a NET-WORK. The *free float* is the amount of time by which the start of an activity may be delayed without interfering with the start of any subsequent activity.

FLOAT FINISH A rather rough concrete finish, obtained by finishing with a wooden float.

FLOAT GLASS Thick sheets of glass made by floating the molten glass on a surface of molten metal, which produces a smooth, polished surface. *See also* PLATE GLASS.

FLOAT VALVE *See* BALL COCK.

FLOATING FLOOR Separation of the floor wearing surface from the rest of the building to provide locally DISCONTINUOUS CONSTRUCTION for insulation against impact sound. A concrete floor may be insulated from the supporting structure by an insulating layer, such as mineral wool (Fig. 40), and a wooden floor may be similarly insulated by a resilient quilt.

FLOATING-POINT ARITHMETIC *See* FIXED-POINT ARITHMETIC.

FLOOR In Europe and Australia the first floor or storey is normally the floor above the ground floor. In America this is normally called the second floor or storey, the ground floor being the first floor. Moreover, in large buildings the level of the ground floor may depend on the door chosen. The term 'floor' is therefore being displaced by *level*, level 1 being the lowest level served by a stair or lift; this is usually below ground.

FLORENCE, DOME OF *See* RENAISSANCE.

FLOW DIAGRAM A diagram which employs both writing and standardised graphical symbols (Fig. 41) to indicate the flow of an operation required to solve a particular problem. It is useful for planning any complicated operations, but is mainly used in conjunction with digital computers. *See also* SYSTEM ANALYSIS.

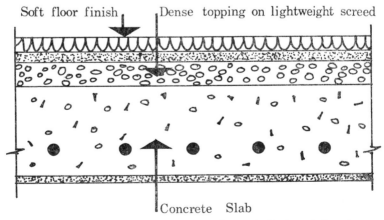

Soft floor finish   Dense topping on lightweight screed

Concrete Slab

FIG. 40. Floating floor. One type of floating floor suitable for reinforced concrete construction.

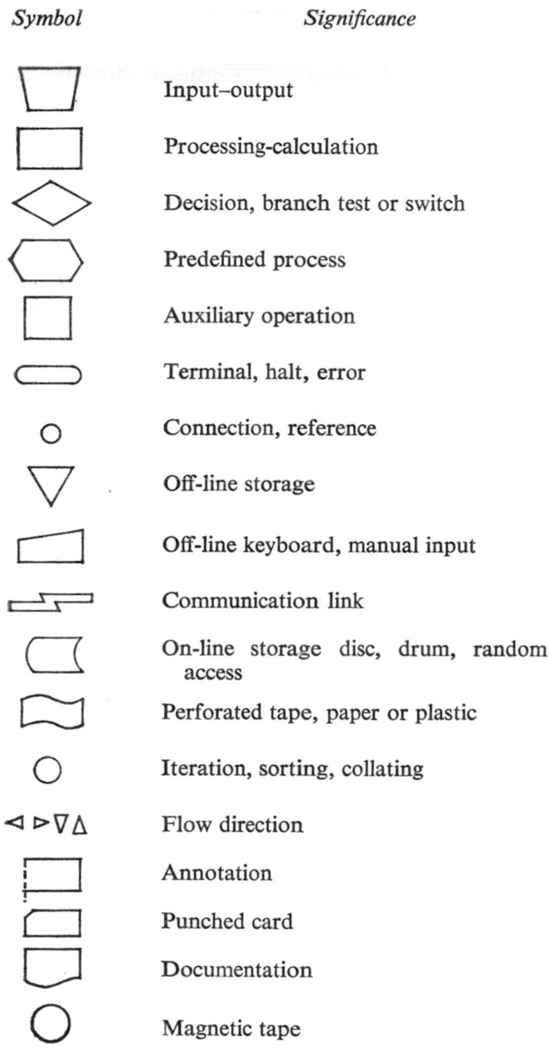

Symbol	Significance
	Input–output
	Processing-calculation
	Decision, branch test or switch
	Predefined process
	Auxiliary operation
	Terminal, halt, error
	Connection, reference
	Off-line storage
	Off-line keyboard, manual input
	Communication link
	On-line storage disc, drum, random access
	Perforated tape, paper or plastic
	Iteration, sorting, collating
	Flow direction
	Annotation
	Punched card
	Documentation
	Magnetic tape

FIG. 41.   Standard flow diagram symbols.

FLOW METER An instrument for measuring the quantity of a fluid, such as water or gas, which flows through a pipe in a unit of time. The most common type is based on the VENTURI TUBE.

FLUIDITY The ability of a material flow; the opposite of *viscosity*.

FLUON *See* POLYTETRAFLUOR-ETHYLENE.

FLUORESCENT PAINT A pigment which converts invisible to visible radiation, and therefore appears brighter than surrounding surfaces. Certain organic dyes, tungstates, borates and silicates have this property; but the brilliance of some of them dulls after a period of time due to weathering. Unlike PHOSPHORESCENT PAINT, it does not exhibit an afterglow.

FLUORESCENT TUBE An electric lamp without a filament, consisting of a tube 1 to 1½ in (25 to 40 mm) diameter and 4 to 8 ft (1·2 to 2·5 m) in length, coated inside with a fluorescent powder. An alternating current passing through the vacuum emits electromagnetic radiation, which is *outside* the visible spectrum. The fluorescent glass walls of the tube change the invisible radiation into light waves. Fluorescent tubes cost more than INCANDESCENT LAMPS, but they consume less power for the same amount of light. They also, if properly arranged, give a more evenly diffused light. Also called a *cold cathode lamp*.

FLUSH SWITCH A switch which has been recessed so that its front is flush with the wall.

FLUSH VALVE A valve supplying a precise quantity of water directly from the water-supply pipes, to flush a WC. Its use is forbidden by most British and Australian water authorities; but the arguments for this prohibition are not supported by scientific evidence. *See* BALL COCK.

FLUTED COLUMN A column decorated with concave channels.

FLUTTER A sustained oscillation due to aerodynamic forces, elastic reactions, and inertia.

FLUX *See* LUMINOUS FLUX and SOLDERING.

FLY ASH The finely divided residue resulting from the combustion of ground or powdered coal, which is transported from the fire box through the boiler by flue gases. It has POZZOLANIC properties, and is sometimes blended with cement for this reason.

FLYING BUTTRESS *See* BUTTRESS.

FLYING SHORE *See* SHORE.

FOAMED CONCRETE *See* CELLULAR CONCRETE.

FOAMED PLASTIC *See* EXPANDED PLASTIC.

FOCUS *See* ELLIPSE and PARABOLIC REFLECTOR.

FOG ROOM *See* MOIST ROOM.

FOIL (*a*) Thin metal SHEET, generally less than 0·005 in (0·1 mm) thick. (*b*) A rounded ornament, once widely used as a window decoration. It is a stylised leaf, commonly divided into three (*trefoil*), four (*quatrefoil*) or five (*cinquefoil*) divisions.

FOIL STRAIN GAUGE ELECTRIC RESISTANCE STRAIN GAUGE made from thin foil printed on a thin lacquer film.

FOLDED-PLATE ROOF A roof structure formed by flat plates, usually of reinforced concrete, joined at various angles (Fig. 42). It has many of the properties of SHELL CONSTRUCTION, but since the component parts are not curved, they do not act in accordance with the MEMBRANE THEORY, and the structure is subject to substantial bending moments. Folded plates can be designed to perform the same function as domes (D),

barrel vaults (A), and northlight shells (B). *Hipped plate roofs* are hip roofs formed by folded plates.

FOLLY A useless and costly structure, which might be thought to show the folly of the client or his architect. The term is particularly applied to sham Gothic or classical ruins built in the landscaped parks of English country houses.

FOOT-POUND The FPS unit of energy or work. The unit of bending moment or twisting moment is generally called pound-foot (or pound-inch). The two units are, however, identical.

FOOTING A foundation for a wall or column.

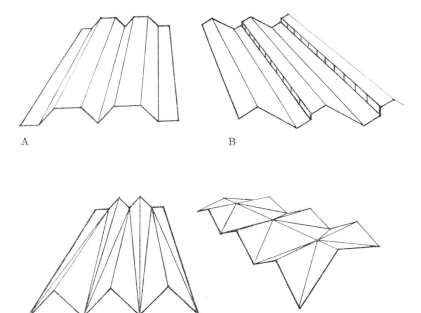

A

B

C

D

FIG. 42.   Folded plate roofs.

FORCE Defined as anything which changes or tends to change the state of rest of a body, or its motion in a straight line. The forces most commonly acting on buildings are the weight of the materials from which they are built, the weight of the contents, and the forces due to snow, wind and earthquakes. *See also* STATICS.

FORCED CIRCULATION Circulation of fluid by pumping, as opposed to *gravity circulation.*

FORCED VIBRATION The vibration imparted to a body by a periodic force. If its period is the same as that of the NATURAL FREQUENCY of vibration of the body, its effect is cumulative, and RESONANCE results.

FORM LINING The lining of concrete FORMWORK can be utilised to impart a smooth or patterned finish to the concrete surface; to absorb moisture in order to obtain a drier consistency at the surface; or to apply a set-retarding chemical which enables the aggregate to be EXPOSED.

FORMALDEHYDE A gas (H.CHO) of pungent odour, readily soluble in water. The *amino-plastics* phenol formaldehyde and urea formaldehyde are condensation products of formaldehyde with phenol or urea; they are THERMOSETTING. *See also* BAKELITE.

FORMWORK The mould for freshly placed concrete (which gives it its shape), as well as the supporting structure and bracing required to support its weight. Also called *shuttering.*

FORTRAN *FORmula TRANslation* the most widely used PROBLEM-ORIENTED LANGUAGE for scientific computer programming, particularly in conjunction with IBM computers. *See also* ALGOL.

FORUM *See* AGORA.

FOUNDATION The material or materials through which the load of a structure is to be transmitted to the earth.

FOUNDATION FAILURE *See* DIFFERENTIAL SETTLEMENT.

FOUNDATION PRESSURE *See* EARTH PRESSURE, BOUSSINECQ PRESSURE BULB and COULOMB'S EQUATION.

FOUR-WAY REINFORCEMENT IN CONCRETE SLABS Reinforcement in FLAT SLABS consisting of bands parallel to the column centre-lines and also to both diagonals. It has been almost entirely replaced by TWO-WAY REINFORCEMENT.

FOURIER SERIES The sum of an infinite number of SINE and cosine waves of successively higher orders, which can be made to represent any actual shape of wave with an accuracy depending on the number of terms used. It is named after the French physicist who developed it in 1822 for his work on heat transfer. Fourier series are important in the analysis of sound waves.

FOURTH-POWER LAW OF RADIATION *See* STEFAN-BOLTZMANN LAW.

FOYER Originally the green-room in French theatres. Now usually the entrance hall of a theatre, in which the audience may congregate during intervals.

FPS UNITS The units of the traditional British system, based on foot, pound and second. *See also* CGS UNITS, US CUSTOMARY UNITS, and *Appendix G.*

FRACTION The ratio of two integers, *e.g.* 2/3. A fraction is a *rational number. See also* APPROXIMATE and IRRATIONAL NUMBERS.

FRACTIONAL HORSEPOWER MOTOR An electric motor with a rating of less than 1 hp.

FRACTURE Failure caused by breaking of the material into two parts. It causes a sudden failure of a structural member, as distinct from the gradual failure caused by PLASTIC DEFORMATION or CRUSHING FAILURE. *See also* CLEAVAGE FRACTURE and GRIFFITH CRACK.

FRAME An assembly of structural members. Most frames are *rectangular, i.e.* they consist of horizontal and vertical members only. Most frames are also *plane, i.e.* they can be designed as two-dimensional structures, connected by additional members at right angles. Frames which cannot be so designed are called SPACE FRAMES. All types of frame may be either ISOSTATIC or HYPERSTATIC. *For solution of plane isostatic frames see* METHOD OF SECTIONS, RECIPROCAL DIAGRAM, RESOLUTION AT THE JOINTS, and TENSION COEFFICIENT.

FRAME CONSTRUCTION Any type of construction in which the building is supported mainly by a frame, and not mainly by load-bearing walls. BALLOON-FRAMED timber houses, BRICK-VENEER houses, steel-framed buildings, and reinforced-concrete frame buildings all belong to this type.

FRASS Dust made by wood boring insects, and one way of identifying them.

FREE-BODY DIAGRAM A diagram obtained by making an imaginary cut through a structure, and considering the STATIC equilibrium of one or both parts separately as, for example in the METHOD OF SECTIONS (Fig. 61). Since the conditions of equilibrium must be satisfied by every stationary structure, the diagram can be drawn even for highly complex hyperstatic structures. It is sometimes possible to obtain a simple answer for the critical forces by this approach, without having to analyse the entire structure.

FREE-FIELD ROOM *See* ANECHOIC CHAMBER.

FREE FLOAT *See* FLOAT.

FREEDOM *See* DEGREES OF FREEDOM.

FREESTONE A building stone which can be freely carved in any direction. Most LIMESTONES and the finer-grained SANDSTONES are in this category.

FRENCH CHALK Finely ground TALC.

FRENCH POLISH SHELLAC dissolved in methylated spirits.

FRENCH ROOF *See* MANSARD ROOF.

FRENCH TRUSS *See* FINK TRUSS.

FRENCH WINDOW or DOOR A long window reaching to floor level, and opening in two leaves like a pair of doors.

FREQUENCY The number of cycles of a periodic phenomenon which occur in a given time interval. Frequency is measured in HERTZ (cycles per second), or cycles per minute, etc. The product of frequency and *wavelength* equals the wave velocity. Thus the wavelength of any kind of electromagnetic RADIATION, multiplied by its frequency, equals the VELOCITY OF LIGHT. Similarly the wavelength of sound, or of any other vibration in air, multiplied by its frequency, equals the VELOCITY OF SOUND.

FREQUENCY ANALYSER *See* SOUND FREQUENCY ANALYSER.

FREQUENCY DISTRIBUTION CURVE A curve relating the magnitude of a test result (or an observed variable characteristic) to its frequency (or the number of times it occurs). If the phenomenon varies only within a narrow range (as, for example, the test results for carefully

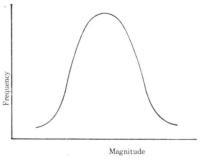

FIG. 43. Frequency distribution curve.

controlled concrete), the curve is narrow and steep. As quality control declines, the curve becomes wider (Fig. 43). *See also* HISTOGRAM, GAUSSIAN CURVE and COEFFICIENT OF VARIATION.

FREQUENCY, RESONANT *See* RESONANCE.

FRESCO Painting on wet lime plaster with pigments mixed with water. The colours dry and set with the plaster. The range of pigments is restricted to a small range which is sufficiently alkali resistant. The plasterer and the painter must work together since the plaster does not absorb colour after it has reached a certain degree of dryness. Later alterations are impossible. The technique is now rarely employed, but it was the medium of the greatest wall and ceiling paintings of the Renaissance (*e.g.* the ceiling of the Sistine Chapel). Hence the term 'fresco painting' is sometimes used, incorrectly, for painting on dry plastered walls and ceilings.

FRESHNESS *See* STUFFINESS.

FRICTION The resistance to motion which is brought into play when one body is moved over another. When the two surfaces are examined under a microscope, it is found that they touch only at a few points; hence the true area of contact is independent of the nominal area of contact. The contact area is, however, increased by the contact pressure. Frictional resistance is therefore proportional to the pressure between the bodies, and independent of the overall surface area of the bodies in contact. The *coefficient of static friction* is defined as W/P where W is the *limiting friction*, or force at which the body just starts to move, and P is the contact pressure. The *coefficient of kinetic friction* (which is lower) is the ratio W/P required to maintain motion against the frictional resistance. *Rolling friction* is the resistance offered when a body rolls over a surface. The *coefficient* of rolling friction is Wr/P, where r is the radius of the rolling body. *See also* ANGLE OF FRICTION.

FRICTION-GRIP BOLT *See* HIGH-TENSILE *bolt*.

FRICTION, INTERNAL *See* ANGLE OF INTERNAL FRICTION.

FRICTION LOSS (*in post-tensioning*) The loss of stress resulting in curved TENDONS due to friction with the concrete or duct lining. Some friction loss occurs also in nominally straight cables due to the unavoidable imperfections of construction, which causes slight bends where the tendons touch the ducts.

FRICTION PILE *See* BEARING PILE.

FRIEZE (*a*) In classical architecture, the middle division of the entablature, between the ARCHITRAVE and the CORNICE. (*b*) In recent buildings, a decorated band along an internal wall, immediately below the cornice. (*c*) In rooms with a picture rail, the part of the wall above that rail, whether decorated or not.

FRIGIDARIUM The cool room in a Roman bath.

FROG A depression on one or both of the larger faces of a brick or block. It provides a key for the mortar, and reduces the weight of the brick. The frog is not necessarily filled with mortar.

FROST HEAVE Swelling of soil due to the expansion of the ground water when it turns into ice. It usually causes uplift, since the soil is restrained in other directions.

FROZEN-STRESS METHOD *See* THREE-DIMENSIONAL PHOTOELASTICITY.

FRP Fibre-reinforced plastic.

FRUSTUM OF A CONE or **PYRAMID** The portion left after cutting off the upper part by a plane parallel to the base.

FULCRUM The point of support, or pivot, of a lever.

FULLER, R. BUCKMINSTER *See* GEODESIC DOME.

FULLY FIXED *See* BUILT-IN.

FUNCTION A term introduced by the German mathematician G. W. Leibnitz, for the relation between two variables. If y is a function of x, and x is known, then y is also known.

FUNCTIONALISM A term widely used in the literature on modern architecture, which has a range of meanings. The dictum 'Form follows function' is generally credited to LOUIS H. SULLIVAN (*Kindergarten Chats*, 1901–2). However, he meant it essentially as a pronouncement on aesthetics. The term functionalism has also been used to describe a technically oriented design which concentrates on the functioning of the building in accordance with its requirements, and treats form as a secondary consideration. Many of the arguments on functionalism in the 1920s and '30s derive from the writings of VIOLLET-LE-DUC via LE CORBUSIER whose book *Vers une architecture* (1923) contained the phrase *A house is a machine for living in*. The movement lost some of its support when it became obvious that efficiently functioning buildings and precisely calculated structures produce beautiful forms only if the architects exercise good aesthetic judgment; the process is not automatic.

FUNGICIDAL PAINT A paint which discourages fungal growth, *e.g.* by the inclusion of a mercury salt. It is particularly used in the tropics.

FUNGUS *See* DRY ROT.

FUNICULAR ARCH An arch which is purely in compression under a series of point loads. It is the reverse of a string carrying the same load system as a suspension cable (Fig. 44). Also called a *linear arch*.

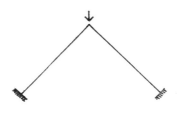

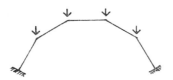

FIG. 44.   Funicular arches.

FUNICULAR POLYGON *See* LINK POLYGON.

FUSE A safety device for preventing the overloading of a circuit (which might start a fire), and for protecting electrical apparatus from excess current. It consists of a short length of metal, which conducts a safe current, but melts when the current is above the specified value, thus breaking the CIRCUIT.

FUSIBLE LINK *See* FIRE DOOR.

FUSION WELDING The WELDING of metals or plastics solely by melting the edges of the pieces to be joined, without mechanical pressure. Additional weld metal may be provided by a filler rod. *See* ARC WELDING.

FUTURISM An Italian movement which intended to 'harmonise the environment of man with the decisive developments in science and technology'. Its architectural manifestations were initiated by an exhibition *Città Nuova*, mostly produced by *Antonio Sant'Elia*, held in Milan in 1914, and by the *Manifesto dell'architettura futurista*, published by *Marinetti* in the same year. Only a few buildings resulted directly from this movement.

G

GABLE The triangular part of the end wall of a building with a sloping roof; it is frequently of a material different from the rest of the wall.

GABLE ROOF A sloping roof with gables. It has inclined slopes on two sides, and vertical gables on the other two (Fig. 45). By contrast, a *hip roof* has four inclined slopes which meet at the *hips*.

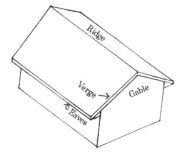

FIG. 45.   Gable roof.

GABLE WALL A wall of which a gable forms a part.

GABLE WINDOW A window built into a gable. The room behind usually has a sloping ceiling. *See also* DORMER WINDOW.

GAGE *See* GAUGE.

GALE A wind of force 8 on the BEAUFORT SCALE. Its speed is approximately 40 mile/h (64 km/h) 30 ft (10 m) above ground. A gale blowing steadily on a vertical face exerts a pressure of about 5 psf (240 N/m²).

GALERIES DES MACHINES A hall built for the Paris International Exhibition of 1889, and since demolished. It consisted of three-pinned portals with a span of 375 ft (115 m). The structural design, by V. Contamin, set a new record.

GALLON (GAL) Liquid measure in FPS UNITS. The Imperial gallon equals 4·546 litres, and the US gallon equals 3·785 litres.

GALTON WHISTLE An instrument for producing ultrasonic vibrations of constant amplitude and frequency.

GALVANIC CELL A cell containing two dissimilar metals and an electrolyte.

GALVANIC CORROSION Electrochemical corrosion.

GALVANIC PROTECTION SACRIFICIAL PROTECTION given to a metal by making it the cathode to a sacrificial anode.

GALVANIC SERIES The ELECTROCHEMICAL SERIES, named after the eighteenth century Italian physicist who discovered the generation of electricity by chemical means.

GALVANISED IRON Normally galvanised steel (not iron) sheet.

GALVANISING The coating of steel or iron with zinc, either by immersion in a bath of zinc covered with flux at a temperature of about 450°C, or by ELECTRODEPOSITION from cold sulphate solutions. The zinc is capable of protecting the iron from atmospheric corrosion even when the coating is scratched, since the zinc is preferentially attacked by carbonic acid, forming a protective coating of basic zinc carbonates. *See also* SACRIFICIAL PROTECTION.

GALVANOMETER Precision instrument for measuring *small* electric currents. A *galvanoscope* is an instrument capable of detecting, but not measuring, small currents. *See also* AMMETER.

GAMMA IRON Unalloyed iron in the temperature range from 1670 to 2560°F (910 to 1405°C). It has a FACE-CENTRED cubic space lattice. *See* Fig. 71 and also ALPHA IRON.

GAMMA RAYS Electromagnetic RADIATION produced by radioactive materials. It is similar to X-RAYS, which are produced by high-voltage apparatus.

GANG NAIL A plate from which a number of spikes have been punched and pushed out at right angles. This can be pressed into two pieces of timber, acting like a gusset plate with compound nails. *See also* NAIL PLATE and TIMBER CONNECTOR.

GANTRY CRANE A portal crane with four legs running on rails. It can pass over railroad tracks, motor vehicles, or even complete houses under construction. This is the most expensive type of crane, used only for large building projects or permanent installations (*e.g.* in PRECAST CONCRETE factories). *See also* TOWER CRANE and DERRICK CRANE.

GANTT CHART A bar chart which shows, for example, the duration of the various processes in a building operation by the length of a bar. It is named after Henry L. Gantt, who popularised its use in the early 1900s. *See also* CRITICAL PATH METHOD.

GAP-GRADED AGGREGATE Aggregate characterised by a particle-size distribution in which certain intermediate sizes are wholly or substantially absent. It is more common to use *continuous grading*. An extreme case of gap grading is NO-FINES CONCRETE.

GARGOYLE A water spout projecting from the gutter of a building, so that it throws the rainwater clear of the wall. It generally terminates in a grotesquely carved animal or human head with an open mouth, through which the water discharges. Neo-Gothic buildings of the nineteenth and twentieth century frequently used gargoyles for decorative purposes only, and avoided the discharge of the water on passers-by by connecting the gutters to downpipes.

GARRET A room in the ATTIC.

GAS CONCRETE Term occasionally used for a CELLULAR CONCRETE made with a gas-forming agent, such as aluminium powder.

GAS CUTTING *See* FLAME CUTTING.

GAS LAWS *See* BOYLE'S LAW and CHARLES' LAW.

GAS WELDING *See* ACETYLENE.

GASKET A piece of material placed around a joint to make it leakproof. An *inflatable gasket* is used around windows in some curtain walls; it is deflated by the maintenance crew, and filled again with air after cleaning the window.

GASLIGHT *See* WELSBACH MANTLE.

GATE VALVE A casting machined to receive a gate which closes the opening. The gate may be lifted and lowered by turning a screw. *See also* GLOBE VALVE.

GAUGE *See* BOURDON GAUGE, DIAL GAUGE, STANDARD WIRE GAUGE. Also spelt *gage*.

GAUGE FACTOR A factor by which the reading of a gauge, *e.g.* an ELECTRIC RESISTANCE STRAIN GAUGE, must be multiplied to give the correct answer. It is usually supplied by the manufacturer.

GAUGE LENGTH The length on a test piece or structural model over which a STRAIN measurement is made.

GAUSSIAN CURVATURE The product of the two principal curvatures at a point on a curved surface, *e.g.* a SHELL. The term is commonly used in differential geometry to classify surfaces (Fig. 46). A SINGLY-CURVED surface has zero Gaussian curvature because the curvature in one direction is zero. A sphere has constant positive Gaussian curvature. A *dome*-type surface has positive Gaussian curvature, because the curvature is the same way in two mutually perpendicular directions, and a *saddle-type* surface has negative Gaussian curvature, because the curvature is upwards in one direction, and downwards at right angles to that direction. A *synclastic* surface is one with a positive Gaussian curvature and an *anticlastic* surface is one with a negative Gaussian curvature. A surface with zero Gaussian curvature is, by definition, DEVELOPABLE.

GAUSSIAN CURVE An exponential curve fitting the normal FREQUENCY DISTRIBUTION (Fig. 43), named after the early nineteenth century German physicist:

$$y = K e^{-\frac{1}{2}t^2}$$

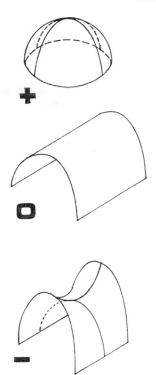

FIG. 46. Gaussian curvature. If two cuts at right angles curve the same way, it is positive; if they curve in opposite directions, it is negative; if one has no curvature, it is nil.

where $t = (x - \mu)/\sigma$ is the *standardised normal variate*, σ = STANDARD DEVIATION, μ = MEAN, x = magnitude of phenomenon, y = frequency of phenomenon, K = a constant, and e = the base of the natural logarithm.

GAY-LUSSAC'S LAW *See* CHARLES' LAW.

GAZEBO A turret on top of a house, a bow window, or a garden house commanding an extensive view. *See also* BELVEDERE.

GEL   The apparently solid material formed from a COLLOIDAL solution which is undisturbed. The major part of mature hydrated PORTLAND CEMENT is in gel form.

GENERATOR   *See* ALTERNATOR and DYNAMO.

GEODESIC DOME   A hemispherical, or approximately hemispherical, LATTICE dome. The concept is due to *Richard Buckminster Fuller*, who has built geodesic domes ranging in diameter from a few feet (to protect radar installations) to 384 ft (117 m) (for a railroad repair shop in Baton Rouge, Louisiana, built in 1958).

GEODESY   The branch of surveying concerned with the mapping of extensive areas, in which allowance must be made for the curvature of the earth's surface (which is neglected in *plane surveying*).

GEODETIC CONSTRUCTION Originally construction of a hyperstatic space frame, particularly in aircraft design, whose members follow GEODETIC curves, each compression member being braced by an ORTHOGONAL tension member. The term now generally implies STRESSED-SKIN CONSTRUCTION, in which the skin of the aircraft or other structure forms a hyperstatic space frame with the ribs, which prevent the skin from BUCKLING.

GEODETIC LINE   The shortest possible line which can be drawn from one point of a curved surface to another. The geodetic lines of a sphere are called *great circles*, because they are the circles with the greatest

diameter which can be drawn on a sphere. They correspond to straight lines on a plane surface. The *equator* and all the *meridians* of *longitude* on the earth's surface are great circles. The parallels of *latitude* (other than the equator) are *small circles*, and travelling along them is not the shortest distance between two points. They correspond to curved lines on a plane surface (Fig. 47).

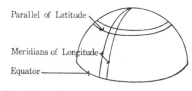

FIG. 47.   Geodetic lines (great circles) and small circles. The great circles are the meridians of longitude and the equator. The small circles are the parallels of latitude (except the equator).

GEOGRAPHY   The science of the earth's surface.

GEOLOGICAL FAULT   A fracture in a rock formation, accompanied by displacement of one part relative to another. The displacement, or *throw*, may be a fraction of an inch, or many feet. Faults are a source of weakness in foundations, since they may cause movement of the entire building, or of one part relative to the other. Movement along major faults is a common cause of EARTHQUAKES.

GEOLOGY   The science which describes the earth's crust. A *geological map* is one which shows the outcrops of all *rocks* and sedimentary deposits, but without the top soil.

GEOMETRIC MEAN   *See* MEAN.

GEORGIAN ARCHITECTURE
The architectural style prevailing in England during the reign of the four Georges, *i.e.* during the eighteenth and early nineteenth century. It extends from the revival of PALLADIANISM by Lord Burlington to the age of Nash and Robert Adam. *See also* COLONIAL ARCHITECTURE.

GERBER BEAM A continuous beam which is made isostatic by the insertion of two hinges in alternate spans (Fig. 48). It thus consists of cantilevered beams alternating with short spans simply supported from the ends of the cantilevers. Gerber beams are useful where continuous structures are assembled from heavy precast concrete elements, or where differential settlement of the foundations may be expected. *See* CANTILEVER BRIDGE.

GESSO A hard white surface, prepared to serve as a basis for painting or bas-relief. It usually consists of plaster of paris, or whiting and glue.

GIBSON, SIR DONALD *See* CLASP.

GIGA *See* BILLION.

GIRDER A synonym for BEAM. It is generally a primary beam which supports secondary beams, as opposed to a JOIST. A *steel girder* may be a single large steel section, a PLATE GIRDER, or a lattice girder (such as a HOWE, PRATT or WARREN girder).

GLARE A condition of vision in which there is discomfort, or a reduction in the ability to see significant objects, or both, due to an unsuitable distribution of range of luminance, or to extreme contrasts in space or time. *See also* BLINDING, DIRECT, DISABILITY, DISCOMFORT and REFLECTED GLARE.

GLASS A hard and brittle amorphous substance, made by fusing silica (sometimes in combination with the oxides of boron or phosphorus) with certain basic oxides (notably those of sodium, potassium, calcium, magnesium and lead) and cooling the product rapidly to prevent crystallisation or devitrification. Most glasses melt between 800°C and 950°C. Heat resisting glass generally contains a high proportion of boric oxide. The brittleness of glass is such that minute surface scratches during manufacture greatly reduce its strength. *See* GRIFFITH CRACK. The two main traditional types of glass are CROWN GLASS and FLINT GLASS. The difference in their chemical composition gives them different refractive indices, and the two materials have been used in conjunction to produce ACHROMATIC lenses. Glasses occur naturally, due to the rapid cooling of molten rock, *e.g.* OBSIDIAN. Glass particles have also been found which are believed to be of cosmic origin, *e.g. australites*. *See also* CYLINDER, FLOAT, PLATE, and SHEET GLASS.

GLASS BRICK A hollow block of glass, which is translucent but not

Fɪɢ. 48.   Gerber beam.

119

transparent. It can be used in brick and block walls, or in glass–concrete construction.

GLASS–CONCRETE CONSTRUC-TION Reinforced concrete into which blocks of glass, capable of transmitting compressive stresses, have been cast. The glass may be in the form of hollow translucent blocks, which provide diffused white light, or it may be coloured, to provide a decorative pattern in the manner of stained glass. *See also* BÉTON TRANSLUCIDE.

GLASS FIBRE *See* FIBREGLASS.

GLASS PAPER An abrasive paper.

GLASS WOOL *See* MINERAL WOOL.

GLAZE (*a*) To install glass panes in window or door frames. It normally implies the traditional method of using PUTTY to hold the glass in position. (*b*) To produce a shiny surface on photographic prints. (*c*) A brilliant glass-like surface given to tiles or bricks. It may be transparent or opaque, white or coloured. *See also* SALT GLAZE. (*d*) A nearly transparent thin coat of colour put on paint to enhance the original colour.

GLAZED TILE A glazed *clay tile.*

GLAZING BAR *See* MULLION.

GLAZING BEAD A strip which presses the glass against the glazing bar. It is an alternative to the use of window PUTTY.

GLC Greater London Council.

GLOBE THERMOMETER Instrument devised by H. M. Vernon

in 1930 as a means of indicating the combined effects of radiation and convection, as they influence the human body. It consists of a hollow copper sphere containing an ordinary thermometer at the centre of the sphere. The temperature of the instrument depends on the environment in which it is placed. If the mean radiant temperature is higher than the air temperature, the temperature recorded in the globe thermometer will be above the air temperature; conversely with the surroundings cooler than the air, the globe thermometer temperature will be below air temperature. The copper globe may be left *bright*, or it may be painted *black*, so that it behaves like a BLACK BODY.

GLOBE VALVE A casting machined to receive a bulbous circular disc which closes a circular opening between the two chambers. The disc is operated by a screw, and it can resist a higher fluid pressure than that of a GATE VALVE.

GLOBOSCOPE A paraboloidal mirror in which a distorted image of all the surrounding buildings is seen in stereographic projection, as observed from above. The image can be viewed, or photographed. It is possible to study sunlight penetration by superimposing a transparent sun-path diagram on a globoscopic view.

GLOSS The reflection of light from a painted surface. It ranges from full gloss (the smoothest mirror-like surface attainable), through semi-gloss, eggshell gloss and eggshell flat, to flat (a matt surface without sheen, even when viewed from an oblique angle).

GLUE Sticky substance for joining two pieces, particularly of wood, by bonding of the contact surfaces. The traditional carpenter's glue is made from bones and other waste products. Other traditional glues are fish glue, casein glue, and vegetable glues such as soya glue and cassava glue. These have now been largely superseded by *synthetic* RESINS, particularly for exterior and other waterproof applications.

GLUED LAMINATES *See* LAMINATING.

GLYCERIN A trihydric alcohol ($CH_2OH.CHOH.CH_2OH$). It is a constituent of many paints and varnishes.

GOING The distance from face of riser to face of riser on a staircase. *See also* RISER.

GOLD A metallic element, one of the few to be found as a metal in the native state, and highly resistant to corrosion. It has consequently been used since ancient times for coins. Its chemical symbol is Au, its atomic number is 79, its atomic weight is 197·2, its specific gravity is 19·3, and its melting point is 1063°C. *See also* ELECTRON and WHITE GOLD.

GOLDEN SECTION A geometric construction used by Euclid to draw the regular pentagon. From it can be derived the golden number $\varphi = \frac{1}{2}(1 + \sqrt{5}) = 1\cdot618$, which was widely claimed in the nineteenth and twentieth centuries to be the key to the beautiful proportions of Greek and other classical architecture. It has never been proved that the golden section was used in Greek architecture, or even that it was known in Pericles' lifetime. However, it was used in the revival of Greek architecture and other classical forms in the nineteenth century. In 1950 Le Corbusier attempted unsuccessfully to revive the golden number in the MODULOR. *See also* THEORY OF PROPORTIONS.

GOODS LIFT Freight elevator.

GOTHIC ARCHITECTURE A style characterised structurally by POINTED ARCHES, RIBBED vaults and flying BUTTRESSES. It originated in France in the late twelfth century. *See* EARLY ENGLISH, DECORATED, FLAMBOYANT, and PERPENDICULAR STYLE. *See also* VIOLLET-LE-DUC'S DICTIONARY.

GOTHIC REVIVAL An architectural movement which flourished particularly in the nineteenth century. Also called *Neo-Gothic*.

GPM Gallons per minute.

GRADE BEAM *or* **SLAB** Reinforced concrete beam or slab which forms the foundation for the superstructure; it is normally placed directly on the ground.

GRADING *See* PARTICLE SIZE ANALYSIS.

GRAFFITO A plaster surface decorated by scoring a pattern while it is still wet. The plaster is normally applied in two or more layers of different colour, and the lower coats are exposed by the graffito work. Also spelt *sgraffito* and *scraffeto*.

GRAIN BOUNDARY The boundary between two crystals. *See also* BUBBLE MODEL.

GRAIN SIZE *See* PARTICLE SIZE.

GRAINING The now obsolete art of painting a surface to look like the grain of wood, marble, etc.

GRAM MOLECULE The quantity of a substance whose mass in grams is equal to its molecular weight.

GRANITE An IGNEOUS rock formed by PLUTONIC INTRUSIONS (consisting of relatively large crystals due to slow cooling). It is of acid composition, its main constituents being QUARTZ, FELSPAR and MICA. Granite is a little heavier than concrete, and generally much stronger (up to 20 000 psi or 140 MN/m²); however, its fire resistance is poor. The term 'granite' is sometimes applied to other high-strength igneous rocks.

GRANOLITHIC CONCRETE Concrete mixed with specially selected aggregate, whose hardness, surface, and particle shape make it suitable for a wearing surface on a heavy-duty floor.

GRAPH PAPER If normal CARTESIAN CO-ORDINATES are used, a printed REFERENCE GRID of closely spaced parallel lines. Graph paper is also available in POLAR CO-ORDINATES, and on a LOGARITHMIC SCALE.

GRAPHICAL ANALYSIS *See* LINK POLYGON, RECIPROCAL DIAGRAM, POLAR DIAGRAMS and WALDRAM DIAGRAM.

GRAPHITE One of the allotropic forms of CARBON. It is present in grey CAST IRON. Colloidal graphite is an excellent lubricant which can be used at high temperatures. Also called *black lead*.

GRATICULE A graduated scale placed in the eyepiece of a microscope.

GRAVEL Naturally occurring deposits of unconsolidated sediment, ranging from about 3 inches to $\frac{3}{16}$ inch (75 to 5 mm), and resulting from the disintegration of rock. Larger samples are called boulders, and smaller material SAND. Most gravel consists of SILICA. Gravel, either whole or crushed, is used as COARSE AGGREGATE for concrete, and as a protective layer on top of built-up roofing.

GRAVITY *See* ACCELERATION, KILOPOND, LBF, NEWTON, and VERTICAL.

GRAVITY CIRCULATION *See* FORCED CIRCULATION.

GRAVITY RETAINING WALL *See* RETAINING WALL.

GRAVITY, SPECIFIC *See* SPECIFIC GRAVITY.

GREAT CIRCLE *See* GEODETIC LINE.

GREEK CROSS A cross with four equal arms. *See also* LATIN CROSS.

GREEK ORDERS, THE *See* ORDERS.

GREEN CONCRETE Concrete which has SET, but not sufficiently HARDENED.

GREEN WOOD Timber which still contains most of the moisture which was present in the living tree; this is reduced by *seasoning*.

GRID *See* REFERENCE GRID and MODULAR GRID.

GRIFFITH CRACK An ideal crack postulated in the theory of the fracture of brittle materials, particularly *glass*. It takes the shape of an elliptical hole, of length 2*c*, and was proposed by A. A. Griffith in 1920. He determined the stress required for elastic crack propagation as

$$\left(\frac{2E\gamma}{\pi c}\right)^{\frac{1}{2}}$$

where E is the modulus of elasticity, π is the circular constant 3·14, and γ is the true surface energy; *i.e.* the work done in creating new surface area by the breaking of atomic bonds.

GRIP LENGTH *See* BOND (*b*) and TRANSFER LENGTH.

GRITSTONE *See* SANDSTONE.

GROIN The curved line at which the soffits of two intersecting vaults meet.

GROIN VAULT A CROSS VAULT.

GROPIUS *See* BAUHAUS and DEUTSCHER WERKBUND.

GROUND FLOOR *See* FLOOR.

GROUT A cement, mortar or concrete slurry which is sufficiently fluid to penetrate into rock fissures, masonry joints or prestressing ducts without segregation of the constituents. A commonly used mixture consists of equal volumes of cement and sand, with an appropriate amount of water.

GROWTH RINGS Rings on the transverse section of a trunk or branch of a tree, which mark successive cycles of growth. Also called *annual rings*.

GUM ARABIC A white powder obtained from certain acacia trees grown mostly in the Sudan and Senegal. It is used for glue, and as a base for transparent paint. *See* TEMPERA PAINTING.

GUM TREE *See* EUCALYPT.

GUM VEIN Local accumulation of natural resin, which occurs as a wide streak in certain hardwoods, particularly the eucalypts. It is a defect which can seriously weaken the timber.

GUN COTTON *See* NITROCELLULOSE.

GUN METAL An alloy containing about 90 per cent copper, 8 per cent tin and 2 per cent zinc.

GUNITE *See* SHOTCRETE.

GUSSET A piece of plate to which the members of a truss are joined, if they are too small or inconveniently shaped for a direct connection.

GUY ROPE A rope which secures or steadies a DERRICK or a temporary structure.

GYMNOSPERM *See* CONIFER.

GYPSUM Calcium sulphate dihydrate ($CaSO_4.2H_2O$). It is a natural mineral which is the raw material of gypsum PLASTER. ALABASTER is a pure variety of gypsum.

H

H Chemical symbol for *hydrogen*.

HAIR Reinforcement for lime and gypsum plaster. Hair from bullocks and goats was once used; however, manila fibre, sisal and asbestos are now more common. *See* FIBROUS PLASTER.

HAIR CRACKS Fine cracks just visible with the naked eye.

HAIR HYGROMETER An instrument based on the experimental relation between the increase in the length of certain animal hairs and the relative humidity of the atmosphere, due to the absorption of moisture by the hair. The principle is commonly used in HYGROGRAPHS.

HALF BAT One half of a brick, cut in two across the length. *See also* QUEEN CLOSER.

HALF-LAP JOINT Joint formed between two pieces of timber, each HALVED so that the surfaces are flush in the assembled joint. It may be an END JOINT (Fig. 49) or a *right-angled*, or END-LAP JOINT.

HALF-OCTAVE-BAND ANA-LYSER *See* SOUND FREQUENCY ANALYSER.

HALF-ROUND A semi-circular moulding.

HALF-TIMBERED Building constructed with a timber frame, the spaces between the frame timbers being filled with brickwork, plaster, or WATTLE AND DAUB. This is still a vernacular form of construction in some rural districts of Europe. It was widely used in TUDOR England, even for large urban buildings. Hence 'half-timbered' suburban houses are intended to convey the spirit of Merrie England; they are usually of brick or brick-veneer construction, with thin vertical and diagonal pieces of timber superimposed.

HALIDES The binary salts (*chlorides, bromides, iodides* and *fluorides*) formed by the union of a *halogen* (chlorine, bromine, iodine or fluorine) with a metal. The best-known halide is common salt, NaCl.

HALVING Cutting away one half of the thickness of each of two pieces of timber, so that the surfaces are flush in the assembled joint (Fig. 49). It may be used for lengthening joints (*half-lap end joint*), for right-angled joints at a corner (*end-lap joint*), and

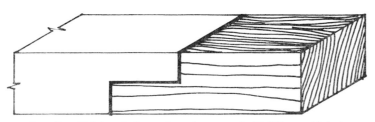

FIG. 49.   Halving timber reduces its strength (Half-lap end joint).

for crossed timbers. Evidently halving seriously reduces the flexural strength of the timber, and it can only be used where the bending moments are low, or where flexural strength is not critical.

HAMMER *See* BALL-PEEN HAMMER and CLAW HAMMER.

HAMMER-BEAM ROOF A medieval timber roof without a tie. The hammer beams are supported on brackets projecting from the wall, but (unlike a conventional tie) they do not meet. Instead, the horizontal reaction is transmitted by arched braces and struts. Because of the opening between the hammer beams, the clear height of the room is greater than in one with KING-POST or queen post roof trusses. The most celebrated example is the roof of Westminster Hall in London (the only surviving part of the old Westminster Palace), which was frequently copied during the nineteenth century Gothic Revival.

HAND *See* RIGHT-HAND.

HAND LEVEL *See* CLINOMETER.

HARD SOLDER *See* SOLDERING.

HARD WATER Water containing calcium or magnesium salts in solution. These are picked up if water passes slowly through limestone or dolomite. The salts react with soap, and thus make washing and laundering difficult. They are also deposited on heating, and thus tend to block the pipes of hot-water systems. Water which is too hard must be treated to remove some of the salt, or *softened*.

HARDBOARD A FIBRE BOARD formed under pressure to a density of 30 to 60 lb/ft^3 (480 to 960 kg/m^3). Most hardboards have one smooth and one textured surface; however, hardboards are also made with decorative and veneered surfaces, and they can be treated to be water-resistant. Also called *compressed fibre board*.

HARDENER A CATALYST which increases the hardening rate of synthetic resins or glues. It is mixed with the resin immediately before use. Also called *accelerator*.

HARDENING OF CONCRETE The stage in the chemical reaction between cement and water when the concrete hardens and gains sufficient strength to bear its own weight and that of the construction loads. *See also* SETTING and AGEING.

HARDENING OF METAL *See* CASE-HARDENING, COLD-WORKING and QUENCHING.

HARDNESS SCALE *See* MOHS' SCALE.

HARDWARE The physical items of computer equipment, *i.e.* the input, processor, store and output, as opposed to the SOFTWARE.

HARDWOOD Timber from trees belonging to the botanical group *angiospermae*, *i.e.* all trees but the CONIFERS, which are called *softwoods*. While some softwoods are moderately hard, some hardwoods are very soft (*e.g.* BALSA WOOD). However, taken as a group, the hardwoods are much harder than the softwoods. Practically all Australian native timbers are hard, and many can only be nailed in their green state. *See also* BOXED HEART.

HARDY CROSS METHOD *See* MOMENT DISTRIBUTION METHOD.

HARMONIC Any *overtone* in a single musical note of complex wave form with a frequency which is an exact multiple of the fundamental or pitch frequency. *See also* SINE WAVE and OCTAVE.

HARMONIC PROPORTIONS The proportions which produce simple harmonies in music, notably the third and the fifth. These proportions were recognised in Ancient Greece, and discussed by Vitruvius. The relation between the harmonic proportions of music and architecture has been traced by Rudolf Wittkower in *Architectural Principles in the Age of Humanism,* London 1949. The important ratio $5/3 = 1.667$ is very close to the GOLDEN SECTION. Although it seems likely that *Palladio* used harmonic proportions, the influence of his *Four Books* accounts indirectly for the interest shown in the golden section in the nineteenth century.

HAUNCH (*a*) The part of an arch near the springing. (*b*) A section of a beam whose depth is increased because of an increase in bending moment.

H-BEAM *See* H-SECTION.

He Chemical symbol for *helium.*

HEAD (*a*) The energy possessed by a fluid due to its elevation above some datum, or due to any other cause, such as its velocity. (*b*) The enlarged part of a bolt. (*c*) The upper end of a column. (*d*) The upper part of a vertical timber. (*e*) The topmost member of a door or window frame.

HEADER A brick, block, or stone laid across the wall, to bond together the STRETCHER bricks. The shorter length, or head, of the brick is visible on the face of the brickwork. *See also* BOND (*a*).

HEADROOM The clear vertical distance between the ground or floor and the lowest point overhead.

HEARING THRESHOLD An empirical curve relating the frequency to the lower limit of sound pressure (in DECIBELS) below which persons with normal hearing are unable to perceive sound. Also called *threshold of audibility. See also* DAMAGE RISK CRITERION and PHON.

HEART The centre of a wooden log. *See* BOXED HEART.

HEAT ABSORBING GLASS *See* ANTI-ACTINIC GLASS.

HEAT GAIN *See* THERMAL CAPACITY.

HEAT GAIN, SOLAR *See* SOLAR HEAT.

HEAT INSULATION *See* THERMAL RESISTANCE.

HEAT OF HYDRATION The quantity of heat liberated or consumed when a substance takes up water.

HEAT PUMP An air-conditioning installation which can act either as a heating or as a cooling unit. The refrigerant soaks up heat from the inside in summer, and from the outside in winter. After a COMPRESSION CYCLE, the heat is released (on the outside in summer, and on the inside in winter).

HEAT-TREATMENT OF METAL *See* ANNEALING, NORMALISING, TEMPERING and QUENCHING.

HECTO (h) Prefix for 100 times, from the Greek word for hundred; *e.g.* 1 hectare = 100 are.

HEEL *See* CANTILEVER RETAINING WALL.

HELD WATER *See* CAPILLARY WATER.

HELIARC PROCESS A process similar to ARGON-ARC WELDING, but employing helium. It is used in the United States where helium is readily available.

HELICAL REINFORCEMENT Small-diameter reinforcement wound around the main, or LONGITUDINAL, REINFORCEMENT of columns. It restrains the lateral expansion of the concrete under compression, and consequently increases the column strength. Also called (incorrectly) SPIRAL REINFORCEMENT. *See also* HOOP REINFORCEMENT.

HELICAL STAIR A stair whose treads are arranged along a helix. Commonly (but incorrectly) called a *spiral stair*.

HELIODON *See* SOLARSCOPE.

HELIUM The lightest of the inert gases, used in HELIARC WELDING. Its chemical symbol is He, its atomic number is 2, its atomic weight is 4·002, and its boiling point is −268·9°C.

HELMHOLTZ RESONATOR A resonant absorber, named after the nineteenth century Prussian scientist. It consists of a narrow neck connected to a larger volume of air which vibrates. This sound energy can be fed back into the room, or it can be partly absorbed within the resonator, which then becomes an absorber. Resonators can be tuned to a particular frequency, which can be adjusted with a movable piston. More commonly, perforated panels are used over an airspace. The resonant frequency (in hertz) of a perforated panel is approximately $170/(md)^{\frac{1}{2}}$, where m is the mass of the panel, in pounds per square foot, and d is the depth of the air space behind it, in inches. *See* ACOUSTIC BOARD.

HEMIHYDRATE A compound containing one-half of a molecule of water to one molecule of another substance. The hemihydrate most widely used in building is PLASTER OF PARIS.

HENCKY–VON MISES FAILURE CRITERION A FAILURE CRITERION widely used for the design of ductile metals, such as structural steel, under combined stresses. It is based on the strain energy absorbed in the distortion of the material.

HEPTAGON *See* POLYGON.

HERRINGBONE In masonry and carpentry, a zigzag pattern or bond.

HERTZ Unit of frequency. 1 hertz (Hz) = 1 cycle per second, 1 kilohertz = 1000 cycles per second. It is named after the German nineteenth century physicist.

HESSIAN Burlap.

HEXAGON *See* POLYGON.

HEXAGONAL CLOSE-PACKED LATTICE *See* CLOSE-PACKED HEXAGONAL LATTICE.

HEXAHEDRON *See* POLYHEDRON.

HIDING POWER The ability of a paint to obscure existing colours and patterns; it is a measure of its opacity. *Covering power* is the hiding power for a given spreading rate.

HI-FI High-fidelity.

HIGH-ALUMINA CEMENT Cement manufactured from BAUXITE and limestone. It is chemically different from PORTLAND CEMENT and gains its strength more rapidly than high-early-strength Portland cement. Also called *calcium aluminate cement*.

HIGH-BOND BAR *See* DEFORMED BAR.

HIGH-CARBON STEEL Carbon steel containing more than 0·5 per cent of carbon. *See also* LOW-CARBON STEEL.

HIGH-EARLY-STRENGTH CEMENT A PORTLAND CEMENT which gains strength more rapidly than the ordinary variety, and costs slightly more. It has a higher SPECIFIC SURFACE through finer grinding, and thus allows the chemical action to proceed more rapidly. Also called *rapid-hardening cement*.

HIGH-FIDELITY Sound reproduction of a superior, but generally undefined quality.

HIGH GLOSS An enamel-like finish on paint or varnish.

HIGH-LEVEL LANGUAGE *See* PROBLEM-ORIENTED LANGUAGE.

HIGH-LIGHT WINDOW A CLEARSTOREY window. The term is generally preferred for modern buildings.

HIGH-PRESSURE STEAM CURING *See* AUTOCLAVE.

HIGH-SPEED STEEL A high-alloy steel used in metal-cutting tools operating at high speeds. It contains 12 to 22 per cent of tungsten and is capable of intense hardening.

HIGH-STRENGTH BOLT *See* HIGH-TENSILE BOLT.

HIGH-STRENGTH STEEL Steel with a high YIELD or PROOF STRESS, generally above 60 000 psi (400 MN/m^2).

HIGH-TENSILE BOLT A bolt tightened to a carefully controlled tension with a calibrated torsion wrench. This ensures on the one hand that the bolt is not damaged by over-tensioning, and on the other hand it provides high frictional forces between the two pieces of steel to be connected. This form of jointing is more convenient than site RIVETING, and more reliable than the use of BLACK BOLTS.

HIGHLIGHTING Emphasising 'high' parts of a painting or surface decoration with a lighter colour, to create a three-dimensional effect, as in a relief.

HILL, SIR LEONARD   *See* KATA-THERMOMETER.

HINGE JOINT   *See* PIN JOINT and PLASTIC HINGE.

HINGELESS FRAME   *See* RIGID FRAME.

HIP *and* **HIP ROOF**   *See* GABLE ROOF.

HIPPED-PLATE ROOF *See* FOLDED-PLATE ROOF.

HIPPODROME   Literally an arena for equestrian performances.

HISTOGRAM   Literally a diagram showing a web-like pattern. Its most common use is as a *frequency histogram,* which shows the relation between the magnitude of an observed variable characteristic and its frequency as a series of bars, or vertical rectangles; each rectangle is drawn so that its height corresponds to the frequency (Fig. 50). *See also* FREQUENCY DISTRIBUTION CURVE.

HITCHCOCK, HENRY-RUSSELL *See* INTERNATIONAL STYLE.

HMSO   HM Stationery Office, London.

HOGGING   Camber.

HOGGING MOMENT   A negative BENDING MOMENT, such as occurs at a support (Fig. 66); it causes a 'hogging' deformation. *See also* SAGGING MOMENT.

HOLE, EFFECT ON STRENGTH *See* STRESS CONCENTRATION.

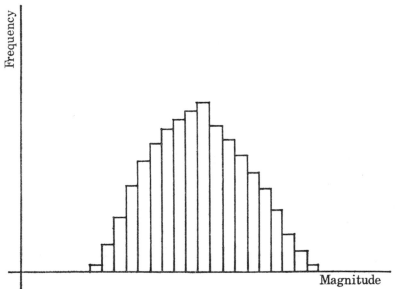

Fɪɢ. 50.   Histogram.

HOLLOW BLOCK An extruded BLOCK of concrete or burnt clay, which consists largely of voids, and is consequently a good insulator, but of limited strength. It is used for partitions and in floors. Also called a *hollow tile*.

HOLLOW-CORE DOOR A door formed of a light timber frame, faced on both sides with hardboard or plywood.

HOLLOW-TILE FLOOR A reinforced concrete floor which is cast over a soffit consisting of hollow blocks (of burnt clay or concrete) between narrow ribs (Fig. 51). Its structural design is like that of a RIBBED SLAB; however, it presents a flat, uniform soffit after plastering, and its thermal insulation is superior to that of a solid slab. Also called a *pot floor*.

HOME INSURANCE BUILDING The first building with a steel frame, erected in Chicago in 1884, and demolished in 1931. It was nine storeys high, and designed by *W. L. B. Jenney.*

HOME UNIT Australian term for an APARTMENT which is sold and not rented.

HOMOGENEOUS Having identical characteristics throughout

HONEYCOMB CORE A structure of air cells, resembling a honeycomb,

placed between plywood panels. It has excellent insulating properties and adequate flexural strength. The honeycomb is commonly made of paper.

HONEYCOMBED CONCRETE Concrete with voids caused by failure of the cement mortar to fill all the spaces between the particles of the coarse aggregate.

HOOKE BODY In rheological terminology, an ELASTIC material. The rheological model of this body, named after the seventeenth century English scientist, is an elastic spring, which extends instantly when pulled, and contracts instantly when released. (Fig. 81).

HOOKED BAR A concrete reinforcing bar whose end is bent to improve its anchorage, generally through 180°. The minimum radius of the hook must be sufficient to avoid stress concentrations in the concrete, and a free length of bar is required at the end of the hook to prevent pullout of the bar. Hooks are rarely required in DEFORMED BARS.

HOOKE'S LAW In 1678 Robert Hooke published his observation that all known ELASTIC materials had deformations which were directly proportional to the applied loads. Today the law is usually stated as 'STRESS is proportional to STRAIN'. (*See* Fig. 93). Hooke's law is the basis of the ELASTIC DESIGN of structures.

FIG. 51.   Hollow-tile floor, or concrete joist construction with tile fillers.

Because of the linear relation between stress and strain, the theory is relatively simple (*see* SUPERPOSITION). The linearity of the stress–strain relationship for the structural materials used in building does not, however, mean that the entire theory of elastic structures is linear; buckling, for example, is a notable exception. The main alternative to the structural theory based on Hooke's law is LIMIT DESIGN.

HOOP FORCES The MEMBRANE forces in a dome which follow the horizontal hoops, or parallels of latitude. They are compressive near the crown in all domes. They are compressive throughout the shell in a *shallow dome*, but HOOP TENSION occurs in the lower portion of a hemi-spherical dome (Fig. 29).

HOOP REINFORCEMENT Closed hoops around the main, or longitudinal, reinforcement of columns. They restrain the BUCKLING of the longitudinal steel, but not the lateral expansion of the concrete. They are consequently not as effective as HELICAL REINFORCEMENT; but since they are simpler to make and fix, they are more commonly used as lateral column reinforcement. Also called *tie*.

HOOP TENSION The tension which occurs in the lower portion of a hemi-spherical DOME (Fig. 29), in a cylindrical or spherical water-tank, or in the *tension ring* which absorbs the horizontal components of the reactions of a shallow dome. Masonry materials are ill-adapted to resist tension. In classical masonry structures hoop tension was sometimes partly absorbed by chains placed around the masonry structure (*e.g.*

St Peter's in Rome). In modern concrete structures it is resisted by reinforcement. Since the reinforcement extends when it resists the tension, it is advisable to PRESTRESS water tanks to ensure against leakage. Tanks can also be built of steel, if adequate provision is made against corrosion.

HORIZON The circle, as seen by an observer, which has an altitude of zero. If the true horizon is obstructed, *e.g.* by mountains or buildings, it may be necessary to use an *artificial horizon*. The point above the observer, which makes an angle of 90° to the horizon, is the *zenith*.

HORIZONTAL *See* VERTICAL.

HORSEPOWER (hp) The performance of 33 000 foot-pounds of work per minute (= 746 watts); originally intended to equal the capacity of a horse when rating the power of a mechanical vehicle. However, a horse can perform better than 1 hp, at least for a short time. The metric horse-power (force de cheval—*ch*, or Pferdestärke—*PS*) equals 4500 kg f m/min, which is 0·986 hp. In the SI SYSTEM the horsepower is replaced by the watt, 1 ch = 735·5 W.

HOT-AIR SEASONING Drying timber in a kiln.

HOT DIPPING *See* GALVANISING.

HOT ROLLING *See* HOT WORKING.

HOT-WIRE ANEMOMETER A remote-reading instrument for measuring low air speeds. An exposed fine resistance wire is heated by the passage of an electric current, and the effect of air movement on its temperature is measured. The probe is small enough to be used for model analysis in a wind tunnel.

HOT WORKING The mechanical working of a metal above the temperature for recrystallisation, as opposed to COLD WORKING.

HOWARD DIAGRAM A polar diagram which provides a graphical solution for the buckling of laterally loaded struts, developed by the British aeronautical engineer, H. B. Howard.

HOWE TRUSS An *isostatic* truss (Fig. 52) consisting of top and bottom chords connected by diagonal *compression* members and vertical tension members, as distinct from a PRATT and a WARREN TRUSS.

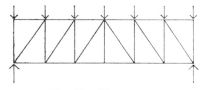

FIG. 52. Howe truss.

hp Horsepower.

H-SECTION An I-BEAM with wide flanges, which is also particularly useful in column design.

HT High tensile.

HUE *See* COLOUR.

HUGGENBERGER TENSO-METER An EXTENSOMETER with a gauge length of 10 to 20 mm, employing a compound lever system giving a magnification of about 1200. Because of its lightness and short gauge length, it is one of the mechanical instruments most commonly used in structural model analysis.

HUGGENBERGER TENSOTAST *See* DEMEC STRAIN GAUGE.

HUMAN TOLERANCE OF NOISE *See* DAMAGE RISK CRITERION, HEARING THRESHOLD and BACKGROUND NOISE.

HUMIDIFIER Component of an AIR CONDITIONING plant which removes suspended dirt by spraying or washing, and raises the moisture content, if necessary. Also called an *air washer*. The spray equipment of an air conditioning plant may be used either for *humidification* or for DEHUMIDIFICATION.

HUMIDITY Water vapour within a given space. The *absolute humidity* is the weight of water vapour per unit volume. *See also* RELATIVE HUMIDITY.

HUMUS The dark coloured, fertile portion of the topsoil, which contains a large proportion of rotting vegetation. It is excellent for growing plants, but a poor foundation material, and it is removed before construction commences.

HUNG CEILING *See* FALSE CEILING.

HUNGRY JOINTS *See* STARVED JOINTS.

HURRICANE A wind of force 12 on the BEAUFORT SCALE. Its speed is approximately 75 mile/h (120 km/h).

HYDRANT A connection to a water main, usually for extinguishing a fire.

HYDRATED LIME Calcium hydroxide $(CaO(OH)_2)$, also called *slaked lime*. It is formed by slaking QUICKLIME, *i.e.* adding water to it. The old-fashioned product was made in a lime maturing pit, and the *lime putty* required some weeks to mature. It is now common practice to use *dry hydrate* (a dry powder supplied in bags) for lime mortar.

HYDRATION Addition of water, particularly to cement, lime and plaster, and the subsequent chemical action.

HYDRAULIC CEMENT An old-fashioned term to distinguish cement proper from lime. Cement sets and hardens under water due to the interaction of the cement with the water. Lime is washed out if submerged for a long period in water, because hydrated lime is soluble. Lime mortar was the normal jointing material for blocks of stone and for brick prior to the nineteenth century, and both lime and cement mortar were used in the nineteenth and early twentieth centuries.

HYDRAULIC FRICTION The loss of HEAD caused by roughness or obstruction in a pipe or channel.

HYDRAULIC GLUE An old-fashioned term for a glue which retains its adhesion under water. Few of the traditional glues satisfied this requirement, but most synthetic RESINS do. *See also* MARINE GLUE.

HYDRAULIC GRADIENT The difference in the water level (in a pipe or in soil) between two points, divided by the shortest path between them. *See* DARCY'S LAW.

HYDRAULIC JACK *See* JACK.

HYDRAULIC LIFT *or* **ELEVATOR** A lift (or elevator) raised by a hydraulic ram underneath it. It requires a high-pressure water supply. Some hydraulic lifts are still in use.

HYDRAULICS The science of the flow of fluids.

HYDROGEN The lightest element known; it exhibits both metallic and non-metallic properties. On burning it forms water (H_2O). Its chemical symbol is H, its atomic number is 1, its atomic weight is 1·008, it has a valency of 1, and its boiling point is $-252\cdot7°C$.

HYDROGRAPH A curve showing the variation of water flow over a period of time.

HYDROMETER An instrument for measuring the specific gravity of a liquid.

HYDROSTATIC PRESSURE The pressure at any point in a liquid, which is at rest. It equals the density of the liquid, multiplied by the depth of the point under consideration. *See* CENTRE OF PRESSURE and RETAINING WALL.

HYGROGRAPH A recording hygrometer, which plots the change in RELATIVE HUMIDITY with time on a clock-driven drum. Many hygrographs are based on the HAIR-HYGROMETER principle.

HYGROMETER An instrument for measuring the humidity in the air. The simplest type is the WET-AND-DRY BULB THERMOMETER.

HYGROSCOPIC MATERIAL A material which readily absorbs moisture.

HYPAR *See* HYPERBOLIC PARABOLOID.

HYPERBOLA The section of a right circular CONE by a plane which intersects the cone on both sides of the apex. Its equation is of the form $x^2/a - y^2/b = 1$, where a and b are constants.

HYPERBOLIC FUNCTIONS A set of six functions *sinh*, *cosh*, *tanh*, *coth*, *sech* and *cosech*, which are analogous to the TRIGONOMETRIC FUNCTIONS, but are derived from the properties of the hyperbola instead of the circle. They can also be interpreted in terms of EXPONENTIAL FUNCTIONS:

$$\sinh x = \tfrac{1}{2}(e^x - e^{-x})$$

and

$$\cosh x = \tfrac{1}{2}(e^x + e^{-x})$$

See also Appendix D.

HYPERBOLIC PARABOLOID A geometric surface which has the equation $z = kxy$, where x, y and z are the Cartesian co-ordinates, and k is a constant (Fig. 53). It has a negative GAUSSIAN CURVATURE, *i.e.* it is a saddle-shaped. It is a RULED surface, and can be generated by moving a straight line over two other straight lines inclined to one another. Frequently abbreviated to *hypar*.

HYPERBOLIC PARABOLOID SHELL A shell in the form of an hyperbolic paraboloid. Since the shape can be generated by two systems of straight lines, its formwork is more readily constructed from straight pieces of timber than that of a dome. Although the theory was worked out by *F. Aimond* in the early 1930s, the practical design owes most to *Felix Candela*, who has built a large number of varied concrete hyperbolic paraboloid shells in the 1950s. (*See* Figs. 53 and 100).

HYPERBOLOID OF REVOLUTION *See* SURFACE OF REVOLUTION and Fig. 95.

HYPERSTATIC STRUCTURE A structure which has more members or

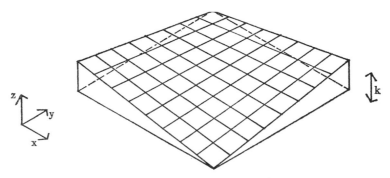

FIG. 53.   Hyperbolic paraboloid.

restraints than can be solved by the use of STATICS alone; also called a *statically indeterminate* structure. A hyperstatic structure has more members or restraints than are required by MÖBIUS' LAW, and those in excess of static requirements are called REDUNDANCIES. If the redundancies are removed, the structure becomes ISOSTATIC. The principal methods for the solution of hyperstatic structures are the MOMENT DISTRIBUTION, SLOPE DEFLECTION, and STRAIN ENERGY METHODS. When electronic digital computers are used, solution is more convenient by the FLEXIBILITY METHOD or by the STIFFNESS METHOD.

HYPO A colloquial term for *sodium thiosulphate*, the material most commonly used for fixing photographic film after developing. It removes the unreduced silver halide emulsion.

HYPOCAUST A central heating system used in Ancient Roman baths and (occasionally) villas. Hot air and gases from a fire were passed through masonry chambers and flues under the floors.

HYPOSTYLE A hall whose roof is supported by rows of columns giving a forest-like appearance.

HYPOTENUSE The longest side of a right-angled triangle, which is opposite the right angle.

HYPSOMETER An instrument in which water is boiled and the boiling temperature measured. It may be used at ground level for calibrating thermometers, or in conjunction with a calibrated thermometer to measure the altitude above ground level, since water has a lower boiling point when the air pressure is reduced.

HYSTERESIS *See* MECHANICAL HYSTERESIS.

I

I The common symbol for the second moment of area (moment of inertia).

IABSE International Association for Bridge and Structural Engineering, Zurich.

I-BEAM A metal section, particularly a rolled steel JOIST, shaped like the letter I. By concentrating the material in the flanges, where the highest flexural stresses occur, and giving the section maximum depth, the material is used in the most economical way. Also called an H-section, if the flanges are relatively wide.

ICE Institution of Civil Engineers, London.

ICELAND SPAR A pure, transparent, and crystalline form of calcium carbonate ($CaCO_3$). It is noteworthy for its double reflection and perfect cleavage, and hence is utilised for producing plane-polarised light in the form of NICOL PRISMS.

ICOSAHEDRON *See* POLY-HEDRON.

ID Internal diameter.

IEAust Institution of Engineers, Australia, Canberra.

IEE Institution of Electrical Engineers, London.

IFBWW International Federation of Building and Woodworkers, Copenhagen.

IFHP International Federation for Housing and Planning, The Hague.

IGNEOUS ROCK Rock formed by the solidification of magma, injected from the earth's interior into its crust, or extruded on its surface. Igneous rocks are classified according to whether they are predominantly acid or basic, and according to their grain size (which depends on the rate of cooling). *See* BASALT, DIORITE, DOLERITE, GRANITE, LAVA, MINOR INTRUSIONS, OBSIDIAN, PLUTONIC INTRUSIONS, PORPHYRY and PUMICE. *See also* METAMORPHIC and SEDIMENTARY ROCK.

ILLUMINATION *See* LIGHTING.

IMAGINARY NUMBER A *real* number multiplied by $\sqrt{-1}$. Thus the square-root of a positive number is real, the square-root of a negative number is imaginary. A *complex* number is the sum of real and imaginary numbers. *See also* IRRATIONAL NUMBER.

IMechE Institution of Mechanical Engineers, London.

IMMEDIATE ACCESS STORE The form of storage provided in the CENTRAL PROCESSOR UNIT of a digital computer.

IMMERSION HEATER An electric resistance heater immersed in a water tank.

IMMERSION VIBRATOR *See* VIBRATED CONCRETE.

IMMISCIBLE Incapable of being mixed to form a homogeneous liquid; *e.g.* water and oil are immiscible.

IMPACT FACTOR A factor by which a DYNAMIC LOAD is multiplied to allow for the vertical impact which it makes on the floor. Impact factors for bridges generally range from 1·0 to 1·6. Impact factors are rarely used in the design of buildings.

IMPACT RESISTANCE Capacity of a material to resist suddenly applied or shock loads. *See* CHARPY TEST and IZOD TEST.

IMPACT SOUND Noise transmitted through building elements by impact, such as footsteps and vibrating bodies; as opposed to AIRBORNE SOUND. Sound insulation against impact sound is best achieved by DISCONTINUOUS CONSTRUCTION.

IMPEDANCE, ACOUSTIC *See* RESONANCE.

IMPEDANCE TUBE An instrument for measuring SOUND ABSORPTION on small samples for normal (90°) sound incidence only. It consists of a long tube, with the test sample at one end. At the other end is a loudspeaker. The incident wave is partly reflected and partly absorbed by the sample, and the reflected wave interferes with the incident wave. The pressure distribution is examined with a travelling microphone which is small enough not to disturb the sound field. From the ratio of the pressures at the maximum and minimum positions, the sound reflection coefficient and the absorption coefficient can be determined. Also called a *standing-wave tube*. *See also* REVERBERATION CHAMBER.

IMPREGNATION OF TIMBER The process of saturating timber with a preservative, such as CREOSOTE OIL.

IMPULSE TURBINE *See* REACTION TURBINE.

IN SITU *See* CAST IN SITU.

IN SITU PILE A concrete pile cast, with or without a casing, in its final location, as distinct from a pile which is precast and subsequently driven.

INCANDESCENT LAMP (*a*) Originally a WELSBACH MANTLE impregnated with the oxides of thorium and cerium, made white-hot by a gas flame. (*b*) A small filament, usually of tungsten, placed inside a glass bulb in an inert gas, made white-hot by the passage of an electric current, *i.e.* the ordinary *filament light bulb*, as opposed to a FLUORESCENT TUBE.

INCISE To cut into a material, *e.g.* masonry or wood. The term implies a shallow cut as in an engraving, rather than a sculptural carving.

INCLINATOR A platform running on an inclined rail for taking supplies, dustbins, etc. from the street to a house at a much higher or lower level, otherwise accessible only by steps.

INCLINOMETER *See* CLINOMETER.

INCOMBUSTIBLE A material which does not burn in a standard test in a furnace (usually lasting $2\frac{1}{2}$ hours). Also called *non-combustible*.

INDENTED WIRE Wire with machine-made surface indentations. Unlike the hot-rolled deformations or DEFORMED BARS, the indentations can be cold-rolled, and therefore placed also on high-tensile wires. They are used to improve the bond, particularly in PRE-TENSIONING tendons.

INDEX OF PLASTICITY *See* PLASTICITY INDEX.

INDIRECT GLARE *See* DIRECT GLARE.

INDIRECT MODEL ANALYSIS Analysis of a structural problem by means of a model which represents an analogy to the behaviour of the structure. There are several methods, the BEGGS' DEFORMETER being a typical example. An indirect model may be considered as an ANALOGUE COMPUTER, although the model bears a physical resemblance to the structure under investigation. In practice the various forms of indirect model analysis have been superseded by DIRECT MODEL ANALYSIS.

INDIVIDUAL-MOULD PRE-TENSIONING *See* PRE-TENSIONING.

INDUSTRIALISED BUILDING Building with factory-made components (which may be small or comprise several rooms in one piece) to reduce the amount of work on the site. Work in the factory is deliberately increased to speed up construction and/or reduce cost. The term 'industrialised building' superseded *prefabrication* (with which it is practically synonymous) in the 1950s, after prefabricated houses had acquired a poor reputation in the late 1940s because of several unsatisfactory designs, which gave 'prefab' the quality of a term of abuse. Industrialised building may be in accordance with a *closed system*, whereby all components for a building are made by one manufacturer, and are generally designed only for one type of building; or it may be in accordance with an *open system*, whereby a building is assembled from stock components, which are interchangeable and may be produced by different manufacturers.

INELASTIC Not ELASTIC. Inelastic deformation is not immediately recovered when the load is removed. It may be permanent, *i.e.* it remains permanently when the stress is removed (*see* PLASTICITY), or it may be recoverable over a period of time (*see* CREEP).

INERT GASES Helium, neon, argon and krypton.

INERT PIGMENT A pigment which does not undergo chemical change.

INERTIA, MOMENT OF *See* MOMENT OF INERTIA.

INFINITE SERIES A regular arrangement of mathematical terms which can be continued indefinitely. The sum of all the terms of a *nonconvergent* infinite series is infinity. The sum of all the terms of a *convergent* series can be ascertained, *e.g.* the series $1 + \frac{1}{2} + \frac{1}{4} + \frac{1}{8} + \cdots$ adds up to 2.

INFLAMMABLE Combustible.

INFLATABLE GASKET *See* GASKET.

INFLECTION *See* CONTRAFLEXURE.

INFLUENCE LINE A diagram showing the effect of moving a load along a beam (frame, arch, etc.). Thus an influence line for bending moment shows the variation of bending moment *at one point* of the beam as the load (usually a concentrated unit load) is moved along the beam; by contrast, the bending moment diagram shows the variation of bending moment along the beam due to *one combination of loads*. The effect of a number of loads can be obtained from the influence line by adding up the effects of the various loads, and the worst load combination can thus be obtained from a single influence line, obviating the need for several bending moment diagrams. Influence lines can also be drawn for shear, torsion, deflection, etc.

INFRA-RED RADIATION Electromagnetic radiation with wavelengths longer than 7600 Å (760 nm) *i.e.* beyond the red end of visible light. It is RADIANT HEAT, and forms part of the radiation received from the sun.

INGLE-NOOK A corner by an open fire, usually with a built-in seat.

INGOT A mass of metal cast into a mould. It is the raw material for rolling and forging.

INHIBITING PIGMENT A pigment which prevents corrosion of a metal surface. *See* PRIMER.

INITIAL PRESTRESS The force applied to the concrete by a TENDON at the time of the prestressing operation. *See* LOSS OF PRESTRESS.

INITIAL SET The onset of stiffening in concrete, cement, plaster, etc., after water has been added. For Portland cement it is measured by the *Vicat test*, which consists of placing a weighted test needle on the hydrated cement pat. The initial set is defined by failure of the initial-test needle to penetrate the pat. The *final set* is defined by failure of the final-test Vicat needle to make a $\frac{1}{2}$ mm indentation on the cement pat.

INITIAL STRESS *See* INITIAL PRESTRESS.

INITIAL-TIME-DELAY GAP *See* INTIMACY.

INJECTION MOULDING The moulding of liquid plastics, liquid metal, or other material by injection into a mould.

INORGANIC FIBRE BOARD Board made from inorganic fibres, such as asbestos or fibreglass.

INPUT The method of transferring information into a digital computer, usually via punched CARDS or paper TAPES. However, a LIGHT PEN may also be used.

INPUT–OUTPUT DEVICE A unit, usually containing an electric typewriter, which is used both to transfer information to the CENTRAL PROCESSOR UNIT of a digital computer and to print out information in a form intelligible to the user. It may be a remote console connected to a central

processor unit by telephone on a TIME-SHARING basis.

INSOLATION Exposure to the sun's rays. *See also* SOLAR HEAT.

INSTRUCTION A word or set of characters in MACHINE LANGUAGE, which directs a digital computer to execute a particular operation. A series of instructions constitutes a PROGRAM.

INSULATING BOARD FIBRE-BOARD of a density not exceeding 25 lb/ft³ (400 kg/m³), specifically designed to give good thermal insulation. *See also* ACOUSTIC BOARD.

INSULATION The prevention of the flow of an electric current, or the retardation of the flow of heat or passage of sound.

INSULATION OF IMPACT SOUNDS *See* SOUND INSULATION and DISCONTINUOUS CONSTRUCTION.

INSULATION, SOUND *See* SOUND INSULATION.

INSULATION, THERMAL *See* INSULATING BOARD and THERMAL RESISTANCE.

INTAGLIO Originally a design INCISED or carved into a material, as opposed to a design carved in RELIEF; now also used for a shallow design *pressed* into a surface.

INTEGER A whole number, which may be positive, negative or zero. The term excludes fractions, square roots, and imaginary numbers.

INTEGRAL The sum of a series of consecutive values of a function.

Geometrically the integral is the area contained under the curve of the function.

INTEGRAL NUMBER *See* INTEGER.

INTEGRAL WATERPROOFING Waterproofing concrete by an admixture to the cement or the mixing water, as opposed to waterproofing applied subsequently to the surface. It should be pointed out, however, that carefully placed concrete is often as waterproof as concrete with an admixture. *See also* SURFACE WATER-PROOFER.

INTERCOLUMNIATION The space between classical columns measured in diameters. The five main ratios mentioned by Vitruvius are $1\frac{1}{2}$ (called *pycnostyle*), 2 (*systyle*), $2\frac{1}{4}$ (*eustyle*, the one most frequently used), 3 (*diastyle*) and 4 (*araeostyle*).

INTERFACE STRENGTH *See* BOND.

INTERFERENCE The effect of superposing two or more trains of waves of equal wavelength. The resultant amplitude is the algebraic sum of the amplitudes of the interfering trains. When two sets of circular waves interfere, a system of hyperbolic stationary nodes and antinodes is formed, which are known as interference fringes in the case of visible light. *See also* MOIRÉ FRINGES and ISOCLINICS.

INTERIOR SPAN A span, other than the end span, in a CONTINUOUS BEAM. If the spans are equal, the bending moments are generally highest in the end span (*see* Fig. 66).

INTERIOR SUPPORT A support, other than the end support, in a CONTINUOUS BEAM. If the spans are equal, the bending moment is generally highest at the first interior support (*see* Fig. 66).

INTERNAL FRICTION *See* ANGLE OF INTERNAL FRICTION.

INTERNAL VIBRATOR *See* VIBRATED CONCRETE.

INTERNALLY RECEIVED COMPONENT *See* DAYLIGHT FACTOR.

INTERNATIONAL STYLE A term coined by *Henry-Russell Hitchcock* to denote the 'modern' style of architecture which came into existence in the 1920s, and had its classical period in the 1930s and the late 1940s. It was first used in a book, which he published with *Philip Johnson* in 1932, called *International Style. See also* BAUHAUS and CIAM.

INTERPOLATE To infer the position of a point on a graph defined by several other known points, by assuming that the curve is smooth. *Extrapolation* is the same process, continuing the graph on either end *beyond* the last known point; it is inherently less accurate than interpolation.

INTIMACY As a visual quality, smallness. As an acoustical quality, the impression created that music is being played in a small hall, or has *presence*. A listener's impression of the size of the hall is determined by the *initial-time-delay gap*, *i.e.* the interval between the sound which arrives directly at the ear and the first reflection which arrives from the walls or ceiling.

INTRADOS The inner curve of an arch.

INVERSE-SQUARE LAW 'As a spherical wave of light or sound travels outwards from a point source, its intensity decreases in inverse proportion to the square of its distance.'

INVERT LEVEL The level of the lowest portion at any given section of the inside surface of a drain, sewer or other liquid-carrying conduit. It determines the hydraulic gradient available for moving the liquid in the conduit.

ION An atom which carries a charge (either positive or negative) because it has had electrons added or removed.

IONIC BOND Atomic bond by attraction between unlike ions.

IONIC ORDER One of the three Greek and five Roman ORDERS. It originated in Asia Minor in the sixth century BC. Its distinctive feature is the column capital which is decorated with a scroll.

IONISATION ANEMOMETER A remote reading instrument for measuring low air speeds which is less sensitive to natural convection than the HOT-WIRE ANEMOMETER. It consists of a sphere coated with radioactive material which is surrounded by a collecting cage and an earthed screen.

IRC *See* DAYLIGHT FACTOR.

IRIDESCENCE The play of the colours of the spectrum on a surface. It is produced by interference of light reflected from the front and back of a very thin film.

IRON A metallic element which exists in the three forms of ALPHA IRON, GAMMA IRON and *delta iron*. Its chemical symbol is Fe, its atomic number is 26, its atomic weight is 55·84, its specific gravity is 7·87, and its melting point is 1535°C. *See also* WROUGHT IRON, STEEL and CAST IRON.

IRON OXIDES Apart from their use as a raw material from which metallic iron is made, the oxides also provide a number of traditional *mineral pigments*, mostly named after the place in Italy from which they were originally obtained. Haematite (Fe_2O_3) provides *Venetian red*. Magnetite (Fe_3O_4) provides purple and black pigments. Ferrous oxide (FeO) is the main ingredient of *Ocher* (yellow), *Sienna* (yellow) and *burnt Sienna*. It is one of the ingredients of *Umber* (dark brown), manganese oxide being the major constituent.

IRRATIONAL NUMBER A real number which cannot be expressed as a fraction of two integers. $\sqrt{2}$ and π are irrational, but not IMAGINARY, numbers. *See also* APPROXIMATE NUMBER and RATIONAL NUMBER.

I-SECTION *See* I-BEAM.

ISI Indian Standards Institution, Delhi.

ISO International Organisation for Standardisation, Geneva.

ISOBARS Curves relating points at the same pressure, *e.g.* atmospheric or barometric pressure.

ISOCHROMATICS Lines of equal colour; in PHOTOELASTIC analysis, the colour fringes which denote lines of equal difference between the principal stresses. When MONOCHROMATIC light is used, the fringes are black and white, but the term 'isochromatic' is still applied; they can be separated from the ISOCLINICS with QUARTER-WAVE PLATES.

ISOCLINICS Lines in a stressed body which connect the points at which the principal stresses have the same direction. In PHOTOELASTIC analysis the light is extinguished when the direction of polarisation is the same as that of the principal stresses, and the isoclinics thus show up as dark lines superimposed on the ISOCHROMATICS.

ISOHYETS Lines drawn on a map through places having equal amounts of rainfall.

ISOLATION *See* RUBBER MOUNT-ING.

ISOMERS Molecules with the same composition, but with different structures.

ISOMETRIC PROJECTION *See* PROJECTION.

ISOSCELES Equal-legged. An isosceles triangle has two equal legs. A triangle with three equal legs is *equilateral*.

ISOSTATIC LINE A line tangential to the direction of one of the principal stresses at every point through which it passes. Also called *stress trajectory* (*see* Fig. 90).

ISOSTATIC STRUCTURE A structure which can be solved by the use of STATICS alone; also called a *statically determinate* structure. An isostatic structure must have the appropriate number of members in accordance with MÖBIUS' LAW. If a structural member or restraint is removed, it becomes a MECHANISM; if one is added, it becomes a HYPER-STATIC structure.

ISOTHERMAL (*a*) A line on a thermodynamic, psychrometric, or meteorological chart connecting points of equal temperature. (*b*) A reaction proceeding at a constant temperature. *See also* ADIABATIC.

ISOTOPES Atoms of the same chemical element with identical chemical properties, but different atomic weights. The *atomic weight* of the entire element is determined by the proportions in which the various isotopes are present. Some isotopes are radioactive; but the term 'isotope' does not imply radioactivity as such, although it is often used that way.

ISOTROPIC Having the same properties on all directions, as opposed to AELOTROPIC. In structural theory, the term usually means having the same strength, modulus of elasticity, and Poisson's ratio in all directions.

IStructE Institution of Structural Engineers, London.

ISWG Imperial standard wire gauge.

ITBTP Institut Technique du Bâtiment et des Travaux Publics, Paris.

IZOD IMPACT TEST An impact test carried out on a notched specimen which is cantilevered from one end. The energy absorbed during fracture is measured with a pendulum.

J

JACK A portable machine for lifting heavy loads through short distances. The smaller jacks are operated by a screw, the larger by a hydraulic ram.

JACK ARCH (*a*) A FLAT ARCH. (*b*) A shallow arch of brick or plain concrete, spanning about 3 ft (1 m), used in the nineteenth century.

JACKBLOCK METHOD A method for erecting a multi-storey prestressed concrete building, while carrying out most of the construction at ground level. The ground-floor slab is cast first, and serves as a casting bed for the other floors. The roof slab is cast next, and jacked up one floor. The top floor and its load-bearing walls are then constructed and jacked up one floor; and so on. The principle is similar to that of the LIFT-SLAB method. However, in the jackblock method the structure is pushed up, instead of being pulled up, and the vertical supports are cast at ground level, as well as the floor slabs.

JAMB The vertical side posts used in the framing of a doorway or window. The outer part of the jamb, which is visible, is called the *reveal*.

JAPANESE LACQUER An extremely durable glossy varnish obtained from tapping the sap of the Japanese varnish tree (*rhus vernicifera*).

JAPANESE MAT *See* TATAMI.

JENA PLANETARIUM The first thin-shell concrete dome designed as a membrane structure, by *F. Dischinger* in 1922. It was a hemisphere with a span of 82 ft (24·9 m). *See also* LEIPZIG MARKET HALLS.

JENNEY, WILLIAM LE BARON *See* CHICAGO SCHOOL and HOME INSURANCE BUILDING.

JIB The lifting arm of a crane. Also called a *boom*.

JOGGLE To give a slight, sharp bend to a steel angle to make it fit as a web stiffener over the angles at the top and bottom of a built-up girder.

JOHANSEN'S METHOD *See* YIELD-LINE THEORY.

JOHNSON, PHILIP *See* INTERNATIONAL STYLE.

JOINT The space between two adjacent components, irrespective of whether it is filled with a jointing material or not.

JOINT FILLER *See* JOINT SEALANT.

JOINT SEALANT Material used to exclude water and foreign solid matter from joints. *See also* MASTIC (*b*).

JOINTLESS FLOORING *See* MAGNESITE FLOORING and SCREED.

JOIST synonym for BEAM. It is generally the member which directly supports the floor or roof, as opposed to a GIRDER. A *steel joist* may be an RSJ or an OPEN-WEB JOIST.

JOULE (J) Unit of energy, named after the nineteenth century English physicist. It is the energy dissipated in 1 second by a current of 1 ampere flowing across a potential difference of 1 volt. In the SI SYSTEM the joule also replaces the erg and the calorie; 1 joule $= 10^7$ erg $= 0.239$ calories.

JOULE'S EQUIVALENT *See* MECHANICAL EQUIVALENT OF HEAT.

JOULE'S LAWS The three laws of the nineteenth century English physicist, J. P. Joule are as follows: (*a*) The intrinsic energy of a given mass of gas is a function of temperature alone; it is independent of the pressure and volume of the gas. (*b*) The molecular heat of a solid compound is equal to the sum of the atomic heats of its component elements in the solid state, *and* (*c*) the heat produced by a current I passing through a conductor of resistance R for a time t is proportional to I^2Rt.

JOURAWSKI'S METHOD *See* RESOLUTION AT THE JOINTS.

JUGENDSTIL The German version of ART NOUVEAU.

K

k (or K) Abbreviation for *kilo*.

K Chemical symbol for *potassium* (kalium, the latinised version of the Arabic *alkali*).

KALSOMINE A cheap white or tinted wash made of whiting and glue

mixed with water. Also spelt *calcimine*.

KAOLIN *See* CHINA CLAY.

KAOLINITE A finely crystalline form of hydrated aluminium silicate $(Al_2Si_2O_5(OH)_4)$.

KATA THERMOMETER Instrument devised by Sir Leonard Hill in 1914 to measure the physiological effect of the environment, and determine the COMFORT ZONE; it can also be employed for measuring the air velocity (*see* ANEMOMETER). It is an alcohol thermometer with a bulb 40 mm long and 18 mm in diameter. In the stem there are graduations 5°F apart. The standard type ranges from 100°F to 95°F, *i.e.* approximately the temperature of the human body. The thermometer is heated slightly above 100°F, and is then allowed to cool in the air. The time required for the kata thermometer to drop from 100° to 95°F is noted, and this gives an indication of the air's cooling power. The measurement can be made with the kata thermometer as a DRY BULB or as a WET BULB thermometer.

KATHODE *See* CATHODE.

KEENE'S CEMENT A hard plaster used for finishing coats. It consists of ANHYDRITE, *i.e.* anhydrous gypsum plaster $(CaSO_4)$, and an ACCELERATOR. It can be applied over Portland cement rendering, but should not be mixed with lime. Also called *Parian plaster*.

KEEP The principal tower of a medieval castle, which contained sufficient accommodation to serve as permanent living quarters in time of siege.

KELVIN SOLID In rheological terminology, a VISCOUS solid. The RHEOLOGICAL MODEL, named after the nineteenth century British physicist, consists of a spring and a dashpot in parallel. To obtain the model for a *visco-elastic solid*, another spring is added in series. *See* CREEP, MAXWELL LIQUID and STRESS–STRAIN DIAGRAM OF CONCRETE.

KENTLEDGE Ballast used to give stability to a crane, provide a reaction for a jack etc. It may consist of scrap iron, concrete, or any heavy building material.

KEY A mechanical bond, *e.g.* between old and new concrete.

KEYSTONE *See* CROWN.

kgf *See* KILOPOND.

KIESELGUHR *See* DIATOMACEOUS EARTH.

KILN A large oven used for the artificial seasoning of timber, for the firing of pottery, for the baking of brick, for the burning of lime, and for the burning of cement clinker.

KILO (k OR K) Prefix for one thousand times, from the Greek word for 1000; *e.g.* 1 kg = 1000 grams.

KILOPOND The unit of force in the conventional (CGS) metric system, as distinct from the SI SYSTEM. It is the gravitational pull on a mass of 1 kilogram. The abbreviations *kp*, *kgf* (kilogram-force) and *kg* are all in use, and have the same meaning if applied to forces. 1 kp = 9·807 NEWTONS.

KILOWATT HOUR *See* WATT HOUR.

KINEMATIC VISCOSITY The ratio of viscosity to density.

KINETIC ENERGY The ENERGY which a body possesses by virtue of its motion.

KINETIC FRICTION *See* FRICTION.

KING CLOSER A closer used to fill an opening in a course larger than a half-brick, but less than a full brick. *See also* QUEEN CLOSER and HALF-BAT.

KING-POST ROOF TRUSS A traditional timber truss consisting of a pair of rafters, held by a horizontal tie beam, a vertical *king post* between the tie-beam and the ridge, and usually also two struts to the rafters from the foot of the king post (Fig. 54a). Contrary to first impression,

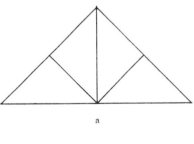

a

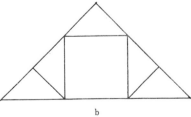

b

FIG. 54. King-post (a) and queen-post (b) roof.

the king-post is usually in tension. *See also* QUEEN POST and HAMMER-BEAM.

KIOSK Originally a summer house in a Turkish or Persian palace garden. Hence a small building serving as a bandstand in a park, or a small shop for selling refreshments or newspapers.

KIP Kilo-pound or 1000 lb. Kips per square inch is abbreviated ksi.

KNAPPED FLINT *See* FLINT.

KNEEBRACE A diagonal member joining the top of a column to the roof truss, used in some structures to improve their resistance to wind loading (Fig. 55).

KNOCKED DOWN Prefabricated, but not assembled.

KNOT A branch or limb embedded in a tree, and cut through in the process of manufacture. It may be a source of weakness (comparable to a hole drilled into the timber), if it is *loose, hollow* or *decayed*. On the other hand, a *sound, tight* or *intergrown* knot may be quite harmless.

KNURL To mill or otherwise roughen a surface to provide a better grip, *e.g.* on the head of a thumb screw.

kp *See* KILOPOND.

KRAFT PAPER A strong brown paper, made from wood pulp and sulphate.

KREMLIN A fortified enclosure in a Russian town.

ksi Kilo-pounds per square inch, or thousands of pounds per square inch.

KTESIPHON ARCH *See* CTESIPHON ARCH.

kVA Kilovolts multiplied by amperes. Used instead of *kilowatts* as a measure of power in alternating current circuits.

K-VALUE *See* THERMAL CONDUCTIVITY.

kW Kilowatt = 1000 watts.

kWh Kilowatt hour.

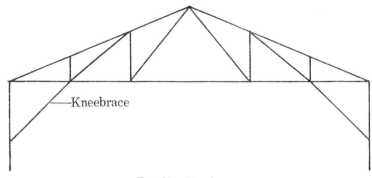

FIG. 55. Kneebrace.

L

LAC *See* SHELLAC.

LACQUER A glossy finish which dries quickly by evaporation of the vehicle. *See also* JAPANESE LACQUER and VARNISH.

LAITANCE A layer containing cement and very fine aggregate brought to the surface of concrete by BLEEDING. It has poor durability and strength, and generally is removed if it forms.

LAKE A pigment which consists of a DYE precipitated on an inorganic base.

LAMBERT The brightness of a surface which reflects 1 lumen/cm^2 is 1 lambert.

LAMELLA ROOF A roof frame consisting of a series of intersecting skew arches, made up of relatively short members, called lamellas, fastened together at an angle so that each is intersected by two similar adjacent members at its midpoint, forming a network of interlocking diamonds. (Fig. 56).

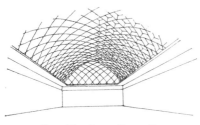

FIG. 56.   Lamella roof.

LAMINAR FLOW *See* STREAMLINE FLOW.

LAMINATED ARCH An arch built of laminated wood; it is usually ISOSTATIC, *i.e.* three-pinned.

LAMINATED PLASTIC A stiff board made from sheets of paper, cloth, linen or silk, soaked with synthetic resin (*e.g.* MELAMINE RESIN) and sandwiched between layers of that resin. One side is usually given a glossy or matte decorative finish. Laminated plastics imitating wood are produced by printing the grain of the wood, from a photograph, on a sheet of paper, which forms the top lamination.

LAMINATED WOOD *See* LAMINATED ARCH and PLYWOOD.

LAMINATING The process of bonding *laminations*, or thin plates, together with adhesive.

LAMP BLACK Carbon black.

LAMP DIMMER *See* THEATRE DIMMER.

LANGUAGE *See* PROBLEM-ORIENTED LANGUAGE.

LANTERN (*a*) A lamp, particularly a light in a street or garden. (*b*) A circular or polygonal turret with windows all around, crowning a DOME or other roof. *See also* MONITOR.

LAP The length by which one piece of material overlaps another.

LAP JOINT *See* BUTT JOINT.

LAPIS LAZULI A decorative variety of calcite, stained a deep blue by three minerals (lazurite, sodalite and hauyne). It is used as a stone veneer, and in powdered form it is the original ULTRAMARINE pigment. The Hebrew word *sappir* in the Bible probably means lapis lazuli.

LAPPING OF REINFORCING STEEL The overlapping of steel bars or fabric to provide transfer of stress from one piece of steel to the next through BOND with the concrete. *See* SPLICE.

LARGE CALORIE *See* CALORIE.

LARGE NUMBERS *See* LAW OF LARGE NUMBERS.

LASER A device which provides a light beam of extremely high intensity and directionality.

LATE WOOD *See* EARLY WOOD.

LATENT HEAT Thermal energy expended in changing the state of a body without changing its temperature, *e.g.* converting water to steam at 100°C. *See also* SENSIBLE HEAT.

LATERAL BUCKLING BUCKLING of a deep narrow beam sideways. It is prevented by limiting the ratio of depth to width.

LATERAL LOADING *See* EARTH-QUAKE LOADING and WIND PRESSURE.

LATERAL REINFORCEMENT Ties, hoops, helical or other secondary reinforcement used in columns, as opposed to the main or longitudinal reinforcement. Lateral reinforcement is also called *transverse reinforcement*.

LATERAL SUPPORT The horizontal propping of a column to reduce its EFFECTIVE LENGTH.

LATEX Originally the viscous, milky fluid which is exuded when the rubber tree (*Hevea Brasiliensis*) is tapped. It is a colloid of caoutchouc dispersed in water, which forms rubber by coagulation. The term is now also applied to artificial emulsions of natural or artificial rubber, or of certain synthetic resins (such as polyvinyl acetate), used in *emulsion paints*.

LATH Originally a sawn strip of wood, $1-1\frac{1}{2}$ in (25–40 mm) wide, $\frac{1}{4}-\frac{3}{8}$ in (6–10 mm) thick, and about 4 ft (1·2 m) long, used particularly as a foundation for gypsum plaster. The term is now applied also to other materials which can serve as a foundation for plastering, *e.g. metal lathing*.

LATHE A machine for turning metal or wood. The piece to be turned is mounted on a face plate, between two centres, and the cutting tool is stationary. The cutting speed is higher for small-diameter pieces and for soft materials (such as wood). *See also* MILLING MACHINE.

LATIN CROSS A cross with three short arms and one long arm. *See also* GREEK CROSS.

LATITUDE *See* GEODETIC LINE.

LATTICE, SPACE *See* SPACE LATTICE.

LATTICE STRUCTURE An open girder (OPEN-WEB JOIST), column, cylindrical shell, dome (GEODESIC DOME) or other structural type, built up from members intersecting diagonally to form a lattice. Lattice structures may be built in any material, but lightness is usually a prime objective.

LATTICE WINDOW *See* LEADED LIGHT.

LAVA The molten IGNEOUS rock material which issues from a volcanic fissure or vent, and consolidates on the surface. *See also* OBSIDIAN and PUMICE.

LAVATORY Literally, a wash basin. By extension a room containing a wash basin and a wc, and a room containing a wc without a wash basin.

LAW OF LARGE NUMBERS 'If n successes are obtained in N trials of an event, then the ratio n/N approaches a fixed probability as N increases.' This is one of the basic postulates of STATISTICS.

LAY The number of helix diameters in which the strand of a cable or wire rope makes one complete turn of $360°$ in its helix. The lay of a cable may be right-hand or left-hand, the former being the more common.

LAY LIGHT A window fixed horizontally in a ceiling.

LAYER, NEUTRAL *See* NEUTRAL AXIS.

LAZY SUSAN *See* REVOLVING SHELF.

L-BEAM A beam whose section has the form of an inverted L. It is usually formed by the combination of an edge beam (or SPANDREL) and the adjacent portion of the floor slab. *See* EFFECTIVE FLANGE WIDTH and T-BEAM.

lbf Symbol for pound as a force, *i.e.* the gravitational pull on a mass of 1 lb. The symbol *lb* is frequently used both for mass and for force.

LE CORBUSIER *See* BÉTON BRUT, CIAM, FUNCTIONALISM, MODULOR and PILOTIS.

LEAD A grey metal which is soft and easily deformed plastically at room temperature; however, at a sufficiently low temperature it shows elastic behaviour. Because of its corrosion resistance and the ease with which it can be formed it was widely used for roofing, flashings and damp-proof courses. Due to its increasing cost it has largely been replaced by other, less easily worked materials. Lead is an excellent insulator against radiation, including X-RAYS, and also a good sound insulator in the form of limp panels. Its symbol is Pb, its atomic number is 82, its atomic weight is $207·21$, its specific gravity is $11·34$, and its melting point is $327·3°C$. *See also* BLACK, RED and WHITE LEAD.

LEAD PAINT *See* WHITE LEAD.

LEAD PRIMER *See* RED LEAD.

LEADED LIGHT A window consisting of relatively small pieces of glass, often diamond shaped, held in lead strips of H-section (called *cames*). Also called a *lattice window*. The method evolved in the Middle Ages, because of the difficulty of making large sheets of glass; it became obsolete in the seventeenth century with the perfection of the broadglass process. Its popularity in the late nineteenth century was due to the Gothic Revival. *See also* MULLION.

LEAF *See* WYTHE.

LEAN CONCRETE A concrete with a low cement content, as opposed to a *rich concrete*.

LEAN-TO ROOF *See* SHED ROOF.

LEAST SQUARES *See* METHOD OF LEAST SQUARES.

LEDGER A main horizontal member of wooden or steel formwork. It is normally supported on the vertical scaffold poles (uprights) and it in turn supports the SOFFIT of the formwork, or the PUTLOGS which support the soffit.

LEFT-HAND *See* RIGHT-HAND.

LEIPZIG MARKET HALLS A series of thin-shell concrete domes, designed as membrane structures, with a span of 248 ft (75 m), the first to exceed the span of the PANTHEON, built in 1927 in Germany. *See also* JENA PLANETARIUM.

LENGTHENING JOINT A joint between two pieces of material, normally timber, which run in the same direction, as opposed to an *angle joint*. Most lengthening joints are END JOINTS.

LEONARDO DA VINCI *See* PARALLELOGRAM OF FORCES.

LEVEL AND LEVEL 1 *See* FLOOR.

LEVEL OF CONTROL *See* STANDARD DEVIATION.

LEVEL, SPIRIT *See* SPIRIT LEVEL.

LEVEL, SURVEYOR'S *See* SURVEYOR'S LEVEL.

LEVER ARM The distance between the resultant tensile force and the resultant compressive force in a structural member subjected to bending. The resistance moment of the section is the product of the resultant tensile or compressive force, and of the lever arm (Fig. 64). Also called *moment arm*.

LEVER PRINCIPLE The principle used for the composition and resolution of parallel forces. If a lever is in equilibrium under the action of a number of parallel forces, then the sum of the moments of these forces about any point is zero. The principle was discovered by *Archimedes of Syracuse* ca. 250 BC, and reputedly used by him to devise machines for throwing missiles in the defence of Syracuse when it was attacked by the Romans.

LFT Latest finish time (of an activity in a NETWORK).

LIBERTY *See* ART NOUVEAU.

LIBRARY SUBROUTINE *See* SUBROUTINE.

LIFT *See* ELEVATOR.

LIFT OF CONCRETE The concrete placed between two successive construction joints. In columns, it is generally the height of one storey.

LIFT SLAB A reinforced or prestressed concrete FLAT PLATE cast at ground level, and jacked up to its correct level after the concrete has hardened. Also, an entire system of floor slabs and roof cast on top of one another at ground level, and hoisted into position by the lift-slab method. *See also* JACKBLOCK METHOD.

LIFTING TACKLE *See* DIFFERENTIAL PULLEY BLOCK and DERRICK.

LIGHT Synonym for window; *see* DEAD, LAY and OPEN LIGHT. *See also* VISIBLE SPECTRUM.

LIGHT ALLOYS Alloys of low specific gravity. They include the aluminium alloys; the others are too expensive for use in buildings at the present time.

LIGHT ANALOGY *See* ACOUSTIC MODELLING.

LIGHT-GAUGE STRUCTURE Structure built up from cold-rolled steel or aluminium sheet. *See also* LOCAL BUCKLING.

LIGHT PEN A form of INPUT device which is used in conjunction with a CATHODE-RAY TUBE to feed graphical information into a digital computer.

LIGHTING *See* ARTIFICIAL LIGHT, COLOUR, DAYLIGHT, GLARE, LUMINOUS FLUX, PSALI and WORKING PLANE.

LIGHTNESS The attribute of visual sensation by which a body is judged to transmit or reflect a greater or smaller proportion of the incident light. This is the property called *value* in the *Munsell Book of Color*.

LIGHTWEIGHT AGGREGATE Concrete aggregate which weighs less than gravel or crushed stone. Natural materials include VERMICULITE and PUMICE. Artificial materials include expanded clay or shale, and products made from FLY ASH or BLAST FURNACE SLAG. The resulting *lightweight concrete* reduces the dead load on the columns, and the consequent saving in space may be more significant than the higher cost of the lightweight aggregate.

LIGHTWEIGHT CONCRETE (*a*) Concrete made with LIGHTWEIGHT AGGREGATE. Usually sand is used as FINE AGGREGATE, but the weight can be reduced further by using small-size lightweight aggregate also for the fines. (*b*) an even lighter material can be obtained by introducing air bubbles. *See* CELLULAR CONCRETE.

LIME A generic term for calcium oxide (CaO) and calcium hydroxide (CaO(OH)$_2$). *See* QUICKLIME and HYDRATED LIME.

LIME CONCRETE The predominant type of concrete (*opus caementitium*) used by the Ancient ROMANS.

LIME MORTAR Mortar made of lime and sand. It was the general medium for laying stone and brick until the nineteenth century, and it was still widely used well into the twentieth century. It has now been largely superseded by Portland cement mortar, which is HYDRAULIC, and also stronger. However, this change has not met with universal approval. Cement mortar is stronger than many types of stone and brick. Thus, if cracking occurs due to foundation settlement, temperature and moisture movement, the crack is liable to pass through the stone or brick, rather than through the mortar joint. This causes irreparable damage, whereas a crack in lime mortar joint can be repaired by repointing. Thus an admixture of lime with cement mortar is favoured by some designers and builders. *See also* MASONRY CEMENT.

LIMESTONE SEDIMENTARY rock containing a large proportion of calcium carbonate ($CaCO_3$). It is formed by the consolidation of calcareous ooze, which may be formed by organisms, by chemical precipitation, or by the weathering of pre-existing limestone. Most limestones are easily carved (*see* FREE-STONE and PORTLAND STONE). Limestone is a raw material for LIME mortar and for PORTLAND CEMENT.

LIMEWASH *See* WHITEWASH.

LIMIT DESIGN Design based on the *limiting loads* at which the structure collapses. PLASTIC HINGES are reached first at the sections subjected to the greatest curvature (Fig. 72). Formation of these hinges allows redistribution of bending moments, and the formation of further hinges with increase in load. The limit is reached when sufficient hinges have formed to turn the STATICALLY IN-DETERMINATE STRUCTURE into a MECH-ANISM. It then collapses (Fig. 57). Limit design, unlike elastic design, gives no information on the deformation of the structure under the working loads, which may be an important consideration. A separate ELASTIC DESIGN is needed, if deformations need to be determined. The WORKING LOADS are derived by dividing the limiting loads by the LOAD FACTOR. Also known as *ultimate strength design* or *plastic design*. The main alternative approach is called WORKING LOAD DESIGN or ELASTIC DESIGN. *See also* YIELD-LINE THEORY.

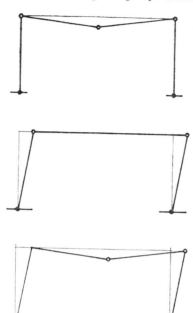

FIG. 57. Limit design. Three failure mechanisms of the failure of a two-pin portal by the formation of plastic hinges, the last two involving side-sway.

LIMIT OF LIQUIDITY, PLASTICITY *See* LIQUID, PLASTIC LIMIT.

LIMIT OF PROPORTIONALITY *See* PROPORTIONAL LIMIT.

LIMITING STRENGTH OF MATERIAL The strength above which a material ceases to contribute to the strength of the structure. It is usually defined, for convenience, in terms of standard tests which approximate to the limiting strength. In structural steel, it is taken as the YIELD STRESS, since the steel suffers very large deformations above that point which are liable to cause structural collapse. In concrete, it is the CYLINDER STRENGTH or (with an appropriate adjustment) the CUBE STRENGTH. The limiting strength is used directly in LIMIT DESIGN. In elastic design MAXIMUM PERMISSIBLE STRESSES are used, which are derived from the limiting strength.

LIMITS OF SIZE *See* TOLERANCE.

LIMONITE Iron ore composed of a mixture of hydrated ferric oxides.

LINE OF THRUST The curve produced by the points through which the resultant thrust passes, *e.g.* in an arch. If tensile stresses are inadmissible, it must lie within the MIDDLE THIRD.

LINEAL FOOT A foot, as distinct from a square foot or a cubic foot.

LINEAR ARCH *See* FUNICULAR ARCH.

LINEAR EQUATION An equation which can be plotted as a straight line. A linear algebraic equation is one of the type $y = ax + b$, where a and b are constants. *See also* DIFFERENTIAL EQUATIONS.

LINEAR PROGRAMMING Programming, *e.g.* by digital computer, of a series of relations which can be expressed as LINEAR EQUATIONS. Since a straight line is defined by any two points on it, this is simpler than programming more complex relations.

LINK POLYGON A graphical construction for solving problems of static equilibrium, based on the POLYGON OF FORCES. It can be used for determining the forces in roof trusses, drawing bending moment diagrams, and determining the line of thrust in an arch. It is also called a *funicular polygon*. Instead of a graphical construction, an experimental solution can be obtained with a freely hanging string, which produces a *string polygon. See also* Fig. 44.

LINOLEUM Floor covering made from jute or similar fabric, impregnated with oxidised linseed oil, resin and a filler, such as cork.

LINSEED OIL A vegetable oil obtained by crushing the seeds of FLAX. When exposed to the air, it thickens and darkens through oxidation, forming a tough skin. It is used for paints and varnishes, for linoleum, and for glazier's putty.

LINTEL (*a*) In classical architecture, the horizontal member which spans between the posts in TRABEATED construction. (*b*) A short beam, particularly one spanning across a door or window opening, and carrying the wall above it.

LIQUID A state of matter in which the shape of a given mass depends on the containing vessel, but its volume is independent thereof; liquids are practically incompressible.

LIQUID LIMIT The water content of a (clayey) soil which marks the boundary between its plastic and liquid state. It is determined by a standard test, in which the wet soil is placed in a cup, and divided by a standard grooving tool. The cup is then raised by a cam and allowed to drop, and this procedure is repeated until the groove is closed over a length of $\frac{1}{2}$ in (13 mm). A curve is plotted relating the number of drops to the water content, and the liquid limit is the water content correspond-ing to 25 drops. *See also* PLASTICITY INDEX.

LIQUID-MEMBRANE CURING *See* CURING (*a*).

LIQUIDUS LINE The line separat-ing the liquid from the liquid–solid phase. In a PHASE DIAGRAM it shows the variation of the composition of an alloy with the temperature when solidification is complete. (Fig. 71).

LITHARGE Lead monoxide (PbO). *See also* RED LEAD.

LITHIUM BROMIDE *See* ABSORP-TION CYCLE.

LITHOPONE An opaque white pigment, which is non-poisonous (*see* WHITE LEAD). It is a co-precipitated mixture of zinc sulphide (ZnS) and barium sulphate (BaSO$_4$).

LITMUS An organic colouring matter used as a reagent to test the alkalinity or acidity of liquids. It turns red in acid solutions, and blue in basic solutions. *See also* pH VALUE.

LIVE LOAD A load which is not permanently applied to a structure. As opposed to a *dead load*, it may or may not be acting at any given time. *See also* SUPERIMPOSED LOAD.

LIVENESS The reverberant quality of an auditorium. A hall which reflects too little sound back to the listener is called dead or *dry*.

ln Continental abbreviation for natural logarithm (log$_e$).

LOAD *See* DEAD LOAD, LIVE LOAD,

EARTHQUAKE LOADING, SUPERIMPOSED LOAD and WIND LOAD.

LOAD-BALANCING Arranging the prestressing cables in concrete so that the load is completely balanced by the prestress, and the member is subject purely to compression. For example, a beam or slab carrying a uniformly distributed load (which produces a parabolic bending moment diagram) requires a parabolic cable for load balancing (Fig. 58). The principle of load balancing can also be extended to torsion, by running the cables diagonally across all four faces of the member, to produce a twisting moment exactly equal and opposite to that applied by the load. *See also* DEFLECTED TENDON.

LOAD-BEARING WALL *See* BEARING WALL.

LOAD-EXTENSION CURVE *See* STRESS–STRAIN DIAGRAM.

LOAD FACTOR Factor used in LIMIT DESIGN to provide a margin of safety against collapse. It includes an allowance for inaccurate assumptions in the loading conditions, for inade-quate control over the quality of the materials, for imperfections in work-manship, and for minor approxima-tions made in the structural theory. It does not allow for arithmetical error. The factors are usually laid down in the building code, and depend on the control which can be exercised over design, materials and construction, and over the use to which the building will be put. The combined load factor is generally of the same order of magnitude as the FACTOR OF SAFETY, but the two factors are not identical. In the first place they are used in different theories,

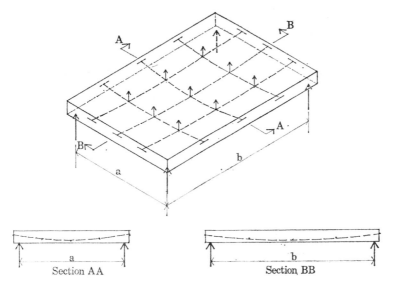

FIG. 58.   Load-balancing in a two-way prestressed concrete slab.

which may, but need not, lead to the same relation between load and stress. Secondly, it is possible to apply a different factor to each type of load; for example, the dead load which can be accurately predicted requires a factor only a little above 1, whereas a much higher factor is required for live loads. This differential factoring of loads is, indeed, one of the main advantages of limit design. On the other hand, the factor of safety can be differentiated between steel (with its higher standard of quality control) and concrete.

LOAD-INDICATING BOLT   A high-tensile bolt embodying a small projection which is compressed as the bolt is tightened. The gap thus indicates the tension in the bolt, and it can be measured with a feeler gauge.

LOADSTONE   *See* MAGNETITE.

LOAM   A soil which contains sand, silt and clay in roughly equal proportions.

LOCAL BUCKLING   Crinkling of a strut or the compression flange of a beam because it is too thin. Local buckling is particularly liable to occur in *thin-walled sections* and LIGHT-GAUGE STRUCTURES, and it can be prevented by the corrugation of excessively long straight runs of sheet.

LOCK NUT   A secondary nut used to prevent the first nut from working loose.

LODESTONE   *See* MAGNETITE.

LOESS   Silt deposited by wind (*aeolian action*); it is found in many parts of the world, covering large areas.

155

log Logarithm. Unless otherwise stated it means $\log_{10}$, *i.e.* a decimal logarithm.

$\log_e$ Natural logarithm.

LOG A rough, unshaped piece of tree trunk.

LOGARITHM A mathematical function. If *a* and *b* are two numbers, and $a = 10^b$, then the decimal logarithm of *a*, $\log_{10} a = b$, and the *anti-logarithm* of *b* is *a*. The suffix 10 is generally omitted for decimal (or common) logarithms. It follows that the logarithm of the product *cd*

$$\log cd = \log c + \log d$$

so that the process of multiplication can be reduced to the simpler operation of adding the logarithms. This is utilised in LOGARITHMIC TABLES and in SLIDE RULES. *See also* NATURAL LOGARITHM.

LOGARITHMIC SCALE A scale plotted to the logarithm of the variable. An increase of one unit of the scale represents a tenfold increase in the quantity. *Logarithmic graph paper* (log log paper) has logarithmic scales for both the ordinate and the abscissa. *Semi-logarithmic paper* (semi-log paper) has one logarithmic and one ordinary scale.

LOGARITHMIC TABLE A table which lists decimal, or common LOGARITHMS. Multiplication of two numbers may be accomplished by looking up their logarithms, adding them, and then looking up the anti-logarithm of the sum.

LOGGIA A gallery or arcade having one or more of its sides open to the air.

LONG COLUMN *See* COLUMN, LONG.

LONG CYLINDRICAL SHELL *See* BEAM THEORY OF SHELLS.

LONG-LINE PRESTRESSING *See* PRE-TENSIONING.

LONG TON *See* TON.

LONGITUDE *See* GEODETIC LINE.

LONGITUDINAL REINFORCE-MENT The main reinforcement parallel to the long dimension of a structural member, as opposed to the *lateral reinforcement*.

LOOP A routine or subroutine forming part of a digital computer PROGRAM which consists of repeating a group of instructions.

LOSS OF PRESTRESS The loss of INITIAL PRESTRESS by the SHRINKAGE of the concrete, and by CREEP in the steel and the concrete. The reduction in the prestressing force due to friction between the tendon and the duct, and the ELASTIC LOSS OF PRESTRESS are often considered separately.

LOUDNESS *See* PHON.

LOUDSPEAKER An electro-acoustic TRANSDUCER for audition by a number of people.

LOUVER *See* LOUVRE.

LOUVRE A ventilator, used particularly in the hot–humid tropics where ventilation is important for COMFORT. It permits the passage of air with as little obstruction as possible, while blocking vision and excluding rain. Window louvres usually consist of horizontal slats of wood or metal, set at an angle across the path of vision. The term 'louvre' is also applied to the grilles covering the openings in the overhead ducts of an air-conditioning, heating or ventilating installation. *See also* SUNSHADING and PUNKAH LOUVRE.

LOW-ALLOY STEEL Steel containing less than 10 per cent of alloying elements.

LOW-CARBON STEEL CARBON STEEL containing less than 0·25 per cent of carbon. *See also* HIGH-CARBON STEEL.

LOW-HEAT CEMENT A PORTLAND CEMENT in which the generation of heat during setting is reduced by a modification of its chemical composition. It is used in dams and other massive concrete structures where the heat cannot easily be dissipated.

LOW-LEVEL LANGUAGE *See* PROBLEM-ORIENTED LANGUAGE.

LOW-PRESSURE STEAM CURING Steam curing at normal atmospheric pressure, as opposed to AUTOCLAVING.

LUCITE *See* ACRYLIC RESINS.

LÜDERS' LINES Lines which appear on the polished surface of a crystal, or a polished metal surface which is polycrystalline, after it has been stressed beyond the elastic limit, first reported by W. Lüders in 1860. They represent the intersection of the surface by planes on which the shear stress has produced plastic slip. Also called *slip lines*.

LUMBER In America, the product derived from LOGS of timber in a sawmill. Elsewhere the word *timber* is used for both the round logs and the rectangular pieces.

LUMEN (lm) Unit of LUMINOUS FLUX. The flux emitted within a unit solid angle of 1 STERADIAN by a point source having a uniform intensity of 1 CANDELA.

LUMINAIRE A light fitting.

LUMINESCENCE The emission of light as a result of any cause other than high temperature, *e.g.* the effect of ultraviolet light on certain chemicals. *See also* ELECTROLUMINESCENT PANEL.

LUMINOSITY The attribute of visual sensation according to which an area appears to emit more or less light. Also called *subjective brightness* or *apparent brightness*.

LUMINOSITY CONTRAST *See* CONTRAST.

LUMINOUS FLUX The quantity of light passing through an opening per unit of time. It is measured in LUMENS. *See also* ABSORPTION FACTOR and TRANSMISSION FACTOR.

LUMINOUS PAINT *See* FLUORESCENT PAINT and PHOSPHORESCENT PAINT.

LUMINOUS TRANSMITTANCE The ratio of transmitted to incident luminous flux.

LUX (lx) Unit of illumination. An illumination of 1 lumen per square metre.

lx Lux.

M

m Abbreviation for *milli*.

M Abbreviation for the basic *module*, or for *mega*.

μ *See* MU.

MACADAM Uniformly sized stones rolled to form a surface, a process developed by the Scottish road-builder J. L. McAdam in the early nineteenth century. *Macadamising* is the process of laying the surface. *Tarmacadam* is a waterproof surface whose stones are bound together by tar, bitumen or asphalt, as distinct from *waterbound macadam*.

MACH NUMBER The ratio of the velocity of an object to the local speed of sound, named after the nineteenth century Austrian physicist. A Mach number below 1 means subsonic flow, and a number above 1 means supersonic flow.

MACHINE LANGUAGE Information, in the form of a program or data, which may be used by a digital computer directly, *i.e.* information in binary code or machine code. *See also* PROBLEM-ORIENTED LANGUAGE.

MACROSCOPIC Visible to the naked eye, as opposed to *microscopic*. Also called *megascopic*.

MADE GROUND OR MADE-UP GROUND Ground built-up with excavated material or refuse, as distinct from the natural, undisturbed soil. Its load-bearing capacity is often very low.

MAGNESIAN LIMESTONE *See* DOLOMITE.

MAGNESITE Carbonate of magnesium ($MgCO_3$).

MAGNESITE FLOORING A composition of OXYCHLORIDE CEMENT with a filler of sawdust, wood flour, sand or ground silica. It is used to cover concrete floors, and is commonly floated in a layer about $1\frac{1}{2}$ in (40 mm) thick. Also called *jointless flooring*.

MAGNESIUM A very light silvery-white metal which burns in air with a brilliantly white light when ignited. It is produced for commercial purposes by electrolysis from seawater. Its alloys weigh less than aluminium alloys and have comparable strength. They are used in aircraft and automobile design, but are too expensive for use in building. The chemical symbol for magnesium is Mg, its atomic number is 12, its atomic weight is 24·32, its specific gravity is 1·74, its melting point is 651°C, and its valency is 2.

MAGNET STEEL A steel capable of retaining its magnetism after removal from an external magnetic field, and therefore usable for *permanent magnets*. It is usually an alloy steel containing tungsten, cobalt or nickel, quenched from about 900°C. By contrast, an *electromagnet* is formed by winding a coil of wire around a core of *soft iron*. This becomes magnetic when an electric current flows through the wire, and loses its magnetism when the current is switched off. It is therefore a *temporary magnet*.

MAGNETITE Ferrous ferric oxide (Fe_3O_4). It is a black iron ore, which is attracted by a magnet, but does not attract particles of iron to itself. The *lodestone* or *loadstone*, which was used by the early navigators as a magnetic compass, is a form of magnetite which exhibits polarity, and points to magnetic North and South when freely suspended. *See also* IRON OXIDE.

MAIHAK STRAIN GAUGE A type of ACOUSTIC STRAIN GAUGE.

MAIN BEAM *See* GIRDER.

MAIN TIE The tension member joining the feet of a roof truss, generally at wall-plate level.

MAINTENANCE FACTOR The ratio of the illumination in a new lighting installation, to that in the same installation after a specified period of use (or sometimes the reciprocal of this ratio).

MAISONETTE British term for a two-level APARTMENT, with the bedrooms at the upper level.

MALLET A hammer made of wood, rubber, leather, or a soft metal, but not of steel.

MAN-HOUR Unit of work. The amount of work done by one man in one hour.

MANGANESE A hard and brittle metallic element. Its chemical symbol is Mn, its atomic number is 25, its atomic weight is 54·93, its specific gravity is 7·39, and its melting point is 1245°C. It is used in alloy steels for toughness, and also in the alloys of aluminium and copper.

MANIFOLD A pipe or chamber with several outlets or inlets for smaller pipes.

MANNERISM The style of architecture current in Italy from Michelangelo (*ca.* 1500) to the end of the sixteenth century, *i.e.* the period between the High Renaissance and the Baroque. The art historian N. Pevsner characterises it as using motifs in deliberate opposition to their original significance or context.

MANOMETER A PRESSURE GAUGE. It commonly takes the form of a U tube filled with water, oil, or mercury (Fig. 101). One limb is connected to the air or other fluid whose pressure is to be measured, and the other is open to the atmosphere or connected to some other standard pressure.

MANSARD ROOF A roof with two slopes on each side. The lower slope is longer and steeper than the upper. It was named after François Mansart, and is also called a *French roof*.

MARBLE A granular crystalline calcareous rock, formed *metamorphically* from limestone. It was greatly esteemed by sculptors, particularly in Ancient Greece and Rome, and later in the Renaissance, because it could be polished to a smooth finish. The white Pentelic marble (of which the Parthenon in Athens was built) and the white Carrara marble (used by Michelangelo) are still quarried. Although marble sculptures are rarely employed today, the material is used in modern architecture because of the ease with which it can be cut and polished by machine into thin slabs. The term 'marble' is also used, incorrectly, for non-metamorphic limestones and for other decorative rocks capable of taking high polish.

MARBLING The now obsolete art of painting a surface to look like marble.

MARGIN OF SAFETY *See* FACTOR OF SAFETY.

MARINE GLUE Waterproof or HYDRAULIC GLUE, usually consisting of about three parts of pitch, two parts of shellac, and one part of rubber. It has been largely superseded by the development of waterproof synthetic RESINS.

MARL A clay which contains between $\frac{1}{3}$ and $\frac{2}{3}$ of calcareous material. It is particularly suitable for making bricks without the addition of other substances.

MARSH GAS *See* METHANE.

MARTENSITE A constituent of steel which appears needle-shaped under the microscope. It is produced when steel is cooled very rapidly from the hardening temperature, *i.e.* at a speed greater than its critical cooling rate, so that the transformation of AUSTENITE occurs at 400°C or lower. It is essentially ALPHA IRON in non-equilibrium condition, formed directly from the under-cooled austenite. It is the hardest of the decomposition products of austenite, and very BRITTLE.

MASONRY Originally work consisting of blocks of carved stone laid in mortar. Now also work consisting of bricks or blocks.

MASONRY CEMENT A cement which has greater plasticity and water retention than can be obtained with Portland cement mortar (*see also* LIME MORTAR). In addition to Portland cement it contains hydrated lime, pulverised limestone, pozzolan, clay, gypsum, or talc. It may also contain an air-entraining additive.

MASS CENTRE The CENTROID.

MASS CONCRETE PLAIN CONCRETE placed in considerable thickness, so that the dissipation of the heat of hydration generated by the cement may become a problem. It often uses COARSE AGGREGATE of very large size.

MASS LAW OF SOUND INSULATION 'For a single wall the average INSULATION against AIRBORNE SOUND is almost entirely determined by its weight per unit area.' The insulation, or difference in sound pressure level, i, to sound waves at normal incidence is given by

$$i = 10 \log \left[1 + \left(\frac{\omega m}{2\rho c} \right)^2 \right] \text{ decibels}$$

where m is the surface density of the wall, ρ is the density of air, c is the velocity of sound in air, and $\omega =$ angular frequency of the sound waves ($2\pi \times$ frequency). In the practical range of frequencies and partition weights, a doubling in the weight of a partition increases the insulation by 5 DECIBELS.

MASS RETAINING WALL A gravity RETAINING WALL (Fig. 80).

MASTIC (*a*) The resin of the mastic tree, a small evergreen tree found near the Mediterranean Sea, which is used in chewing gum and, dissolved in alcohol, as a varnish. (*b*) A jointing compound which dries on the surface, but remains permanently plastic underneath. The term covers a variety of compounds used in DRY CONSTRUCTION for sealing joints between precast concrete facing panels, curtain walls, windows, pipes, etc. It is usually inserted into the joint with a pressure gun. (*c*) A thick adhesive, consisting of asphalt, bitumen or pitch, and a filler, such as sand. It is used for bedding woodblock floors, bedding and pointing window frames, and for laying and repairing flat roofs.

MATCHED VENEER *See* WOOD VENEER.

MATERIALS SCIENCE Application of the physics and chemistry of the internal structure of materials to the interpretation of their engineering behaviour. Two important aspects are the prediction of the strength and the electrical properties of metals, alloys, and polymers from theoretical considerations. Materials science has been less successful in predicting the behaviour of concrete and timber. *See also* STRENGTH OF MATERIALS.

MATHEMATICAL MODEL A mathematical formulation which describes the known behaviour of a structure or other physical process. *See also* MODEL ANALYSIS.

MATRIX (*a*) A rectangular array of numbers or mathematical terms, used in the solution of simultaneous equations. Because large matrices are easily handled by electronic *digital computers*, a substantial proportion of mathematical theory (*e.g.* for structural design) is now written in *matrix algebra*. (*b*) The cement which binds together the aggregate of concrete. (*c*) The principal constituent of an alloy or mechanical mixture, in which the other constituents are embedded. (*d*) The mould from which printers' type is cast.

MATT *See* GLOSS *and* DIFFUSER.

MAUSOLEUM (*a*) The tomb of Mausolus, King of Caria, built by his wife at Halicarnassus in the fourth century BC. (*b*) Any large or magnificent tomb. (*c*) A large room or building of gloomy, tomb-like appearance.

MAXIMUM PERMISSIBLE STRESS The greatest stress permissible in a structural member under the action of the WORKING LOADS. It is usually defined as (LIMITING STRENGTH OF MATERIAL)/(FACTOR OF SAFETY).

MAXWELL, CLERK *See* PHOTO-ELASTICITY and RECIPROCAL THEOREM.

MAXWELL DIAGRAM *See* RECIPROCAL DIAGRAM.

MAXWELL LIQUID In rheological terminology, a visco-elastic liquid. The rheological model, named after the nineteenth century British physicist, consists of a spring and dashpot in series. *See also* NEWTON LIQUID and HOOKE BODY.

MEAN The *algebraic mean* is the average of all values (*i.e.* their sum, divided by the number of values), considering their positive or negative sign; it may be either positive or negative. The *arithmetic mean* is the average of all values, neglecting their sign, *i.e.* taking them all as positive. The *geometric mean* of *n* values is the *n*th root of their product. *See also* MEDIAN and MODE.

MECHANICAL ADVANTAGE The ratio of the load raised by a machine, such as a lifting tackle or a lever, to the force required to operate it. The *velocity ratio* is the ratio of the distance through which the operating force has to be moved, to the distance moved by the load. In an ideal machine the two ratios would be equal; however, due to friction the mechanical advantage is less. The *efficiency* of the machine is

$$\text{efficiency} = \frac{\text{mechanical advantage}}{\text{velocity ratio}}$$

MECHANICAL ANALYSIS OF PARTICLES *See* PARTICLE-SIZE ANALYSIS.

MECHANICAL BOND *See* KEY.

MECHANICAL CORE *See* SERVICES CORE.

MECHANICAL EQUIVALENT OF HEAT In FPS units, 1 BThU = 778 ft lb (In metric units, 1 calorie = $4 \cdot 18 \times 10^7$ ergs = $4 \cdot 18$ joules). The equivalent was first determined by the nineteenth century English physicist, *J. P. Joule*. This is also known as the first law of THERMODYNAMICS.

MECHANICAL HYSTERESIS The dissipation of energy as heat during a stress cycle. It shows as a loop when the stress–strain diagram is plotted for successive loading and unloading. If the ascending and descending branches coincide, then there is no hysteresis.

MECHANICAL STRAIN GAUGE A gauge for measuring strains by levers or other mechanical means (*e.g.* the HUGGENBERGER TENSOMETER),

as opposed to an electrical, an acoustical or an optical gauge.

MECHANISM A frame which is capable of movement. In terms of structural mechanics, a mechanism may be considered as an ISOSTATIC STRUCTURE from which one or more members have been removed. The removal of each member gives a degree of freedom, just as the addition of each member adds one redundancy. *See also* PANTOGRAPH.

MEDIAN The middle item of a group of observations, arranged in order of magnitude. For example, if there are 21 observations, it is the 11th, and if there are 20 it is the mean of the 10th and 11th. *See also* MEAN.

MEDIUM The liquid part of a paint, in which the pigment is suspended. After the paint has hardened it becomes the binder of the paint film. Also called *vehicle*.

MEGA (M) Prefix for one million, from the Greek word for large; *e.g.* 1 MW = 1 megawatt = 1×10^6 watt.

MEGASCOPIC *See* MACROSCOPIC.

MELAMINE RESIN A material used in LAMINATED PLASTICS. It is produced by the reaction of melamine and formaldehyde with a suitable catalyst. Melamine resin-bonded laminates are characterised by extreme hardness.

MELT The molten portion of the raw materials in a furnace.

MEMBRANE (*a*) A thin, sheet-like structure. (*b*) A thin, impervious sheet.

MEMBRANE ANALOGY An analogy between the mathematical equations for the elastic *torsion* function and the transverse deformation of a stretched membrane, proposed by the German physicist L. Prandtl in 1903. It is utilised as an ANALOGUE COMPUTER for the torsional strength of complex shapes, such as RSJs. The apparatus consists of a rubber membrane stretched over an opening, shaped like the cross section under investigation. The deflected membrane is contoured, and its slope and volume determined. *See also* SANDHEAP ANALOGY.

MEMBRANE CURING *See* CURING (*a*).

MEMBRANE FORCES *See* MEMBRANE THEORY.

MEMBRANE SHELL *See* MEMBRANE THEORY.

MEMBRANE THEORY Theory for the design of *thin* shell structures, based on the assumption that all forces act *within* the membrane, as shown in Fig. 59. Three distinct internal forces are possible in a membrane: N_x and N_y, which are direct forces and may be either tensile or compressive, and the membrane shear V. It is rarely possible to achieve equilibrium with membrane forces alone. A *thick* shell is one which has sufficient thickness to accommodate moments and transverse shear forces; but 'thick' is a relative term, since it may denote a few inches in a span of more than a hundred feet (relatively thinner than an egg shell). The forces and moments in a thick shell are shown in *Fig.* 60.

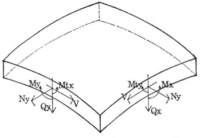

FIG. 60.   Forces and moments in a thick shell.

Eight distinct forces and moments are possible:

(*a*) the membrane forces N_x, N_y and V.
(*b*) shear forces Q_x and Q_y acting across the *x*- and *y*-sections.
(*c*) bending moments M_x and M_y.
(*d*) torsional moments M_{tx} and M_{ty}.

The object of thin-shell design is to limit non-membrane forces and bending moments to a small region near the support of the shell. Two conditions have to be satisfied in the design of shells: (1) the internal forces must balance the external forces at any point, in accordance with the normal laws of statical equilibrium. This problem is *isostatic*, and readily soluble. (2) the deformation of the shell must be compatible with the restraints imposed on its own supports. The problems posed by the

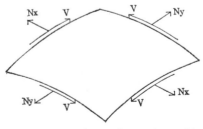

FIG. 59.   Membrane forces in a thin shell.

BOUNDARY CONDITIONS are *hyperstatic*, and empirical rules or simplifications are employed if precise solutions are difficult to obtain. *See also* HOOP and MERIDIANAL FORCES.

MEMORY A term sometimes used for the *store* of a digital computer.

MENDELSOHN, ERICH *See* ORGANIC ARCHITECTURE.

MER The smallest repetitive unit in a POLYMER.

MERCALLI SCALE Scale for classifying the magnitude of an earthquake devised by G. Mercalli and modified in 1931 by H. O. Wood and F. Neumann. It is a 12-point scale based on the observed symptoms. Number 1 on the scale is not felt, number 6 causes some structural damage, and number 9 destroys weak masonry structures and damages frame buildings; number 12 causes nearly total destruction. *See also* RICHTER SCALE.

MERCURY A white metallic element, also called *quicksilver*, with a melting point of $-38.5°C$, and a boiling point of $+356.7°C$. Its chemical symbol is Hg, its atomic number is 80, its atomic weight is 200·61, and its specific gravity is 13·56. It is a solvent for most metals, and the resulting alloys are called AMALGAMS. These played a prominent part in medieval alchemy. Mercury is used in *mercury-vapour lamps*, and other electrical applications. Mercury BAROMETERS are used for measuring air pressure, which is consequently often expressed in millimetres of mercury. At a temperature of 0°C, 750 mm of mercury = 1 atmosphere = 14·7 psi = 101·3 kN/m^2.

MERIDIAN *See* GEODETIC LINE.

MERIDIANAL FORCES The MEMBRANE forces in a dome which follow the lines of the meridians, or circles of longitude (Figs. 29 and 47). Unlike the HOOP FORCES, they are always compressive.

MESH The open spaces of a net; hence the net itself. Wire mesh is used in *sieves* and *screens* to separate materials, such as sand, into sizes. The larger sizes are specified by the size of the opening; the smaller sizes by a standard number. The Test Sieve Numbers standardised by the ASTM and the BSI are similar, but not quite the same.

MESH REINFORCEMENT Welded-wire fabric used as reinforcement for concrete, particularly in slabs. *See also* EXPANDED METAL.

METABOLISM The chemical and physical processes continuously going on in living organisms, whereby food is assimilated and energy released. The *basal metabolism* of a person is his heat production when he is completely at rest in warm surroundings. It is generally taken to be a function of the surface area of his body, although some physiologists claim that it is more nearly proportional to body weight.

METAL (*a*) An element which is malleable, conducts heat and electricity, and can replace the hydrogen of an acid, and so form a base. Metals are electropositive, and their atoms readily lose their ELECTRONS. When untarnished, metals have a characteristic 'metallic' lustre. (*b*) Broken stone used as a COARSE AGGREGATE for concrete, tarmacadam etc. *Blue* *metal* is crushed from any hard igneous or sedimentary rock of a bluish colour, *e.g.* dolerite. (*c*) The liquid glass during the process of manufacture.

METAL COATING *See* GALVANISING, SHERARDISING and CADMIUM PLATING.

METAL LATH *See* LATH and EXPANDED METAL.

METAMORPHIC ROCK IGNEOUS or SEDIMENTARY rock which has been changed by chemical or physical action in the earth's crust into a distinctly new type. *Marble* is metamorphic limestone, and *slate* is metamorphic shale.

METER (*a*) A measuring instrument, particularly for fluids, *e.g.* flow meter. (*b*) American spelling of METRE.

METHANE The name of the short-est hydrocarbon chain, CH_4. It is a colourless gas which occurs in *natural gas*, and is given off in marshes as *marsh gas* and in coal mines as *fire damp*.

METHOD OF LEAST SQUARES A method for fitting the best linear equation to a set of experimental results by making the squares of the residuals a minimum. The 'squaring' is necessarily to eliminate the effect of positive and negative signs, as in the STANDARD DEVIATION.

METHOD OF SECTIONS A method for the design of isostatic trusses, proposed by the German engineer A. Ritter in 1862. It is particularly useful if the magnitude of the force in only one member is required. If the truss shown in Fig. 61 is cut across the main tie, the force in it, T_1, can be obtained in one operation by taking moments about A. The two other unknown forces pass through A, and therefore have

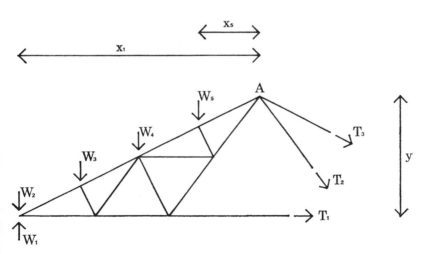

FIG. 61. Methods of sections. Taking moments about A, the force in the tie T_1 is given by $T_1 \cdot y = \Sigma \, W \cdot x$.

no moment about it. *See also* FREE-BODY DIAGRAM, RECIPROCAL DIAGRAM and RESOLUTION AT THE JOINTS.

METHYL METHACRYLATE *See* ACRYLIC RESINS.

METHYLATED SPIRITS Industrial ethyl alcohol to which some methyl alcohol (wood spirit) has been added to make it poisonous, and therefore undrinkable.

METRE The unit of length in the metric system, introduced by the French Academy in 1791. 1 m = 39·37 in. *See* CGS UNITS and SI UNITS.

METRIC SYSTEM *See* CGS UNITS and SI UNITS.

METRICATION The act of converting any other unit to its metric equivalent.

MEWS Originally the Royal Stables at Charing Cross in London, built on the site where the royal hawks were formerly mewed (*i.e.* caged). Hence a row of stables with living accommodation for the staff above, built at the back of London town houses; these have now been mostly converted into fashionable houses.

MEZZANINE An intermediate storey of lower height. It is usually a gallery between the main floor and the floor above it.

MF Melamine FORMALDEHYDE.

Mg Chemical symbol for *Magnesium*.

MHO Unit of conductance. It is the reciprocal of the OHM.

MICA A group of minerals found in igneous rocks, characterised by perfect cleavage, so that they can be split into thin sheets. Different types vary in chemical composition. Some micas are excellent electrical or thermal insulators, and some are transparent.

MICHELANGELO BUONAROTTI *See* MANNERISM.

MICRO (μ) Prefix for one millionth, from the Greek word for small; *e.g.* 1 μs = 1 microsecond = 1×10^{-6} second.

MICRO-CONCRETE Concrete for use in small-scale structural models. Its aggregate has been scaled down.

MICRO-CRACK A crack which is too small to be seen with the naked eye, but can be detected, for example, by ULTRASONICS. The propagation of micro-cracks is responsible for the failure of brittle materials such as GLASS and concrete, but stable microcracks can introduce pseudoplastic behaviour leading to considerable redistribution of stress and an increase in the ultimate load. *See* STRESS–STRAIN DIAGRAM OF CONCRETE.

MICROGRID A rectangular grid, say of fine rubber thread, cast into a transparent MODEL. It shows up the elastic distortion due to membrane stresses, but it is not sufficiently sensitive for quantitative measurements.

MICROMETER An instrument for the precision measuring of length which consists of a G-shaped frame, one of whose legs is a round bar accurately threaded to 100 divisions per inch. A nut running on it is marked to 0·001 in. *See also* VERNIER.

166

MICROMETRE In the SI SYSTEM $1\ \mu\text{m} = 10^{-6}$ m. Previously called a *micron*.

MICRON One-millionth of a metre, and one-thousandth millimetre; in the SI system called a *micrometre*.

MICROPHONE *See* CONDENSER, CRYSTAL, MOVING COIL, OMNI-DIRECTIONAL and UNIDIRECTIONAL MICROPHONE.

MICROSCOPIC Visible with a microscope only, as opposed to *macroscopic*.

MIDDLE STRIP The portion of a FLAT SLAB or FLAT PLATE between the COLUMN STRIPS. Most building codes define the middle strip as the middle half of the slab (Fig. 31).

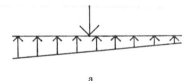

a

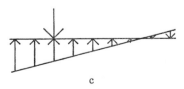

b

c

FIG. 62. Middle-third rule.

MIDDLE-THIRD RULE 'Provided the resultant force lies within the middle third, no tension is developed in a wall or foundation.' Fig. 62 shows (*a*) the THRUST within the middle third, which gives rise to compressive stresses only. The limiting condition (*b*) occurs when the compressive stress drops to zero at one end. The stress distribution is now triangular, and the thrust (which is in equilibrium) must be at the centroid of the triangle, *i.e.* one third from the edge. If the thrust falls outside the middle third (*c*), tensile stresses develop, or else the joint opens up. The middle-third rule is particularly important for the design of RETAINING WALLS, masonry arches and Gothic buttresses.

MIES VAN DER ROHE *See* BAUHAUS and DEUTSCHER WERKBUND.

MIL One thousandth of an inch = 0·001 in.

MILD STEEL Plain carbon steel with a carbon content of 0·1 to 0·2 per cent. It is ductile, has a yield stress of 30 000 to 35 000 psi (200–250 MN/m^2), and is frequently used for structural steel sections and concrete reinforcing bars. (*See* Fig. 93).

MILK OF LIME SLAKED LIME in water.

MILL SCALE The oxide layer formed on structural steel sections and reinforcing bars during hot rolling. If it is loose, it must be removed before concreting or painting, to ensure proper adhesion.

MILLI (m) Prefix for one thousandth, from the Latin word for

1000; *e.g.* 1 millimetre = 0·001 metres.

MILLIBAR (mb) Unit of pressure used in meteorology and climatology. 1 mb = 100 NEWTONS per square metre = 0·750 mm of mercury (measured with a mercury barometer at 0°C). *See also* ATM and BAROMETER.

MILLING MACHINE A machine which removes shavings from a surface by pushing it on a moving table past a *rotating* cutter. Because both the work and the tool move, it can be used for making complex parts, such as gears. A *planing machine* is one which removes shavings from a surface by pushing it on a moving table past a stationary tool. A *shaping machine* is one which removes shavings from a surface fixed to a stationary table with a tool which moves backwards and forwards. *See also* LATHE.

MINERAL FLAX Fiberised asbestos.

MINERAL PIGMENT *See* PIGMENT.

MINERAL WOOL An aggregate of fine filaments, which can be formed into a flexible and resilient mat. It is produced by blowing air or steam through molten blast-furnace slag (*slag wool*) through molten rock (*rock wool*), or through molten glass (*glass wool*). It has excellent thermal insulating properties, particularly in loose form, and it is vermin-proof and rot-proof.

MINIMAL SURFACE The shortest SURFACE which can be formed between a given set of boundaries. It can be obtained experimentally by forming the boundaries out of wire, and then dipping them into a solution (*e.g.* detergent or latex) which contracts to a minimal surface by surface tension. The minimal surface is always a SADDLE.

MINOR INTRUSIONS IGNEOUS ROCKS formed by intrusions in fissures, etc., and consequently much finer grained than PLUTONIC INTRUSIONS.

MIRROR *See* REFLECTION.

MIRROR, SOUND *See* ECHO and WHISPERING GALLERY.

MITER *See* MITRE.

MITRE BOX A U-shaped open box with cuts at 45°, which facilitate the making of 90° mitred joints. The piece of timber to be mitred is placed in the box, and the cuts guide the saw to cut at an angle of 45°. Also called a *mitre block*.

MITRE JOINT A timber joint formed by fitting together two pieces of board on a line bisecting their junction. *Mitring* is the operation of cutting the boards and fitting them together at an angle, usually a right angle.

MITRE SQUARE A *square* for setting out timber. It has one edge of the handle set at 45°, so that it can be used for laying out right-angled mitre joints.

MIX PROPORTIONS The ratio in which the various components of a composite material, such as concrete, mortar or plaster, are mixed together. Mix proportions may be specified by volume, which merely requires the filling of a bucket or gauge box, or by weight which is more accurate for materials subject to BULKING. *See also* WATER-CEMENT RATIO, and WEIGH-BATCHER.

MIXER *See* BATCH MIXER.

MKS UNITS *See* CGS UNITS.

Mn Chemical symbol for *manganese*.

MÖBIUS' LAW A rule defining the number of members required for an ISOSTATIC STRUCTURE, published by A. F. Möbius in 1837. The simplest plane PIN-JOINTED frame consists of three members, and it has three joints. Each additional joint requires two additional members. Consequently the number of members required is

$$n = 2j - 3$$

where *j* is the number of joints (*see*, for example, Fig. 38, where $n = 27$ and $j = 15$). If the number is greater, the frame is HYPERSTATIC; if it is less, it is a MECHANISM. A RIGID JOINT adds a redundancy to the FRAME, and an additional pin joint, or one member less is required. An additional pin joint removes a redundancy, and an additional member or rigid joint is required to maintain the correct number of members. The simplest SPACE FRAME is a tetrahedron, which has six members and four joints. Each additional joint requires three additional members, so that

$$n = 3j - 6$$

See also GEODESIC DOME and SCHWEDLER DOME.

MODE The most common item of a group of observations. If these are plotted as a bar chart, or HISTOGRAM, the mode is the observation corresponding to the longest bar. *See also* MEAN.

MODEL ANALYSIS Analysis of structural and other (*e.g.* lighting, ventilation or acoustic) problems by means of models. *See* DIRECT and INDIRECT MODEL ANALYSIS. *See also* ACOUSTIC MODELLING, ARTIFICIAL SKY, PERISCOPE, SOLARSCOPE and WIND-TUNNEL.

MODEL, MATHEMATICAL *See* MATHEMATICAL MODEL.

MODELSCOPE A PERISCOPE for viewing models.

MODERN ARCHITECTURE *See* CIAM *and* INTERNATIONAL STYLE.

MODULAR· COORDINATION Design of building components to conform to a dimensional standard, based on a modular system.

MODULAR DEVIATION The difference between actual size and modular size.

MODULAR GRID A REFERENCE GRID in which the grid lines are spaced at exact multiples of the MODULE. Drawing paper with a pre-printed modular grid can save much work in setting out the plan and elevation of a building, and in stating dimensions on the drawing.

MODULAR LINE, PLANE A grid-line, grid-plane in a MODULAR reference GRID.

MODULAR RATIO (*of reinforced concrete*) The ratio of the MODULUS OF ELASTICITY of the reinforcing steel to the EFFECTIVE MODULUS OF ELASTICITY OF CONCRETE.

MODULAR SIZE   *See* MODULE.

MODULAR SPACE A space bounded by MODULAR PLANES.

MODULE A unit of length particularly specified for modular co-ordination. In most metric countries the *basic module* (M) is 10 cm or 100 mm; in countries using the FPS system it is 4 in, which is only 1·6 per cent more than 100 mm. A *modular size* is a dimension which is a multiple of the basic module, *i.e.* nM, where n is an integer. The modular size nM is normally considered to include allowances for joints and TOLERANCES. In practice, all multiples of M are not equally useful, and many modular standards therefore specify a system of *preferred sizes*, *e.g.* multiples of $3M$ (300 mm or 1 ft), in preference to multiples of $2M$ (200 mm or 8 in). A system of preferred sizes may also consist of a number pattern which becomes more widely spaced with increasing size, as in the FIBONACCI SERIES, or in various empirical number patterns proposed, for example, by the British Ministry of Works.

MODULOR System of modular co-ordination advocated by Le Corbusier in a book of that name in 1950. It is based on 183 mm (6 ft, *i.e.* the height of a rather tall man), which is mainly subdivided in accordance with the *Golden Section*. Preferred multiples are based on the *Fibonacci series*. The system does not produce the simple numerical relationships required for industrialised building.

MODULUS, BULK *See* BULK MODULUS.

MODULUS OF ELASTICITY The ratio of direct STRESS to STRAIN in an elastic material obeying HOOKE'S LAW. It is the (hypothetical) stress which would produce a unit strain. It is also called *Young's modulus*, after Thomas Young, a British scientist who introduced the concept in 1807. *See also* MODULUS OF RIGIDITY, SECANT MODULUS and Fig. 83.

MODULUS OF RIGIDITY The ratio of SHEAR STRESS to SHEAR STRAIN in a material obeying HOOKE'S LAW. It is the (hypothetical) shear stress which would produce a unit shear strain. The modulus of rigidity, G, is related to the MODULUS OF ELASTICITY, E, by the equation

$$E = 2G(1 + \sigma)$$

where σ is POISSON'S RATIO.

MODULUS OF RUPTURE The nominal stress

$$f = \frac{M}{S}$$

(where M is the ultimate BENDING MOMENT, and S the SECTION MODULUS) at which a beam breaks. It is a common test for the tensile strength of brittle materials (such as concrete), whose flexural strength is controlled by tension. The term is misleading, because of the association of 'modulus' with the modulus of elasticity. It should also be noted that stress distribution at rupture is no longer elastic, and consequently the true tensile stress is lower than M/S.

MODULUS OF SECTION *See* SECTION MODULUS and PLASTIC MODULUS.

MOHR CIRCLE A graphic construction devised by the nineteenth century German engineering professor, Otto Mohr, which enables the stresses acting on a cross-section oriented in any desired direction to be determined if the PRINCIPAL STRESSES are known (Fig. 63). It can be used either for two-dimensional or, with an additional construction, for three-dimensional stress problems. *See also* ELLIPSOID OF STRESS.

MOHR FAILURE CRITERION A FAILURE CRITERION which assumes that failure occurs when the MOHR CIRCLE touches a limiting envelope. If the envelope is an inclined straight line, we get COULOMB'S EQUATION, also known as the *internal friction theory*, which gives good results for granular materials, such as sand, gravel or concrete failing in shear. *See also* COWAN CRITERION.

MOHR'S THEOREM A theorem proposed by Professor Otto Mohr at Dresden Technical University in

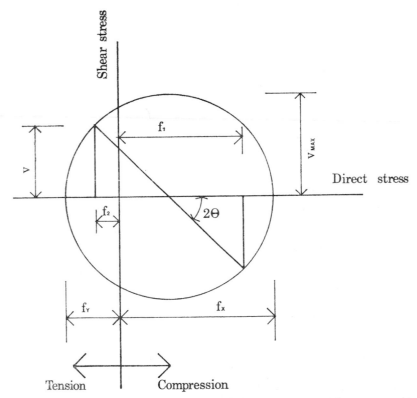

FIG. 63. Mohr circle for a plane stress problem. The principal stresses are f_x (compressive) and f_y (tensile). The maximum shear stress, which occurs at an angle of 45° to the PRINCIPAL PLANES, is v_{max}. At some angle θ, the direct stresses are f_1 and f_2, and the shear stresses are v.

171

1868 for determining the slope and deflection of a beam. It states that the slope and deflection bear the same relation to the bending moment, as the shear force and bending moment respectively do to the load. Consequently the slope and deflection can be determined from a fictitious load, which equals the bending moment divided by *EI* (where *E* is Young's modulus and *I* is the second moment of area); the resulting method is known as the *moment-area method*.

MOHS' SCALE A hardness scale defined by Friedrich Mohs, a German mineralogist, in 1812. It consists of a comparison with 10 standard minerals: (1) talc, (2) gypsum, (3) calcite, (4) fluorite, (5) apatite, (6) orthoclase, (7) quartz, (8) topaz, (9) corundum, and (10) diamond. A material of number 8 hardness is scratched by corundum, but scratches quartz.

MOIRÉ FRINGES Patterns produced by interference between two series of lines, *viz.* an undistorted parallel grid, and a similar grid distorted by a loaded MODEL. This can be achieved either by reflection from the surface of the model, in a manner similar to the photostress method, or by refraction, *i.e.* the light passing through the model. The patterns resemble those of PHOTO-ELASTICITY, but the technique measures slope, and consequently *flexural curvature*. (The isochromatics give the difference between the two principal stresses.)

MOIST ROOM A room in which the relative humidity is kept above 98 per cent, and the temperature is kept constant (usually 23°C ± 2°). It is used for curing test specimens of building materials (particularly concrete) under standard conditions. Also called a *fog room*.

MOISTURE BARRIER Either a *vapour barrier* or a *damp-proof course*.

MOISTURE CONTENT The weight of water in a material, such as a soil, divided by the weight of the solids.

MOISTURE EQUIVALENT, FIELD *See* FIELD MOISTURE EQUIVALENT.

MOISTURE GRADIENT The variation in moisture content between the outside and the inside of a piece of material, particularly wood.

MOISTURE METER Instrument for determining the moisture content of timber by measuring the electrical resistance between two points on its surface.

MOISTURE MOVEMENT (*a*) The movement of moisture through a porous material. (*b*) The effect of moisture movement on the dimensions of the material. In concrete it causes mainly SHRINKAGE. In timber it may cause shrinkage, swelling or distortion, particularly if it is inadequately SEASONED. *See also* CREEP (*b*).

MOL A unit weight of an element or a chemical compound, equal to its molecular weight in grams.

MOLDING *See* MOULDING.

MOLECULAR HEAT The product of the SPECIFIC HEAT of a substance and its molecular weight.

172

MOMENT   *See* BENDING MOMENT, COUPLE, TWISTING MOMENT, MOMENT OF A FORCE.

MOMENT–AREA METHOD Determination of slope from the area, and deflection from the first moment of the area of the bending-moment diagram, using MOHR'S THEOREM.

MOMENT ARM   *See* LEVER ARM.

MOMENT DISTRIBUTION METHOD   A technique devised by the American engineer, Hardy Cross, (who also produced the *column analogy*) in 1930 for the solution of the bending moments in rigid frames and continuous beams by successive approximations. Every span is initially assumed fully restrained at each support. The supports are then released in turn and the out-of-balance moments distributed to the members joined at the support in proportion to their STIFFNESS. This causes CARRY-OVER MOMENTS at the far ends of these members. The joints are released in turn, and the remaining out-of-balance moments are distributed until the residual moments are negligible. The *relaxation method,* devised by *Sir Richard Southwell,*

is a generalised form of the moment distribution method.

MOMENT OF A FORCE   The moment of a force about a given point, or briefly 'the moment', is the turning effect, measured by the product of the force and its perpendicular distance from the point. *See also* COUPLE.

MOMENT OF AN AREA   The sum of the products obtained by multiplying each element of area, dA, by its distance from the reference axis, y. It is therefore

$$\Sigma \, y \, dA \quad \text{or} \quad \int y \, dA$$

The CENTROIDAL AXIS is the axis about which the moment of an area vanishes.

MOMENT OF INERTIA   A measure of the resistance offered by a body to angular acceleration. It is the SECOND MOMENT OF AREA of the body, multiplied by its density. The term 'moment of inertia' is also commonly used as a synonym for 'second moment of area'. *See also* POLAR MOMENT OF INERTIA.

MOMENT OF RESISTANCE   The moment produced by the internal forces in a beam (Fig. 64). For

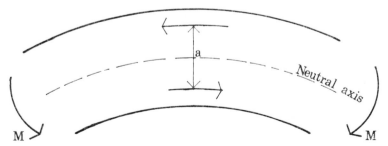

FIG. 64.   The external bending moment must equal the internal resistance moment.

equilibrium, it must equal the BEND-
ING MOMENT due to the external
forces.

MOMENTUM　The product of the
mass of a body and its velocity.

MONDRIAN, PIET　*See* DE STIJL.

MONITOR　A series of windows on
both sides of a single storey factory
roof, which admit daylight, and
sometimes also provide ventilation.
The monitor is a linear version of the
LANTERN used in classical domes.
Monitor frames (Fig. 65) and roof
trusses are shaped to include the
monitor as an integral part of the
steel frame. *See also* NORTHLIGHT
ROOF.

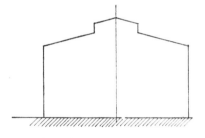

FIG. 65.　Monitor frame.

MONOCHROMATIC LIGHT
Light of a single colour. The visible
spectrum of light ranges from a wavelength of about 4000 Å for violet-
blue to about 7000 Å for red (400–
700 nm). For certain measurements
and photographic applications, this
range of wavelengths is unacceptable.
Monochromatic light can be produced by a filter (which absorbs the
other colours), or by a monochromatic light source, which produces light
of one wavelength only. A sodium
lamp, or a bunsen burner in which

the yellow sodium colour is produced
by adding a little common salt, are
commonly used, although this normally produces two wavelengths,
5890 and 5896 Å.

MONOLITH　A single block of
stone or concrete. Hence *monolithic
construction*, which is concrete cast
with no joints other than construction
joints.

MONOMER　A molecule containing a single mer, as opposed to a
POLYMER.

MONTAUK BUILDING　*See* SKY-
SCRAPER.

MONTE CARLO METHOD　A
procedure which utilises the techni-
ques of statistical sampling to obtain
an approximate solution of a mathematical or physical problem.

MONTMORILLONITE　A
mineral, consisting of finely divided
hydrous aluminium or magnesium
silicate, characterised by a sheet-like
molecular structure. Its expansion
and contraction on wetting and drying are exceptionally high. It is the
main constituent of BENTONITE.

MOON–SPENCER SKY　The standard for an overcast sky adopted by
the CIE, originally proposed by
P. Moon and D. E. Spencer. It is a
completely overcast sky, without any
direct sunlight, whose *luminance* is a
maximum at the zenith. At some
altitude θ its luminance is $\frac{1}{3}(1 + 2$
$\sin \theta)$ times the luminance at the
zenith. This standard has been used
in artificial skies for model analysis,
particularly for the temperate zone.
It gives the worst lighting conditions
on a dull day; however, it is not

suitable for designing windows for hot–arid climates.

MORRIS, WILLIAM *See* ARTS AND CRAFTS and DEUTSCHER WERK-BUND.

MORTAR A mixture of sand and a cementing medium, such as PORT-LAND CEMENT or LIME.

MORTISE A rectangular slot cut into one piece of timber, into which a tenon, or tongue, from another piece is fitted to form a *mortise-and-tenon joint*. A *mortise lock* is one which fits into a mortise in the edge of a door, as opposed to a RIM LOCK.

MOSAIC Inlaid surface decoration for floors and walls formed by small pieces (or *tesserae*) of tile, glass or marble. Present-day mosaics are generally laid in geometric patterns, but representational designs were common in earlier days.

MOTEL A hotel which provides guests with ready access to their motor-cars.

MOULD Formwork. Also spelt *mold.*

MOULDING A strip of wood, stone, metal or plastic, either plane or curved, used to cover a joint, or used for decoration. Hand-carved mouldings played an important part in classical architecture. Mass-produced mouldings can now be bought in many different shapes and sizes.

MOVEMENT *See* MOISTURE MOVE-MENT and TEMPERATURE MOVEMENT.

MOVING-COIL MICROPHONE A microphone in which the dia-phragm exposed to the sound pres-sure is connected directly to a mov-able coil. The movement of the coil in the magnetic gap produces a voltage.

MOVING FORMS *See* CLIMBING FORMWORK.

MOVING STAIR An escalator.

MPH Miles per hour.

MU (μ) Abbreviation for *micro.*

MUFFLER *See* TRANSMISSION LOSS.

MULLION A vertical dividing member of a window, capable of supporting weight. The correspond-ing horizontal member is a *transom.* Windows may be further subdivided by *glazing bars.*

MULTI-ELEMENT PRESTRESS-ING Assembly of an integrated structural member from several indi-vidual units by means of prestressing.

MUMFORD, LEWIS *See* RADBURN PLAN.

MUNSELL BOOK OF COLORS *See* COLOUR.

MUNTIN *See* SASH BAR.

MURIATIC ACID Hydrochloric acid (HCl).

MUSHROOM CONSTRUCTION *See* FLAT SLAB.

MUSLIM DOME *See* SQUARE DOME.

MUTHESIUS, HERMANN *See* DEUTSCHER WERKBUND.

N

n Abbreviation for *nano*.

N (*a*) Chemical symbol for *nitrogen*. (*b*) Abbreviation for *newton*.

Na Chemical symbol for *sodium* (natrium).

NADAI'S SANDHEAP ANALOGY *See* SANDHEAP ANALOGY.

NAIL-GLUED ROOF TRUSS A glued roof truss with plywood gusset plates. The nails hold the truss together until the glue dries; thereafter the strength of the joints is presumed to depend on the glue alone.

NAIL PLATE A metal gusset plate containing holes through which nails are driven into two or more pieces of timber. It may be straight, T-shaped or angle-shaped. *See also* GANG-NAIL and TIMBER CONNECTOR.

NAILABLE CONCRETE Concrete into which nails can be driven. It usually contains LIGHTWEIGHT AGGREGATE, often with the addition of SAWDUST.

NAILER A strip of wood attached to steel, built into a brick wall, or cast into concrete, so that other elements can be fixed with nails. Also called *nailing strip*.

NANO Prefix for one thousandth of a millionth (from the Greek word for dwarf); *e.g.* 1 nm = 1 nanometre = 1×10^{-9} metres.

NAPERIAN LOGARITHM An alternative name for NATURAL LOGARITHM, named after the sixteenth century mathematician, John Napier, who invented logarithms.

NAPPE A sheet of water flowing over a weir.

NATURAL CEMENT A cement obtained by grinding calcined argillaceous limestone burned at a temperature no higher than is necessary to drive off the carbon dioxide. In contrast to PORTLAND CEMENT, it is produced from a natural mixture of argillaceous and calcareous material. Natural cements were more common before the manufacture of Portland cement was perfected. The earliest example is the *Roman Cement* patented by J. Parker in England in 1796.

NATURAL FREQUENCY The frequency of a system acted upon by no external forces, except the original excitation. *See also* FORCED VIBRATION.

NATURAL LOGARITHM A logarithm to the base $e = 2.718\,28$ ..., also called a *Naperian logarithm*. This is of practical importance, because the solution of the integral

$$\int \frac{\mathrm{d}x}{x} = \log_e x$$

Although tables of natural logarithms are obtainable, calculations are usually performed by conversion to common logarithms, for which tables are more readily available.

$$\log_e a = \log_{10} a \times \log_e 10$$

The natural logarithm of 10, $\log_e 10 = 2.3026$. *See also* EXPONENTIAL FUNCTION.

NATURAL SEASONING The drying of timber by stacking it so that it is exposed to the flow of air, but is sheltered from rain and sun. It is an older method than KILN seasoning, and it takes much longer to achieve the same result. Since sufficient time is not always allowed, the extent of the seasoning is open to doubt. *See also* AIR DRYING.

NATURAL STONE Stone which has been quarried and cut, but not crushed into chips and reconstituted as CAST STONE.

NAUTICAL ALMANAC A book of tables, published annually, of the daily movements of the sun, moon and stars. Since 1960 the *British Nautical Almanac* and the *American Ephemeris* have been published in the same form.

NAUTICAL MILE One international nautical mile = 1852 metres. One British Admiralty and US Coast Survey nautical mile = 6080 ft = 1853·2 metres.

NAVE The body of a church, usually separated from the AISLES by lines of columns. *See also* TRANSEPT.

NAVIER'S THEOREM The simple theory of bending. In 1826 L. M. H. Navier, professor at the *Ecole des Ponts et Chaussées* in Paris, published the solution in the now accepted form:

$$\frac{M}{I} = \frac{f}{y} = \frac{E}{R}$$

where M is the bending moment at the section; I is the second moment of area of the section; f is the extreme fibre stress; y is the extreme fibre distance; E is Young's modulus; and

R is the radius of curvature at the section. The equation is often quoted in two different forms:

$$M = FS$$

where F is the MAXIMUM PERMISSIBLE STRESS for the material in the beam, and S is the SECTION MODULUS of the beam, and

$$R = \frac{EI}{M}$$

which is used for determining the SLOPE and the DEFLECTION of the beam.

NBRI National Building Research Institute, Pretoria.

NBS National Bureau of Standards, Washington.

NC *Noise criterion*. A series of standard curves defining critical noise levels.

NEAR-CRITICAL ACTIVITY *See* ACTIVITY.

NEAT CEMENT Cement used without sand, as opposed to cement mortar.

NECKING The contraction in area which occurs when a ductile material, such as mild steel, fails in tension.

NEEDLE A short, stout timber or steel beam which is passed horizontally through a wall, to support the end of a shoring timber.

NEGATIVE BENDING MOMENT A bending moment which causes hogging, or convex, curvature. The sign convention normally employed for the structure of buildings is shown in Fig. 66 (an exactly opposite

STATICALLY DETERMINATE BEAMS

negative b.m.   positive b.m.   negative b.m.

STATICALLY INDETERMINATE BEAMS

FIG. 66.   Sign convention for positive and negative bending moments.

sign convention is also in use for certain types of engineering structures). Bending moments which cause sagging, or concave curvature, are taken as *positive*. In continuous beams and frames, the negative moments occur over the columns, and the positive moments near mid-span.

NEGATIVE REINFORCEMENT Reinforcement placed in concrete beams to resist the NEGATIVE BENDING MOMENT, *i.e.* normally near the top face.

NEO-CLASSICISM A style which emerged in the eighteenth century as a reaction to the late Baroque and Rococo. English PALLADIANISM is an early phase, which was followed in

the middle of the century by a more direct appeal to archaeological precedent.

NEO-GOTHIC *See* GOTHIC REVIVAL.

NEON A colourless, odourless and inert gas. Its chemical symbol is Ne, its atomic number is 10, its atomic weight is 20·183, and its boiling point is −245·9°C. A discharge of electricity through neon at low pressures produces an intense orange–red glow, which is used for neon signs and glow discharge lamps.

NEOPRENE An oil-resistant synthetic rubber.

178

NERVI, P. L. *See* FERROCEMENTO.

NETWORK A connected series of ARROWS representing a building project (Fig. 67). It forms the basis of both CPM and PERT. A network must have one starting point and one terminal point. However, it may have many branches between, and become so complex that the CRITICAL PATH can be determined only with the aid of a DIGITAL COMPUTER. *See also* ACTIVITY, CRASHED TIME, DUMMY ACTIVITY, EVENT and FLOAT.

NEUTRAL AXIS The line in the cross section of a beam where the flexural stress changes from tension to compression, *i.e.* the line where the direct stress is nil. In sections subject to pure bending, the neutral axis passes through the CENTROID. Also called *neutral layer (of the beam)*.

NEUTRON Sub-atomic particle of the same mass as a PROTON, but having no electric charge.

NEWEL An upright post supporting the hand rail at the top and bottom of a staircase, and also at the turn on a landing.

NEWTON (*abbreviated N*) In the SI system of units, the unit of force and weight; it is named after the seventeenth century English mathematician. 1 newton is the force which, applied to a mass of 1 kilogram, produces an acceleration of 1 metre per second per second. Unlike the units of force used in the CGS and FPS units, it is independent of the earth's gravity. The newton is a small unit (roughly the weight of a small apple) for practical purposes, and it is frequently necessary to use the kilo-newton (kN) = 1000 N, or the mega-newton (MN) = 1 000 000 N. *See also* LBF and KILOPOND.

NEWTON LIQUID In rheological terminology, a VISCOUS liquid. The rheological model, named after the seventeenth century Englishman, is a dashpot, which extends indefinitely

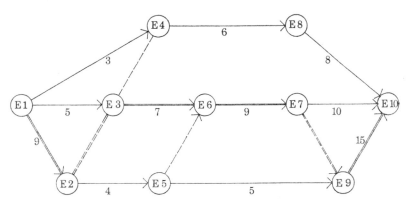

FIG. 67. Network. The events are indicated by circles, and the activities by lines which link the EVENTS together. ARROWS indicate the sequence of activities, and dotted lines indicate DUMMY ACTIVITIES. The number on the arrow is the time duration of the activity; dummies have no time duration. The CRITICAL PATH is indicated by the double line.

when pulled or pushed, but does not deform instantly (Fig. 81).

NICHE *See* ALCOVE.

NICOL PRISM A device for obtaining plane-polarised light, *e.g.* for PHOTOELASTICITY. It consists of a crystal of ICELAND SPAR, cut and cemented together in such a way that a plane-polarised ray of light is freely transmitted, while an ordinary ray of light is reflected out at the side of the crystal.

NITROCELLULOSE Nitric acid ester of cellulose, formed by the action of a mixture of nitric and sulphuric acid on cellulose. Nitrocellulose with a low nitrogen content is not explosive, and it is used as a solvent of lacquer. Nitrocellulose with a high nitrogen content is the explosive *gun cotton.*

NITROGEN A colourless and odourless gas. It constitutes about 78 per cent by volume of the atmosphere. Its chemical symbol is N, its atomic number is 7, its atomic weight 14·008, it has a valency of 3 or 5, and its boiling point is $-195.8°C$.

NO-FINES CONCRETE Concrete made without sand. It therefore contains a high proportion of communicating pores, which provide thermal insulation and drainage.

NO-SKY LINE A line separating all points on the WORKING PLANE at which the sky is directly visible from those at which no section of the sky is directly visible.

NO-SLUMP CONCRETE Concrete with a SLUMP of 1 inch or less.

NOBLE METALS Metals, such as silver, gold and platinum, which have a relatively high positive electrode potential, and consequently do not readily enter into combination with non-metals. They are resistant to most types of corrosion, and specifically to atmospheric oxidation. *See* ELECTROCHEMICAL SERIES.

NOGGING (*a*) Horizontal short timbers which stiffen the vertical studs of a framed partition. (*b*) Brick infilling in the spaces between the studs of a framed partition.

NOISE *See also under* ACOUSTIC and SOUND.

NOISE ABSORPTION *See* SOUND REDUCTION FACTOR and DISCONTINUOUS CONSTRUCTION.

NOISE, BACKGROUND *See* BACKGROUND NOISE.

NOISE, FAN *See* FAN NOISE.

NOISE REDUCTION COEFFICIENT The arithmetic average of the SOUND ABSORPTION COEFFICIENTS of a material at the frequencies of 250, 500, 1000 and 2000 hertz.

NOMINAL SIZE OF TIMBER The size of timber before it is DRESSED. Sizes of dressed timber are generally given as nominal sizes, and the actual size is $\frac{3}{16}$ to $\frac{1}{2}$ inch (5–13 mm) smaller.

NOMOGRAM A diagram used for the evaluation of an equation. Its simplest form consists of three straight lines, each graduated for one variable. By joining any two of them with a straight edge, the third can be read off. However, more complicated forms can be constructed, using more lines, including curved and inclined lines. One nomogram serves the same purpose as a *series* of graphs; it is more compact, but does not show the trend of variations as easily. The use of nomograms has declined with the increasing use of computer printouts.

NONAGON *See* POLYGON.

NON-BEARING WALL *See* BEARING WALL.

NON-COMBUSTIBLE *See* INCOMBUSTIBLE.

NON-CRITICAL ACTIVITY *See* ACTIVITY.

NON-DESTRUCTIVE TESTING Testing which does not destroy the testpiece. It is often performed on the actual structure. Non-destructive testing is particularly important when doubts about strength, composition, etc., arise after the building has been wholly or partly completed, so that the removal of a test-piece might seriously weaken or disfigure it. It is, however, rarely as reliable as conventional methods which result in the destruction of the test-piece by breaking it, or analysing it chemically, etc. X-RAYS, ULTRASONIC WAVES, magnetic particles and BRITTLE COATINGS have been employed for non-destructive strength tests. Tests based on elastic deformation, as in the STRESS-GRADING OF TIMBER, in the testing of concrete with the SCHMIDT HAMMER, or in EXPERIMENTAL STRESS ANALYSIS with strain gauges, are also non-destructive.

NON-DEVELOPABLE *See* DEVELOPABLE SURFACE.

NORMAL DIRECTION Direction perpendicular to the cross section under consideration.

NORMAL FREQUENCY DISTRIBUTION CURVE *See* GAUSSIAN CURVE.

NORMAL STRESS A direct stress, as opposed to a shear stress.

NORMALISING Heating steel to a temperature about 50°C above the transformation range, followed by cooling in still air at room temperature, so that moderately rapid cooling occurs. The object is to eliminate internal stresses, refine the grain size and render the structure of the metal more uniform.

NORMAN STYLE A Romanesque style of architecture developed in England after the Norman conquest in 1066. Like the Anglo-Saxon style, it is characterised by massiveness, which gave way to much lighter construction in the Gothic era.

NORTH-LIGHT ROOF A sloping factory roof having one gentle slope without glazing, and a glazed roof face pointing north. In the temperate zone this has a slope to admit more daylight, but in the subtropics it is usually vertical to exclude direct sunlight. Also called a *sawtooth roof*. In the southern hemisphere a *southlight roof* is used instead. *See also* MONITOR.

NORTHLIGHT SHELL A shell designed as a NORTHLIGHT (or in the Southern Hemisphere, southlight) ROOF. It may take the form of a parabolic *conoid*, with the parabolas facing north, or a cylindrical shell which has been cut (Fig. 68).

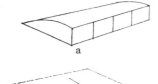

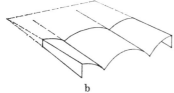

FIG. 68.   Northlight (Southlight) shells.
(a) Cylindrical shell, (b) conoid.

NOTCH EFFECT A locally increased stress in a section due to a
notch, or sharp change in section. It
is normally employed to test IMPACT
RESISTANCE.

NOTE *See* OCTAVE.

NPL National Physical Laboratory,
Teddington, England.

NRC (*a*) Noise reduction coefficient.
(*b*) National Research Council (USA
or Canada).

N-TRUSS *See* PRATT TRUSS.

NUMBER, APPROXIMATE *or*
EXACT *See* APPROXIMATE NUMBER.

NZIA New Zealand Institute of
Architects, Wellington.

NZIE New Zealand Institution of
Engineers, Wellington.

O

O Chemical symbol for *oxygen*.

OAKUM *See* CAULKING.

OBELISK A tapering shaft of
stone, of rectangular section with a
pyramidal apex, usually monolithic.

**OBLIQUE PARALLEL PROJEC-
TION** *See* PROJECTION.

OBSIDIAN A natural GLASS of
granitic composition, originating as
a LAVA. It is generally black with a
vitreous lustre. It fractures with a
sharp, hard edge, and has therefore
been used in building tools by civilisa-
tions which did not know iron, *e.g.*
the Aztecs of Mexico (*see also* FLINT).

OCCUPANCY *See* FIRE LOAD.

OCHRE *See* IRON OXIDE.

OCTAGON *See* POLYGON.

OCTAHEDRON *See* POLYHEDRON.

OCTAVE A range of eight notes
on the diatonic (*i.e.* the conventional)
musical scale. The FREQUENCY of the
octave is precisely twice that of the
base note, and it is the first HARMONIC
of the base note.

OCTAVE-BAND An acoustic frequency range of two.

OCTAVE-BAND ANALYSER *See*
SOUND-FREQUENCY ANALYSER.

OD Outside diameter.

182

ODEON   Originally a public building for musical performances.

OEDOMETER   A machine for determining the CONSOLIDATION characteristics of COHESIVE SOILS. An *undisturbed* sample is loaded between two porous stone plates which allow free passage of water in and out of the sample, and the settlement is measured.

OFF-WHITE   White, with the addition of a small amount of another colour, but insufficient to identify any colour other than white. Also called *broken white*.

OGEE   A doubly-curved line, made up of a convex curve passing without a break into a concave curve, like an S.

OGEE ARCH   A pointed arch formed by two ogee curves (Fig. 69). It was used in Saracen architecture, and became popular in Europe in the

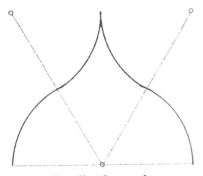

FIG. 69.   Ogee arch.

late Middle Ages. It is not based on sound structural principles, and cannot be used for large spans. *See also* POINTED ARCH.

OGEE MOULDING   A moulding of ogee shape. If the concave part is uppermost, it is called a *cyma recta*. If the convex part is uppermost, it is called a *cyma reversa*.

OHM'S LAW   Law governing the flow of a steady current in an electric circuit, enunciated by the German physicist G. S. Ohm in 1827. It states that the voltage drop produced by the current is proportional to the magnitude of the current. The resistance is defined as the ratio of current to voltage, and its unit is 1 *ohm* (Ω); when the steady current in the conductor is 1 ampere and the voltage difference between two points along it is 1 volt, the resistance is 1 ohm.

OIL–ALKYD PAINT   *See* OLEO-RESINOUS PAINT.

OIL PAINT   Paint with a binder of DRYING OIL, as opposed to water paint.

OIL STAIN   A thin oil paint with very little pigment, used for staining timber. *See also* SPIRIT STAIN.

OLD ENGLISH BOND   *See* BOND (*a*).

OLEORESINOUS PAINT   A paint whose vehicle consists of a mixture of drying oil and resin (such as phenolic or alkyd resin), combined by a cooking process. *See also* ALKYD PAINT and OIL PAINT.

OMNIDIRECTIONAL MICRO-PHONE   A microphone which picks up sound from all directions.

ON GRADE   At ground level or supported directly on the ground.

ONYX A banded variety of silica, consisting of very small quartz crystals. The bands are straight, not curved as in AGATE, and as a result it has been widely used for CAMEOS.

OOLITIC LIMESTONE A limestone formed by the agglomeration of oolites. These are tiny spherical concretions (less than 2 mm in diameter) of calcium carbonate, usually showing a concentric-layered or radiating fibrous structure. The structure resembles the roe of fish cemented together. *See also* PORTLAND STONE.

OPACITY The opposite of *transparency*. In painting, it denotes HIDING POWER of a paint.

OPAL Amorphous hydrous silica ($SiO_2.nH_2O$). Apart from the gemstone, there are less precious varieties, *e.g.* in opaline cherts, which find application as building materials.

OPEN-FRAME GIRDER *See* VIERENDEEL GIRDER.

OPEN-HEARTH PROCESS Steel made in an open hearth from pig iron, scrap iron and limestone. The process was developed in 1858 by Sir William Siemens, and is also known as the Siemens Martin process. *See also* BESSEMER PROCESS.

OPEN LIGHT A window which can be opened, as opposed to a *dead light*.

OPEN SYSTEM *See* INDUSTRIALISED BUILDING.

OPEN-WEB JOIST A lattice JOIST welded from light steel sections, which directly supports the roof or floor. It is mass produced to certain standard lengths.

OPEN-WEB STUD Stud made of open-web steel which permits the passage of plumbing and electrical conduits.

OPTICAL AXIS *See* COLLIMATION LINE.

OPTICAL DENSITY *See* TRANSMISSION FACTOR.

OPTIMISATION The process of determining the optimum conditions, generally with the aid of a DIGITAL COMPUTER. The most common optimisation is for cost, *i.e.* finding the design with the lowest cost. However, other factors can be optimised, such as construction time, weight or maintenance.

ORANGERY A garden building with large windows on the South side (in the Northern Hemisphere), originally intended for growing oranges in a climate which is otherwise too cold in winter.

ORCHESTRA (*a*) Originally the large semi-circular space in a Greek theatre, in front of the stage, where the chorus danced and sang. (*b*) Generally that part of the theatre assigned to the musical performers. (*c*) The company of musicians, particularly the instrumental players.

ORDERS, THE The arrangement of a column and its entablature in classical architecture. The three *Greek* Orders were the DORIC, the IONIC and the CORINTHIAN. The five *Roman* Orders were the TUSCAN, the Doric (which differs from Greek Doric), the Ionic, the Corinthian and the COMPOSITE.

ORDINATE The y-axis, or vertical axis, of a co-ordinate system (Fig. 13).

OREGON *or* **OREGON PINE** *See* DOUGLAS FIR.

ORGANIC ARCHITECTURE Architecture whose appearance has a character similar to that of a natural organism, and gives the same impression of unity. The concept dates from classical times; it is, for example, mentioned by VITRUVIUS, VASARI and MICHELANGELO. In modern times *Erich Mendelsohn* and *Frank Lloyd Wright*, in particular, have described their own works as organic.

ORGANIC CHEMISTRY The study of those compounds of carbon which form chains or rings. Many organic materials are POLYMERS.

ORGANIC CLAY *or* **SILT** A clay or silt which contains the remains of plants or animals. It is normally recognisable by smell when the soil is moulded. Its bearing capacity is generally very low.

ORIEL WINDOW A projecting window on an upper floor, CORBELLED from the wall. *See also* BOW WINDOW.

ORIENTATION The arrangement of a building in relation to the North point.

ORIGIN The point of intersection of Cartesian co-ordinates, *i.e.* the zero point for the x-, the y-, and (in three-dimensions) the z-axes (Fig. 13).

ORTHOGONAL CURVES Curves crossing one another at right angles.

ORTHOGRAPHIC PROJECTION *See* PROJECTION.

ORTHOTROPIC Having physical properties which vary at right angles, *e.g.* the strength of timber along and across the grain. It is a special case of AELOTROPIC.

OSMOSIS The diffusion of a solvent or of a dilute liquid through a skin, permeable only in one direction, into a more concentrated solution.

OUD, JACOBUS JOHANNES *See* DE STIJL.

OUTPUT The method of transferring information from the CENTRAL PROCESSOR UNIT of a digital computer to a device which makes it intelligible to the user. This is usually a typewriter, but it may be a graphical display unit. *See also* DATA MEDIUM and COMPUTER GRAPHICS.

OVAL Any plane figure resembling the longitudinal section of an egg, except an ELLIPSE. The term is commonly used for closed curves resembling an ellipse, but not precisely conforming to its mathematical definition.

OVEN-DRY SOIL Soil dried in an oven at 105°C.

OVEN-DRY TIMBER Timber which does not lose moisture in a ventilated oven heated to 100°C.

OVER-REINFORCED *See* BALANCED DESIGN.

OVERCAST SKY *See* MOON-SPENCER SKY.

OVERSTRETCHING Stressing TENDONS above the INITIAL PRESTRESS. This serves to overcome frictional losses, to temporarily overstress the steel to reduce the CREEP of steel, and to counteract LOSS OF PRESTRESS caused by the subsequent prestressing of other tendons.

OVERTONE *See* HARMONIC.

OXIDATION The chemical combination of an element with oxygen. A fire is caused by rapid oxidation. Corrosion of some materials, *e.g.* steel, is caused by slow oxidation. The opposite process is called *reduction*.

OXY-ACETYLENE WELDING and **CUTTING** *See* ACETYLENE.

OXYCHLORIDE CEMENT A composition of magnesium chloride $(MgCl_2.6H_2O)$ and magnesia (MgO), also called *Sorel's cement*. The magnesia is derived from MAGNESITE; hence the floor finish in which this cement is used, is called *magnesite flooring*.

OXYGEN A colourless and odourless gas which supports combustion. It is the most abundant of all chemical elements, since it forms 21 per cent by volume of the atmosphere, 89 per cent of the weight of water, and almost 50 per cent of the weight of the rocks in the earth's crust. Its chemical symbol is O, its atomic number is 8, its atomic weight is 16·000, its valency is 2 or 4, and its boiling point $-183°$C.

OZONE An unstable form of oxygen, which contains three atoms to the molecule (O_3), instead of two as in ordinary oxygen (O_2).

P

φ *See* PHI.

π *See* PI.

Pa Abbreviation for PASCAL.

PA Polyamide, more commonly called nylon.

PACKING *See* SPACE LATTICE.

PADDLE-WHEEL FAN A CENTRIFUGAL FAN.

PADSTONE A block of stone or concrete, built into a wall to distribute the pressure from a concentrated load.

PAINT *See* PIGMENT *and* VEHICLE. *See also* ALKYD PAINT, ANTI-CORROSIVE PAINT, CEMENT PAINT, CHLORINATED RUBBER, DISTEMPER, FIRE-RETARDANT PAINT, FRESCO, GRAINING, KALSOMINE, LACQUER, LATEX PAINT, MARBLING, OIL PAINT, PLASTIC PAINT, PLASTICS, PRIMER, STAIN, TEMPERA PAINTING, VARNISH and WATER-PAINT.

PAINT REMOVER A liquid which softens paint or varnish so that it can be scraped or brushed off.

PALLADIANISM A style derived from the publications (notably *Quattro libri dell' architettura*, published in 1570) and, to a lesser extent, measured drawings of the buildings of the sixteenth century Italian architect, Andrea Palladio. It was introduced into England by Inigo Jones in the sixteenth century, and revived by Colen Campbell and Lord Burlington in the eighteenth century (*see* GEORGIAN ARCHITECTURE). Hence it influenced the COLONIAL architecture of America and Australia. *See also* HARMONIC PROPORTIONS.

PALLET A lifting tray used for stacking materials with a fork-lift truck.

PAN HEAD A screw or rivet head shaped like a truncated cone.

PANEL, DROP *See* DROP PANEL.

PANEL HEATING A system of heating in which the heating units are concealed in special panels, or built into the walls or ceiling.

PANELLED DOOR A wooden door built with a framed surround, with the spaces between the framing members filled with panels of a thinner material. It has less moisture movement than a door built of parallel planks, and it was used for the better quality buildings before the development of plywood (which has even less moisture movement). 'Panelled' doors used in modern neo-Colonial buildings are often plywood doors with mouldings attached.

PANTHEON A domed building erected in Rome in AD 123, allegedly, by the Emperor Hadrian, and still standing. Unlike most classical domes, it was built mainly of concrete. With a span of 142 ft 6 in (43·5 m) it held the record until the nineteenth century. There is also a Pantheon in Paris (St Geneviève by Soufflot AD 1755), and there was one in London (by James Wyatt, AD 1772); however, these are entirely different in construction.

PANTOGRAPH A mechanism, consisting of a jointed parallelogram with projecting sides, used at one time for the copying of illustrations. By altering the location of the pen or pencil the scale can be reduced or increased.

PAPER *See* KRAFT PAPER and WALL PAPER.

PAPER TAPE *See* TAPE.

PARABOLA A curve produced by plotting the equation $y = ax^2 + b$, where a and b are constants. It is also the shape made by cutting a right circular CONE parallel to one edge. A *cubic parabola* has the equation $y = ax^3 + b$. *See also* ELLIPSE and HYPERBOLA.

PARABOLIC ARCH An arch whose curvature is parabolic. The bending moment diagram for a uniformly distributed load is parabolic, and consequently a parabolic arch carrying a uniformly distributed load is free from bending stresses. The shape of flat parabolic and flat CATENARY ARCHES is similar.

PARABOLIC CONOID *See* SURFACE OF TRANSLATION.

PARABOLIC REFLECTOR A mirror whose cross-section forms a parabola. If the lamp is placed at the focus of the parabola, it produces a parallel beam. This is due to the parabola's property (Fig. 70) that any parallel ray, such as A, meeting the parabola at B, makes the same angle with the tangent at B as a ray coming from the focus F. Hence all rays reflected by the mirror surface from F are parallel. A *spherical reflector* is often used because it is cheaper, but the beam is only approximately parallel, even if a small segment of the sphere is used.

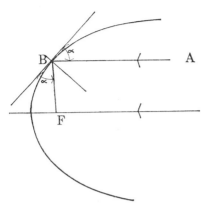

FIG. 70.   Parabolic reflector.

PARABOLOID   *See* SURFACE OF REVOLUTION.

PARABOLOID, ELLIPTICAL
See SURFACE OF TRANSLATION.

PARABOLOID, HYPERBOLIC
See HYPERBOLIC PARABOLOID.

PARALLEL   Lying alongside and always the same distance apart. Thus two parallel straight lines meet only at infinity.

PARALLEL OF LATITUDE   *See* GEODETIC LINE.

PARALLELEPIPED   A solid bounded by six parallelograms, any two opposite ones being parallel to one another. In a *right* parallelepiped all edges are parallel or perpendicular to the base. The CUBE is a special case of a right parallelepiped. All other parallelepipeds are classified as *oblique*.

PARALLELOGRAM   *See* QUADRILATERAL.

PARALLELOGRAM OF FORCES
A figure illustrating a theorem of statics. 'If two forces acting at one point be represented in magnitude and direction by two sides of a parallelogram, their resultant is represented by the diagonal drawn from that point.' In structural design two sides of the parallelogram are usually omitted, leaving the other two sides and the diagonal forming a *triangle of forces. (See* Fig. 78). The parallelogram of forces was first used by *Leonardo da Vinci* in the fifteenth century, and formally published by *Stevinus of Bruges* in 1586.

PARAMETER   A variable in a mathematical relation, which is kept constant for a particular investigation, but which can be made to vary in different cases. Variation of the parameter thus produces a family of curves and surfaces.

PARAPET   The portion of a wall which extends above roof level.

PARIAN PLASTER   *See* KEENE'S CEMENT.

PARIS, PLASTER OF   *See* PLASTER OF PARIS.

PARIS WHITE   *See* WHITING.

PARQUETRY   Small pieces of wood, sometimes of different species, fitted together to form a geometrical design. They are usually glued to a floor.

PARTIAL DIFFERENTIAL EQUATION   *See* DIFFERENTIAL EQUATION.

PARTIAL PRESTRESSING Pre-stressing to a lower level than full prestressing (which eliminates all possibility of cracking under *working loads*). In partially prestressed members tensile stresses exist in the pre-compressed tensile zone of the concrete at the working load.

PARTIALLY FIXED *See* SEMI-RIGID JOINT.

PARTICLE BOARD A FIBRE BOARD formed with only a small amount of pressure (unlike HARD-BOARD), or by EXTRUSION. The binder is usually urea or phenol resin, and only a small amount (4–12 per cent) is required. It can be joined and veneered like plywood, but is cheaper and has better insulation. Also called *chip board*.

PARTICLE-SIZE ANALYSIS Determination of the proportion of particles of each size in a granular mixture, such as soil or aggregate. When the result is plotted on semi-logarithmic graph paper, the particle *grading curve* is obtained. The larger particles are separated by *sieving* or *screening*. For those particles too fine to pass through a sieve (74 μm) the specific surface is determined, usually by STOKES' LAW; or a TURBIDIMETER may be used. *See also* GRAVEL, SAND, SILT and CLAY.

PARTITION *See* BEARING WALL.

PARTY WALL *See* COMMON WALL.

PASCAL (Pa) Unit of pressure in SI UNITS. It equals 1 NEWTON per square metre, and is named after the seventeenth century French mathematician.

PASSIVE EARTH PRESSURE *See* EARTH PRESSURE.

PATENT GLAZING Any system of dry glazing, *i.e.* without the use of putty.

PATINA *See* VERDIGRIS.

PATIO An outdoor paved court forming part of the living area, partially or wholly surrounded by parts of the house.

PATTERN STAINING Discoloration of plaster ceilings of composite construction, caused by the different thermal conductances of the backing. The air circulates more freely over the warmer parts, and deposits more dust on them.

PAVEMENT LIGHT A window of GLASS BRICKS built into a pavement surface, to admit natural light to a space below ground level.

PAVILION (*a*) Originally a large tent. (*b*) A light ornamental structure, roofed but only partially enclosed, in a garden or sports ground. (*c*) A projecting sub-division of a building, often elaborately decorated.

PAWL *See* RATCHET.

PAXTON, SIR JOSEPH *See* CRYSTAL PALACE.

Pb Chemical symbol for *lead* (plumbum).

PC Prime cost.

PE Polyethylene.

PEA GRAVEL Screened gravel, from which particles larger than 10 mm and smaller than 5 mm have been removed by sieving. *See also* SINGLE-SIZED AGGREGATE.

PEARLITE *See* PERLITE.

PEAT Dead, gelatinous, compressible vegetable matter preserved by humic acid in the ground. It is unsuitable as a foundation material. However, it can be used as a fuel after drying.

PEBBLE DASH An external plaster which has been surfaced with small stones, thrown on while the plaster is still wet.

PEDESTAL The base of a classical column or super-structure. In modern construction, a short column whose height does not exceed three times its least lateral dimension.

PEDIMENT In classical architecture, a low-pitched GABLE above a portico.

PEEN *See* BALL-PEEN HAMMER and CLAW HAMMER.

PEG-BOARD A *perforated hardboard*. A pattern of holes is drilled during manufacture. These may serve a decorative or acoustic purpose, or they may, with special hooks, be used to support shelves and other fixtures.

PELMET A built-in head to a window for hiding the curtain rail.

PELTON WHEEL An impulse turbine, consisting of a wheel carrying buckets on its perimeter which are struck by a fast-flowing water jet.

PENDANT (*a*) A hanging ornament. (*b*) A hanging chandelier. (*c*) A decorated boss in stone, stucco, or timber, elongated so that it hangs down. It was used particularly in late Gothic vaulting.

PENDENTIVE *See* SQUARE DOME.

PENETRATION TEST Test of soil in place, which may be either STATIC or dynamic. The information obtained supplements that collected from BOREHOLE SAMPLES.

PENETROMETER An instrument used for conducting a penetration test.

PENTAGON *See* POLYGON.

PENTHOUSE A room, apartment, or separate dwelling built on the roof of a building. The term is used both for a room at the top of a high building which covers the elevator shaft, water tank, etc., and for a luxury apartment on the top storey.

PERFECT FRAME A synonym for an ISOSTATIC frame.

PERFORATED HARD BOARD *See* PEG BOARD.

PERGOLA A structure of intersecting beams, carried on posts, which is open to the sky. It is often used as a garden feature, covered with climbing plants which form a roof.

PERIMETER AIR-CONDITIONING *or* **HEATING** A system which feeds air through registers located along the outer walls, supplied through ducts from a central plenum chamber.

PERIMETER GROUTING Injection of grout at low pressure around the periphery of an area. When the area is subsequently grouted at a higher pressure, the grout injection is confined by the perimeter, with consequent saving in grout.

PERIODIC TABLE OF THE ELEMENTS A classification table of the chemical elements, arranged in ascending order of atomic weight (the ATOMIC NUMBER), which demonstrates that the physical and chemical properties of an element and its compounds vary with the atomic number (*see* Appendix E). The table is arranged in nine vertical columns, and the properties show periodicity in accordance with these nine groups. The system was initiated by the Russian chemist Dimitri Mendeléev in 1869.

PERISCOPE An apparatus consisting essentially of two prisms or two inclined mirrors which enable an observer to obtain a view of objects at a different level. Although it is best known for its use in submarines to see objects above while remaining below water, the process can be reversed. The observer can insert a periscope into a model, and thus obtain from above the view which a person would get if he could stand inside the model (*model scope*). The periscope can be used visually, or in front of a camera.

PERISTYLE A row of columns *surrounding* a building or court.

PERLITE (*a*) A volcanic glass, usually with a higher water content than OBSIDIAN. It can be expanded by heating. *Expanded perlite* is used as an insulating material, and as LIGHTWEIGHT AGGREGATE. (*b*) A EUTECTOID composed of alternate laminae of iron and iron carbide, formed at about 720°C when steel is slowly cooled. The etched section has a pearly appearance. The proportion of perlite in carbon steels increases with the carbon content. Also spelt *pearlite*.

PERMAFROST Permanently frozen ground, which exists in the Northern parts of Russia, Canada and Alaska. Buildings erected on it must be specially insulated to prevent the heat generated by the building from melting the frozen soil and turning it into mud.

PERMANENT DEFORMATION *See* INELASTIC.

PERMANENT MAGNET *See* MAGNET STEEL.

PERMEABILITY The rate of diffusion of a gas or liquid under pressure through a material such as soil or concrete.

PERMEAMETER A laboratory instrument for measuring the coefficient of permeability, using DARCY'S LAW. The head of water has to be kept constant for highly permeable materials, such as sand or gravel; however, for the slowly permeable clays and silts it can be allowed to drop from an initial head (*falling-head permeameter*).

PERMISSIBLE STRESS *See* MAXIMUM PERMISSIBLE STRESS.

PERPENDICULAR A line or plane which meets another at right angles.

PERPENDICULAR STYLE The last phase of the Gothic style in England, following the *Decorated style*. It began in the middle of the fourteenth century and was in use up to the Reformation. The Chancel of Gloucester Cathedral is an early example; Kings College Chapel in Cambridge and Henry VII Chapel in Westminster Abbey are late examples. The style is characterised by intersecting horizontals and verticals.

PERPENDS The vertical joints on the face of brickwork.

PERRY–ROBERTSON FORMULA A formula for the design of columns, which is more sophisticated than the RANKINE FORMULA. It is the work of British engineering professor J. Perry (who derived it), and A. Robertson (who checked it experimentally in 1924), and is used in the British and Australian steel codes. Perry's formula applies to struts which are initially curved, and it is assumed that every slender strut has an unavoidable slight curvature, which is defined by the empirical constant in the formula. *See also* SECANT COLUMN FORMULA.

PERSPECTIVE *See* PROJECTION.

PERSPEX *See* ACRYLIC RESINS.

PERT Originally *Program Evaluation Research Technique*, now interpreted as *Performance Evaluation and Review Technique*. This form of network analysis, developed mainly for military and aerospace work, differs in terminology, rather than in substance, from the CRITICAL PATH METHOD, developed for the construction industry.

PETERSBURG STANDARD *See* BOARD FOOT.

PETROGRAD STANDARD *See* BOARD FOOT.

PETROGRAPHIC MICROSCOPE A microscope fitted with a pair of NICOL PRISMS, one serving as a polariser and the other as an analyser. The characteristic colours produced by polarised light help to identify the minerals, particularly in igneous rocks.

PETROGRAPHY Descriptive petrology.

PETROLOGY The science of rocks.

PEWTER *See* TIN.

PF Phenol formaldehyde.

PFA Pulverised fuel ash, *i.e.* FLY ASH.

pH VALUE The logarithm to the base 10 of the reciprocal of the concentration of hydrogen ions in an aqueous solution, in gram molecules per litre. Water has a pH value of 7; basic solutions are higher than 7, and acid solutions lower than 7. It is normally measured with an electrical instrument, and used particularly to express small differences in the alkalinity or acidity of neutral solutions.

PHAROS An island off Alexandria, where King Ptolemy Philadelphus built a famous lighthouse, which was the largest tower of Antiquity. Hence any lighthouse of the classical period.

PHASE (*a*) A physically and chemically homogeneous portion of an alloy system, as shown in a PHASE DIAGRAM. (*b*) One of the windings or circuits of a polyphase electrical apparatus; also the recurring sequence of the electric wave. *See also* THREE-PHASE.

PHASE DIAGRAM A diagram which shows the temperature as ordinate, and the composition range of an alloy as the abscissa. Although its main use is in metallurgy, it can be used to show the variation of the

phases in non-metallic solutions and solid solutions (Fig. 71). The phase diagram can be used to illustrate the temperatures at which alloys made of any proportion of two or three elements exist in the liquid and the solid state, the temperatures at which transformations occur, the manner in which solubility changes with temperature, and other features of the behaviour of an alloy system. *It is also called a constitutional diagram, an equilibrium diagram, or an alloy diagram. See also* LIQUIDUS and SOLIDUS LINES.

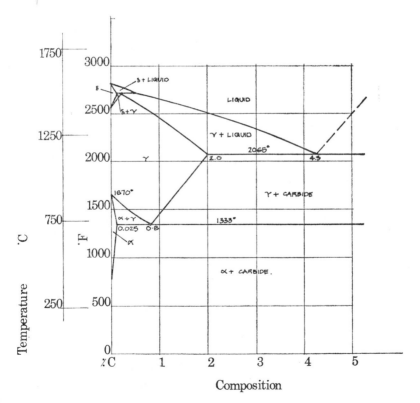

FIG. 71. Phase diagram of iron and carbon. This is the most important of all phase diagrams, since it covers WROUGHT IRON, CARBON STEEL, and CAST IRON.

PHENOL FORMALDEHYDE *See*
FORMALDEHYDE.

PHI (φ, $\varnothing$) (*a*) Greek letter. (*b*)
Diameter, particularly of plain rein-
forcing bars. Deformed bars are
denoted by number, not diameter.
(*c*) Symbol commonly used for the
letter O in ALPHANUMERIC printouts.
0 denotes the number zero. (*d*)
Symbol for the GOLDEN SECTION.

PHON A unit of *loudness*. It is
based on equal loudness contours,
determined experimentally by Fletcher and Munson in 1930. Observers
were asked to match the loudness of
a pure tone at a frequency of 1
kilohertz with that of a tone at
another frequency. The *threshold of
audibility* (2×10^{-4} dynes/cm^2 at
1000 Hz, or 2×10^{-5} N/m^2 at 1
kHz) corresponds to 0 phon. As the
loudness increases, the phon scale
increases proportionately to the
DECIBEL scale, *i.e.* it is a logarithmic
scale. Thus, if the loudness increases
from X to Y, then the gain is $10 \log_{10}$
X/Y phons. *See also* SONE and
HEARING THRESHOLD.

PHOSPHORESCENT PAINT A
paint which emits visible light for
some minutes or hours after visible
or ultra-violet light has fallen on it.
It usually contains calcium sulphide,
strontium sulphide, or zinc sulphide.
See also FLUORESCENT PAINT.

PHOTOELASTIC MATERIALS
A material which has the property,
when stressed, of breaking up light
into two components polarised in the
directions of the principal stresses. It
is possessed by several glasses, ther-
moplastics, and casting resins. The
choice of the material for a model to
be analysed by PHOTOELASTICITY
depends largely on the ease with
which it can be fabricated to the
required shape, and on the cost.
Bakelite, Perspex, Plexiglas, Catalin
and Araldite have been used for plane
photoelasticity. Certain phenol-for-
maldehyde plastics have the property
of freezing the photoelastic stress
patterns at 75°C, and these can be
used for THREE-DIMENSIONAL PHOTO-
ELASTICITY. Thin sheets and liquid
coatings have been developed for
the PHOTOSTRESS METHOD.

PHOTOELASTICITY The
property of certain transparent materials to break up the incident light
into two components polarised in
the directions of the principal stresses.
Light waves are ordinarily oriented
at random. By inserting a NICOL
PRISM or POLARISING FILTER, only
light in one plane is transmitted. A
second Nicol prism or polarising
filter is oriented at right angles to the
first, and a model is placed between
them. Since the two filters are at right
angles, no light is transmitted when
the model is unstressed. When the
model is stressed, coloured fringes
(ISOCHROMATICS) appear which connect points of equal difference
between the two principal stresses
(*see also* MOIRÉ FRINGES). Although
the colours give a clear visual picture,
and often make fascinating patterns,
they make precise measurement
difficult, and MONOCHROMATIC light
is used for numerical analysis. The
photoelastic effect was discovered in
1816 by *Sir David Brewster*, explained
in 1850 by *Clerk Maxwell*, and first
applied to model analysis by *E. G.
Coker* in the 1920s. Apart from the
visual image conveyed, photoelastic
model analysis is particularly useful

for examining STRESS CONCENTRATIONS. *See also* ISOCLINICS, PHOTOELASTIC MATERIALS, PHOTOSTRESS METHOD, POLARISCOPE, QUARTERWAVE PLATE and THREE-DIMENSIONAL PHOTO-ELASTICITY.

PHOTOELECTRIC CELL A device whose electrical state is altered by exposure to visible light. For example, SELENIUM becomes an electrical conductor when illuminated, and it is an insulator in the dark. *See also* EXPOSURE METER and PHOTOMETER.

PHOTOMETER An instrument for comparing the luminous intensities of two sources of light. There are numerous types. For architectural models PHOTOELECTRIC CELLS are commonly used as photometers because they are small and easily moved into confined spaces. They require regular calibration for accuracy.

PHOTOMICROGRAPHY The production of photographic negatives and prints of very small objects, obtained by attaching a camera to a microscope.

PHOTON A QUANTUM of light.

PHOTOSTRESS METHOD PHOTOELASTIC analysis by means of photoelastic coatings, which may be applied to a model in sheet form or as a solution backed by a reflective surface. The method was developed by the French engineer, F. Zandman, in the 1950s for determining surface strains. It has an advantage over the ordinary polariscope in not requiring a transparent model; however, the sensitivity is lower.

PI (π) The circular constant 3·1416 . . . .

PI-THEOREM (π-THEOREM) A theorem published by E. Buckingham in 1914, which establishes the number of dimensionless ratios (πs) required for dimensional similarity between two physical phenomena (*e.g.* a structure and its model). If the two phenomena are determined by r parameters, which can be expressed in terms of n primary dimensions (length, mass, time, etc.), then the number of πs is (r–n).

PIANO NOBILE The floor of a large house which contains the reception rooms. It is usually above ground level, and its ceiling height is greater than that of the other floors.

PICKUP *See* VIBRATION PICKUP.

PICTORIAL PROJECTION *See* PROJECTION.

PICTURE RAIL A moulding fixed to an interior wall. Pictures may be suspended from it by means of metal hooks which fit over the top of the moulding.

PICTURE WINDOW A large window whose bottom ledge is less than waist high.

PICTURESQUE *See* COTTAGE ORNÉ and FOLLY.

PIER A massive compression member, less SLENDER than a column or strut.

PIEZO-ELECTRIC EFFECT An electric charge, set up when certain crystals (such as quartz) are expanded or compressed. Discovered by Jacques and Pierre Curie in 1880, it can be utilised for measuring very small forces in confined locations. The effect may also be utilised for NON-DESTRUCTIVE TESTING, since such crystals can be made to vibrate at ULTRASONIC frequencies by an alternating current, and thus act as ultrasonic transducers.

PIG A mass of metal, such as cast iron, lead or copper, cast into a simple shape, which is subsequently remelted for purification, alloying, or processing. The term originated with the now obsolete method of running the liquid metal from the blast furnace into a channel in a bed of sand, called a *sow*. From there it ran into smaller lateral channels, called pigs.

PIGMENT *See* BLANC FIXE, CADMIUM YELLOW, CARBON BLACK, CHINESE WHITE, IRON OXIDE, LAKE, LITHOPONE, PRUSSIAN BLUE, RED LEAD, RED OXIDE, ULTRAMARINE, VERMILION, WHITING and WHITE LEAD. *See also* FLUORESCENT and PHOSPHORESCENT PIGMENT.

PILASTER A column built into a wall, and projecting slightly from it. In classical architecture it was usually decorated in conformity with one of the ORDERS.

PILE A long slender column of timber, concrete, or steel embedded in the foundation. It may be driven, jacked, jetted, or (in the case of concrete) cast in place. *See also* BEARING PILE and SHEET PILE.

PILE CAP (*a*) A protective cap fitted over the head of a pile during driving. (*b*) A structural member

designed to distribute the load from a column or wall to a group of piles.

PILE HAMMER A hammer for driving piles into the soil. Drop hammers, which depend purely on gravity, go back to Roman times, but are still commonly used because of their reliability. Power-operated double-acting hammers, however, are faster since they can deliver up to 300 blows per minute.

PILE HEAD The top of a pile. Since the penetration of piles driven to REFUSAL cannot be accurately predicted, the heads are often cut off after driving.

PILE SHOE A point of cast steel or cast iron at the foot of a driven pile of timber or precast concrete.

PILLAR A vertical compression member. In classical architecture a *column* was required to be circular in plan, and to conform to one of the orders. A vertical compression member which did not do so, was called a pillar.

PILOTIS A French term for a column or stilt, which supports the building and leaves the ground space open. It has been employed by *Le Corbusier*, and the term 'pilotis' implies a column of the type used in his buildings.

PIN JOINT A joint between two or more members of a structure which transmits no moment, as opposed to a RIGID or SEMI-RIGID JOINT. Pin joints are so called, because in the mid-nineteenth century they frequently consisted of pins pushed through a hole in each of the members to be joined. True pin joints are today rare, except for very large spans where complete certainty of freedom to rotate is required. Normally 'pin joint' denotes a flexible joint, or a joint at the end of a flexible member, which transmits only a negligible moment. Pin joints may be deliberately introduced into structures to render them ISOSTATIC, or to reduce the number of REDUNDANCIES (Fig. 73). *See also* PLASTIC HINGE and ARTICULATED STRUCTURE.

PINNACLE A vertical pointed structure rising above a roof or BUTTRESS. Its weight helps to turn the horizontal thrust of the roof towards the vertical. *See also* FINIAL.

PIPE COLUMN A column made from steel tubing. It is frequently filled with concrete to increase its stiffness and strength.

PISÉ DE TERRE Wall of unburnt clay or chalk, rammed in a damp condition into formwork without reinforcement, except sometimes straw. This is a vernacular form of construction in several arid regions. With a protective coating it can also be used in areas of moderate rainfall. *See also* COB and ADOBE.

PITCH OF A ROOF The angle of a sloping roof, usually defined by the *ratio* of rise to span.

PITH A soft core in the centre of wooden log.

PITOT TUBE An open-ended tube facing in the direction of motion of a fluid, so that the pressure recorded by it equals the total head due to the velocity of the fluid.

PLAIN BAR *or* **WIRE** A reinforcing bar or wire without deformations to improve bond, as opposed to a DEFORMED BAR or an INDENTED WIRE. Bars with deformations not conforming to the building code are also treated as plain bars for assessing the maximum permissible BOND stress.

PLAIN CONCRETE Concrete which is neither reinforced nor prestressed.

PLAN *See* PROJECTION.

PLANE FRAME *See* FRAME and SPACE FRAME.

PLANE OF SATURATION The water table in soil.

PLANIMETER An instrument for measuring areas by mechanical means. It is a simple type of integrator.

PLANING MACHINE *See* MILLING MACHINE.

PLANNING GRID A network of horizontal and perpendicular lines, to assist the designer with a layout plan. It is usually a MODULAR GRID.

PLASTER Any pasty material of mortar-like consistency, used for covering the walls or ceilings of a building. The traditional plasters based on lime or gypsum are now rare, and Portland cement, mixed with sand and water, is the common material for plastering.

PLASTER OF PARIS GYPSUM which has been heated to drive off some of its water ($CaSO_4.\frac{1}{2}H_2O$). When mixed with water it sets rapidly with formation of heat, and expands in the process. Hence it is particularly useful for making accurate casts. Also known as *hemihydrate* plaster. The setting is too rapid for many building applications, and *retarded hemihydrate plaster* contains a retarder, usually *keratin*. If plaster is to be used in conjunction with iron or steel reinforcement, it is necessary to add about 5 per cent of hydrated lime to prevent corrosion. *See also* FIBROUS PLASTER and KEENE'S CEMENT.

PLASTERBOARD A building board made of a core of gypsum or anhydrite plaster, faced with two sheets of heavy paper.

PLASTIC CRACKING Cracking which occurs on the surface of fresh concrete soon after it is placed. It is often confused with SHRINKAGE CRACKING; however, plastic cracks can be filled in by trowelling, while shrinkage cracks occur during the HARDENING STAGE.

PLASTIC DEFORMATION Continuous permanent deformation in metals, which occurs above a critical stress, the YIELD or PROOF STRESS. The ELASTIC deformation results from straining of the crystal lattice. Plastic deformation normally occurs by slipping action across the crystal planes when the interatomic forces become too high. It depends on the ability of the metal to sustain distortion of the crystal structure without fracture. *See also* Fig. 28, VISCOUS FLOW, BRITTLENESS, and SAINT VENANT BODY.

PLASTIC DESIGN Design based on the formation of PLASTIC HINGES. *See* LIMIT DESIGN.

PLASTIC FLOW A widely used

term which is synonymous with CREEP. It is, however, misleading, since plastic deformation is not recovered when the stress is removed, while creep is viscous deformation, largely recoverable over a period of time. *See also* STRESS–STRAIN DIAGRAM OF CONCRETE and Fig. 91.

PLASTIC HINGE After structural steel has reached its LIMITING STRENGTH, which is the YIELD STRESS, it continues to deform at a constant stress until its deformation is several times as much as the total elastic deformation. It is therefore reasonable to consider it as a PLASTIC MATERIAL (Figs. 72 and 93). Consequently a hinge forms, which can be rotated without further increase in the bending moment. When sufficient hinges have formed to turn

a

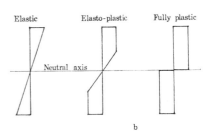

b

FIG. 72. Plastic stress distribution. (a) Idealised plastic stress–strain diagram, *i.e.* a linear elastic relation followed by a constant plastic stress. (b) Stress distribution changing with increasing load from elastic through elasto-plastic to fully plastic.

the structure into a mechanism, it collapses (*see* LIMIT DESIGN and Fig. 57). Plastic hinges also form in reinforced concrete, even though concrete does not exhibit plasticity. The yielding of the steel reinforcement in tension, and the crushing of the concrete in compression may allow sufficient rotation at a hinge to enable the structure to become a mechanism in accordance with the rules of limit design. However, the ultimate angle of rotation of a reinforced concrete 'plastic hinge' is always less than that of structural steel. (*See* Figs. 92 and 103). *See also* PIN JOINT.

PLASTIC LIMIT The water content at which a damp clayey soil just begins to crumble when rolled into a thread approximately $\frac{1}{8}$ in (3 mm) diameter. *See also* PLASTICITY INDEX.

PLASTIC MATERIAL A term which is often confusing in discussions between architects and engineers. In materials science and in rheology, plasticity denotes the ability of a material to deform at a constant stress without fracture, *see* SAINT-VENANT BODY. For structural purposes this must be the LIMITING STRENGTH OF THE MATERIAL. Thus structural steel is the plastic material *par excellence*, from an engineer's point of view. In sculpture, a plastic material is one which can be freely moulded; unlike stone or steel, which must be finished by cutting. From this point of view, concrete is the ideal plastic material of architecture because it can be cast into any mould chosen by the designer. However, as an engineering material concrete is BRITTLE, which is the very opposite of plastic.

PLASTIC MODULUS The SECTION MODULUS used in LIMIT DESIGN, which is tabulated in some section tables. It is based on the uniform stress distribution assumed to exist at a PLASTIC HINGE. For most RSJs it is approximately 15 per cent more than the elastic section modulus. The *shape factor* is defined as the ratio (plastic section modulus)/(elastic section modulus), and its value for structural steel sections is about 1·15. Including the shape factor in the load factor simplifies the use of elastic section moduli for limit design. *See* Fig. 72.

PLASTIC MORTAR A mortar of a consistency which allows it to be readily deformed during bricklaying, etc., without disintegrating. Plasticity is sometimes improved by additives, called plasticisers.

PLASTIC PAINT (*a*) A vague term for a paint whose medium is a *plastic* *i.e.* a synthetic resin. Or (*b*) A *texture paint*, *i.e.* a paint which can be used plastically for relief modelling.

PLASTIC STRESS DISTRIBUTION Stress distribution in a flexural section based on the assumption of fully-plastic stress distribution (Fig. 72). *See also* LIMIT DESIGN, WHITNEY STRESS BLOCK, STRESS–STRAIN DIAGRAM OF MILD STEEL and OF CONCRETE.

PLASTIC TILE A tile made from plastic, commonly PVC, as opposed to a ceramic tile.

PLASTIC WOOD A paste of wood flour, synthetic resin, and a volatile solvent. It is used for filling holes and cracks in timber. Its surface can be painted about an hour after application, so that the stopping of holes is frequently undertaken by the painter, and not by the carpenter.

PLASTICISER (*a*) An admixture to mortar or concrete which increases its workability. However, some plasticisers also reduce the strength. (*b*) A non-volatile substance mixed with the medium of a paint, lacquer or varnish to improve the flexibility of the hardened film.

PLASTICITY *See* PLASTIC MATERIAL.

PLASTICITY INDEX The numerical difference between the LIQUID LIMIT and the PLASTIC LIMIT. It is indicative of the range of water content through which a soil remains plastic.

PLASTICS A generic term for organic substances, mostly synthetic and formed by condensation or polymerisation, which become plastic under heat or pressure. They can then be shaped by moulding or extrusion. They are also used for laminates, paints, lacquers and glues. The main distinction is between THERMOPLASTIC and THERMOSETTING plastics. *See also* ABS PLASTIC, ACRYLIC RESIN, ALKYD RESIN, BAKELITE, CHLORINATED RUBBER, ELASTOMER, EPOXY RESIN, EXPANDED PLASTICS, FIBREGLASS, FORMALDEHYDE, LATEX PAINT, MELAMINE RESIN, POLYETHYLENE, POLYPROPYLENE, · POLYSTYRENE, POLYSULPHIDE, POLYTETRAFLUORETHYLENE, POLYVINYL ACETATE, POLYVINYL CHLORIDE, RESIN and SILICONE.

PLATE *See* SHEET.

PLATE, FLAT *See* FLAT PLATE, and also SLAB.

PLATE GIRDER A steel girder built up from vertical web plates and horizontal flange plates joined with angles. Connection may be by riveting or welding.

PLATE GLASS Glass of better quality than ordinary SHEET GLASS. It is usually thicker, with a smoother surface free of blemishes. It is made by pouring the molten material on a flat, heated iron table with a raised rim, and rolled to the required thickness with a heavy iron roller. *See also* FLOAT GLASS.

PLATED BEAM A plate girder.

PLATEN (*a*) The plate in a printing machine which presses the paper against the inked type. (*b*) A hot steel plate used in presses which make plywood with thermosetting glues. (*c*) A smooth steel plate used to compress a specimen in a testing machine.

PLENUM CHAMBER An air compartment maintained under a pressure slightly above atmospheric, and connected to one or more distributing ducts.

PLENUM SYSTEM A method of air conditioning whereby the air forced into the building is at a pressure slightly above atmospheric.

PLEXIGLAS *See* ACRYLIC RESINS.

PLINTH In classical architecture, the projecting base of a wall or column PEDESTAL, moulded or chamfered at the top. The term is now used for a slight widening at the base of a wall or column.

PLOTTER *See* DIGITAL PLOTTER.

PLUM A random-shaped stone weighing 100 lb (50 kg) or more, which is dropped into MASS CONCRETE to economise on cement and reduce the generation of heat. *Cyclopean concrete* is mass concrete containing a large number of plums.

PLUMB (*a*) To determine the direction of the vertical with a line weighted with lead (*plumbum*). (*b*) Any other method of lining up a building element in the vertical direction. (*c*) Vertical.

PLUMBAGO *Black lead*, an obsolete term for GRAPHITE.

PLUMBER'S SOLDER *See* SOLDER.

PLUTONIC INTRUSIONS IGNEOUS ROCKS which have cooled slowly at great depth below the earth's surface, and are therefore coarse-grained.

PLY A thickness of material, used for building up several layers, as in plywood, built-up roofing, etc.

PLYWOOD Material consisting of two or more PLIES of wood, with the grain of adjacent plies usually at right angles to one another. The outer plies are often veneers of decorative timber, while thicker and cheaper timber may be used inside. Plywood overcomes the inherent weakness of timber across the grain by lamination at right angles. However, a similar effect can be achieved by orienting the wood particles at random in reconstituted timber. *See* HARDBOARD.

PMMA Polymethyl methacrylate.

PNEUMATIC LOADING *See* AIR-BAG LOADING.

PNEUMATIC STRUCTURE A structure held up by a slight excess of internal air pressure above the pressure in the atmosphere outside. It must be sufficient to balance the weight of the roof membrane, and must be maintained by air compressors.

PNEUMATIC TOOL A tool worked by compressed air.

PNEUMATICALLY APPLIED MORTAR *See* SHOTCRETE.

POINT BEARING PILE An end-BEARING PILE.

POINT LOAD A concentrated load, as opposed to a distributed load.

POINT OF CONTRAFLEXURE *See* CONTRAFLEXURE.

POINT OF INFLECTION Point of contraflexure.

POINT SOURCE A source of radiant energy of negligible dimensions compared with the distance between source and receptor.

POINTED ARCH An arch produced by two curves (usually circular curves with a radius equal to the span) meeting in a point at the top. It is closer to the CATENARY ARCH, and therefore more efficient than the CIRCULAR ARCH. Its use in GOTHIC cathedrals greatly contributed to the lightness of their structure. *See also* OGEE ARCH, *and* CIRCLE.

POINTING (*a*) Pressing surface mortar into a RAKED JOINT. Pointed mortar joints are not as durable as joints made with the original bedding mortar; however, the practice of pointing joints in white, black or coloured mortar was once common. (*b*) The finishing operation on a mortar joint, without the addition of surface mortar.

POISSON'S RATIO The ratio of lateral unit STRAIN to longitudinal unit strain, when a piece of material is subjected to a uniform and uniaxial longitudinal stress. For a material which does not contract at all laterally, it would be zero. For a fully plastic material (whose volume remains constant under deformation) it is $\frac{1}{2}$. For many metals, Poisson's ratio is about $\frac{1}{3}$; however, for steel it is about $\frac{1}{4}$. For concrete it is about $\frac{1}{6}$. S. D. Poisson, a French mathematician, proved in 1829 that an elastic axial elongation must be accompanied by a lateral contraction. *See also* MODULUS OF RIGIDITY.

POKER VIBRATOR *See* VIBRATED CONCRETE.

POLAR CO-ORDINATES A system of co-ordinates based on the radial distance *r* from a reference point and the angle θ with a reference axis, in place of the conventional CARTESIAN co-ordinates *x* and *y*. It is useful for problems framed in terms of circular functions. Polar GRAPH PAPER greatly facilitates a graphical solution.

POLAR DIAGRAMS *See* MOHR CIRCLE and HOWARD DIAGRAM.

POLAR MOMENT OF INERTIA Moment of inertia about an axis *normal* to the plane of the section or area. The term 'moment of inertia' without prefix implies a moment about an axis *in* the plane of the section.

POLARISCOPE An instrument for showing phenomena connected with polarised light. If used for analysing models by PHOTOELASTICITY, it consists of two POLARISING FILTERS or NICOL PRISMS, a MONOCHROMATIC light source, and a holder or testing device for the model. QUARTER-WAVE PLATES are frequently added.

POLARISED LIGHT Light whose waves are confined to a single plane. Normal light, whose waves vibrate in space, can be polarised in a plane by passing it through a NICOL PRISM or a POLARISING FILTER, or by reflecting it from a glass plate at a particular angle.

POLARISING FILTER A filter for producing POLARISED LIGHT. The term is more specifically used for sheets of cellulose nitrate or acetate with natural or synthetic crystals embedded in them which transmit light polarised in one plane only.

POLARISING MICROSCOPE A microscope equipped with NICOL PRISMS, which allows minerals and rocks to be examined under polarised light. The characteristic colours of certain materials in polarised light help identification.

POLES The two points where the meridians of longitude intersect.

POLYESTER RESIN *See* FIBRE-GLASS.

POLYETHYLENE (PE) A low-cost thermoplastic polymer of ethylene. It is an ELASTOMER, which is completely waterproof, and is therefore widely used for bags and protective wrapping. It forms a cheap waterproof membrane, but must be protected against puncture. Also called *polythene*.

POLYGON A many-sided plane figure. The term often implies a figure with more than four sides. A five-sided figure is a *pentagon*, a six-sided figure a *hexagon*, a seven-sided figure a *heptagon*, an eight-sided figure an *octagon*, a nine-sided figure a *nonagon*, a ten-sided figure a *decagon*, and a twelve-sided figure a *dodecagon*. A *regular* polygon is one which is equiangular and equilateral. The included angle of a regular (commonly called equilateral) *triangle* is 60°; that of a regular *quadrilateral* (square) is 90°; that of a regular pentagon is 108°; that of a regular hexagon is 120°; that of a regular heptagon is $128\frac{1}{2}$°; that of a regular octagon is 135°; that of a regular nonagon is 140°; that of a regular decagon is 144°; and that of a regular dodecagon is 150°.

POLYGON OF FORCES A figure analogous to the triangle of forces, representing the statical equilibrium of more than three forces (*see* Fig. 78).

POLYHEDRON A solid figure bounded by plane surfaces. There are only five *regular* polyhedra, bounded by identical polygons. The *tetrahedron* is bounded by four equilateral triangles; it has four vertices and six edges. The *cube* (or *hexahedron*) is bounded by six squares; it has eight vertices and twelve edges. The *octa-hedron* is bounded by eight equilateral triangles; it has six vertices and twelve edges. The *dodecahedron* is bounded by 12 regular pentagons; it has 20 vertices and 30 edges. The *icosahedron* is bounded by 20 equilateral triangles; it has 12 vertices and 30 edges. In addition there are a large number of semi-regular polyhedra which are bounded by two or more types of regular polygon, but are otherwise symmetrical. These are important for the design of *geodesic space frames*, because of their versatility. *See also* SPACE FRAME and SPACE LATTICE.

POLYMERISATION The combination of several molecules to form a more complex molecule, having the same empirical chemical formula as the simpler ones. It is often a reversible process (*see* DE-POLYMERISATION). Some of the most successful plastics used in buildings are produced by polymerisation (which is also called CURING). The resulting material is called a *polymer*.

POLYMETHYL METHACRY-LATE (PMMA) *See* ACRYLIC RESINS.

POLYMORPHISM The existence of more than one *crystal structure* for a single composition.

POLYNOMIAL An expression consisting of many terms, but all of the type ax^n, where a is a constant, x is the variable, and n is a positive integer.

POLYPHASE *See* THREE-PHASE and PHASE.

POLYPROPYLENE (PP) A low-cost thermoplastic material, whose properties are similar to those of high-density polyethylene.

POLYSTYRENE (PS) A *thermoplastic* material formed by the polymerisation of styrene (C_6H_5. $CH:CH_2$). It is resistant to moisture, strong alkalis, several acids, and alcohol, but softens at 140°F (60°C). In transparent form it is brilliantly clear. Expanded polystyrene is one of the EXPANDED PLASTICS used as an insulating material.

POLYSULPHIDE A thermosetting resin, used as a building sealant. It is usually polymerised by mixing it with a catalyst immediately before application, and then poured in place.

POLYTETRAFLUORETHYLENE (PTFE) A crystalline, linear polymer, unique among organic compounds in chemical inertness. It is resistant to all alkalis and acids, even to AQUA REGIA. It does not have a melting point, but undergoes a phase transformation at 620°F (330°C), with a sharp drop in strength. It is marketed under the trade names *Teflon* and *Fluon*.

POLYTHENE *See* POLYETHYLENE.

POLYURETHANE *See* EXPANDED PLASTICS.

POLYVINYL ACETATE (PVA) A thermoplastic material formed by the polymerisation of vinyl acetate (CH_3. $COOCH:CH_2$). It is the binding agent in many emulsion paints.

POLYVINYL CHLORIDE (PVC) A low-cost thermoplastic material formed by the polymerisation of vinyl chloride ($CH_2:CHCl$). It is a

rubbery material, used for insulating electrical cables. It is practically incombustible, and resistant to water, oil and many chemicals. Thus it can be used for cold water pipes, as a flooring material, as a waterproof membrane, and as an expansion joint.

PONDING Accumulation of water on a *flat roof* due to an insufficient slope or inadequate drainage. It may cause excessive loads, producing additional and progressive deflection.

POPULATION In STATISTICAL terminology, the totality of all possible values of a particular characteristic for a UNIVERSE. A universe could have several populations associated with it. For example, we could measure the exact dimensions of the concrete cylinders made on a specific building site (population 1), and also test their compressive strength (population 2); or we could interview the same group of persons in an office, and ascertain their response to noise (population 1), their response to the thermal conditions (population 2), and their response to the level of illumination (population 3).

PORCELAIN Glazed pottery made from CHINA CLAY, used for fine table ware, for electric insulators, and for dielectrics.

PORCELAIN ENAMEL *See* VITREOUS ENAMEL.

PORE-WATER PRESSURE The pressure of water in a saturated soil.

POROSITY The ratio of the volume of voids to the total volume of a sample of soil. It equals $v/(1 + v)$, where v is the VOIDS RATIO.

PORPHYRY A generic term for IGNEOUS rocks which contain a few large crystals in a fine-grained ground-mass.

PORTAL A monumental door or gateway. Hence a frame consisting of two verticals and a member which may be horizontal, sloping or arched. RIGID and *two-pinned* portals are hyperstatic. *Three-pinned* portals are isostatic. (Fig. 73).

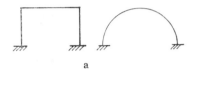

a

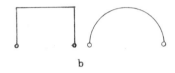

b

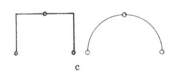

c

FIG. 73. Portal and arch. (a) Rigid portal and arch (*three* REDUNDANCIES). (b) Two-pin portal and arch (*one redundancy*). (c) Three-pin portal and arch (ISOSTATIC).

PORTAL CRANE *See* GANTRY CRANE.

PORTCULLIS A gate sliding in vertical grooves in the jambs of a doorway. It was used to defend fortifications and could be lowered quickly in an emergency.

PORTICO A roofed space, open or partly enclosed, which forms the entrance to a building. It is usually the centrepiece of a classical façade. The roof is often supported by columns at regular intervals, surmounted by a PEDIMENT.

PORTLAND BLAST-FURNACE SLAG CEMENT Cement consisting of a mixture of PORTLAND CEMENT and of BLAST-FURNACE SLAG in specified proportions. It can be produced by mixing the Portland cement clinker with granulated blast furnace slag before grinding, or by blending the ground cement with finely granulated slag.

PORTLAND CEMENT The most common form of cement. It is made by burning together chalk or limestone and clay or shale, and grinding the resulting clinker into a fine powder. The result is a complex mixture of calcium silicates (*see* DICALCIUM *and* TRICALCIUM SILICATE) and calcium aluminates, which sets into a hard paste when it comes into contact with water. Portland cement, mixed with sand and AGGREGATE, forms CONCRETE. The name is due to J. Aspden who patented the first artificial cement (*see also* NATURAL CEMENT) in England in 1824. He claimed that Portland cement mortar, plastered on brickwork and painted cream, was a substitute for PORTLAND STONE.

PORTLAND-POZZOLAN CEMENT A uniform blend of PORTLAND CEMENT with finely ground POZZOLANA, or a cement made by grinding a mixture of Portland cement clinker and pozzolana. The proportion of pozzolana requires careful control.

PORTLAND STONE An OOLITIC LIMESTONE, quarried on the Isle of Portland, off the coast of Southern England. Because of its light creamy colour and the ease with which it can be carved, it has been a particularly popular stone for monumental and residential buildings in London.

POSITIVE BENDING MOMENT *See* NEGATIVE BENDING MOMENT (Fig. 66).

POSITIVE REINFORCEMENT Reinforcement placed in concrete beams to resist the POSITIVE BENDING MOMENT, *i.e.* near the bottom face.

POST-AND-BEAM CONSTRUCTION A system of construction in which posts and beams are the main load-bearing members. *See also* TRABEATED.

POST-HOLE AUGER *See* AUGER (*b*).

POST-STRESSING Obsolete term for *post-tensioning*.

POST-TENSIONING PRESTRESSED CONCRETE in which the TENDONS are tensioned *after* the concrete has hardened, as opposed to PRE-TENSIONING.

POSTULATE A simple proposition of a self-evident nature, which requires no proof, and generally cannot be proved. Some major scientific innovations have resulted from questioning the truth of a long-accepted postulate. Also called AXIOM.

POT FLOOR *See* HOLLOW-TILE FLOOR.

POT LIFE Time interval after mixing during which a liquid material is usable, *e.g.* an adhesive mixed from a powdered resin and a liquid HARDENER. *See also* SHELF LIFE.

POTASSIUM A very reactive ALKALI metal. Its chemical symbol is K, its atomic number is 19, its atomic weight is 39·096, and its specific gravity is 0·86. It has valency of 1, a melting point of $+62 \cdot 5°C$, and a boiling point of 762°C.

POTENTIAL, ELECTRIC *See* VOLT.

POTENTIAL ENERGY *See* ENERGY.

POUND *See* LBF *and* FPS UNITS.

POUND-FEET, POUND-INCHES *See* FOOT-POUND.

POWDER METALLURGY The technique of agglomerating metal powders into engineering components.

POWER *See* HORSEPOWER.

POWER EARTH AUGER *See* AUGER.

POZZOLANA (also **Pozzolan, Pozzuolana**) (*a*) A volcanic dust, first discovered on the slopes of Mount Vesuvius near Pozzuoli, called *pulvis puteolanus* by VITRUVIUS. When mixed with lime mortar, it produces a waterproof cement (or HYDRAULIC CEMENT). Pozzolana was used by the builders of Ancient Rome for high-quality work, but the secret of hydraulic cement was lost after the fall of the ROMAN Empire. (*b*) A natural deposit, usually of volcanic origin, which has pozzolanic properties. (*c*) An artificial substance, usually a siliceous, or a siliceous and aluminous material, with pozzolanic properties. Although it possesses little cementitious value by itself, it reacts, in finely divided form and in the presence of water, with slaked lime ($Ca(OH)_2$) to form a HYDRAULIC CEMENT. The reaction takes place at ordinary temperatures, whereas the manufacture of PORTLAND CEMENT requires burning of the silica and the lime.

PP Polypropylene.

PRANDTL'S MEMBRANE ANA-LOGY *See* MEMBRANE ANALOGY.

PRATT TRUSS An *isostatic* truss, also called *N-truss*, consisting of top and bottom chords, regularly spaced vertical compression members, and diagonal *tension* members, as distinct from a HOWE and a WARREN TRUSS. It is used for medium to long spans in buildings and for small bridges (Fig. 74).

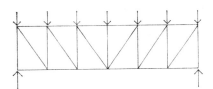

FIG. 74. Flat Pratt truss.

PRE-BORING *See* BORED PILE.

PRECAST CONCRETE Concrete cast and cured, and subsequently placed in its final location, as opposed to CAST-IN-PLACE concrete. Precast concrete can be more carefully controlled, particularly if it is made in a factory, and most of the shrinkage occurs before it is placed in position. However, transportation to the final location is an added expense, and it is difficult to produce a MONOLITHIC structure from precast elements.

PRECAST PILE A concrete pile which is cast and subsequently driven, as opposed to a pile which is cast in place in a BORED hole.

PRECAST STONE *See* CAST STONE.

PRECIPITATION (*a*) Separation and deposition of a substance in solid form from solution in a liquid. (*b*) Condensation and deposition of moisture from water vapour held in the air or in clouds. (*c*) That which is so deposited, *i.e.* a collective term for dew, rain, hail and snow.

PRECIPITATION HARDENING *See* AGE HARDENING.

PREFABRICATION *See* INDUS-TRIALISED BUILDING.

PREFERRED ANGLE For stairs: 30–35°. For ladders: 75–90°. For ramps: up to 15°.

PREMATURE STIFFENING *See* FALSE SET.

PRESENCE *See* INTIMACY.

PRESERVATIVE A substance which inhibits decay, infection or attack by fungi, insects, marine borers, etc. (as distinct from corrosion by chemicals), *e.g.* in timber. *See also* PRIMER.

PRESSED BRICK *See* WIRECUT BRICK.

PRESSURE Force per unit area.

PRESSURE, ATMOSPHERIC *See* ATM.

PRESSURE BULB *See* BOUSSINECQ PRESSURE BULB.

PRESSURE GAUGE *See* BOURDON GAUGE and MANOMETER.

PRESTRESSED CONCRETE Concrete which is precompressed in the zone where tensile stresses occur under load; consequently cracking of the concrete due to tension is avoided (Fig. 79). The prestressing can be accomplished by jacking the concrete against a rigid abutment; but the usual technique is to tension TENDONS of high-tensile steel. Prestressing is classified as PRE-TENSIONED or POST-TENSIONED, depending on whether the tendons are tensioned before or after the concrete has hardened. *See also* REINFORCED CONCRETE and PARTIAL PRESTRESSING.

PRESTRESSED SHELL A shell containing some prestressing TENDONS in addition to normal reinforcement, *e.g.* in the edge beams of *cylindrical shells* or the tie beams of *domes*. It is uncommon to prestress the membrane surface of the shell.

PRESTRESSING CABLE *See* TENDON.

PRE-TENSIONING PRESTRESSED CONCRETE in which the TENDONS are tensioned *before* the concrete has hardened, and generally before it is cast, as opposed to POST-TENSIONING. Pre-tensioning may be carried out

individually for each mould, or a *long line* of wire may be tensioned against fixed anchorages, the concrete units being cast around the wires which are flame-cut after the concrete has hardened.

PRIMARY BEAM *See* GIRDER.

PRIMARY COLOURS (*a*) The colours of three *pigments* from which nearly all reflected colours can be produced. These colours are usually identified as red, yellow and blue; or sometimes as magenta, yellow and cyan. (*b*) In the *psychophysical* sense primary colours are defined as the colours of three *lights*, by whose additive combination all other coloured lights can be produced. These are usually identified as red, green and blue; or as red, green and violet.

PRIME COST SUM (PC) A sum entered in a BILL OF QUANTITIES by the architect or consulting engineer. Its original purpose was to specify the quality of the item, and not allow any choice to the contractors tendering for it. However, prime cost items are also inserted for parts of a building which have not yet been fully designed, and therefore cannot be priced.

PRIME MOVER A machine which converts natural energy into mechanical power, *e.g.* a water turbine or an internal-combustion engine.

PRIMER or PRIMING COAT (*a*) Ground coat of paint applied to timber and other materials as a PRESERVATIVE and as a filler for the pores, which serves as a base for the further coat(s) of paint. (*b*) *Anti-corrosive* paint applied to steel.

PRIMING Filling a SIPHON or a pump with water so that it can be operated.

PRINCIPAL PLANES Three mutually perpendicular planes on which the stresses are purely normal tension or compression. The stresses on these planes are the PRINCIPAL STRESSES. When one of the principal stresses is nil, the space problem reduces to one of *plane stress*. When two of the principal stresses are nil, the condition is one of *uni-axial stress*.

PRINCIPAL RAFTER *See* RAFTER.

PRINCIPAL STRESSES The stresses acting across the PRINCIPAL PLANES. They are the greatest and smallest direct (tensile or compressive) stresses. The greatest shear stresses occur at an angle of 45° to the principal planes. *See* Fig. 63 and MOHR CIRCLE.

PRINCIPLE OF ARCHIMEDES *See* BUOYANCY.

PRINCIPLE OF SUPERPOSITION *See* SUPERPOSITION.

PRISM A polyhedron consisting of two parallel and equal faces (the *bases*), connected by parallelograms. The best known example is the *right triangular prism*, consisting of two right-angled triangles connected by rectangles, which is used in optics. The *cube* is the only prism which is also a regular POLYHEDRON.

PRISMOID A solid which has two parallel polygonal faces. The volume of a prismoid is $\frac{1}{6}h(A_b + 4A_m + A_t)$ where h is the height of the prismoid, A_b and A_t the areas of the polygons on the parallel top and bottom faces, and A_m is the area at mid-height. This formula is used for the computation of the volume of excavation of earth-works by the *prismoidal rule*, also known as SIMPSON'S RULE.

PROBABILITY A measure of the likelihood of the occurrence of a chance event. If the event can occur in N mutually exclusive and equally likely ways, and if n of these possess a characteristic E, then the probability that the event has this characteristic E is the fraction n/N.

PROBLEM-ORIENTED LAN-GUAGE A computer programming language which is oriented to the requirements of a computer user, and thus allows him to write a source program in terms which are in everyday use. However, the choice of words has to be carefully defined even in a problem-oriented language. It is a *high-level language*, as opposed to the older *low-level languages* which are closely related to the internal binary language of a digital computer. *See also* ALGOL, FORTRAN and MACHINE LANGUAGE.

PRODUCT OF INERTIA The sum of the products obtained by multiplying each element of an area dA by its co-ordinates with respect to two mutually perpendicular axes in the plane of the area; *i.e.* $\int xy \, dA$.

PROFILE A sectional drawing, usually vertical.

PROGRAM A series of INSTRUCTIONS necessary for the solution of a problem with a digital computer. The same program can be re-used with different data, if the problem remains the same. *See also* COMPILING ROUTINE, FLOW DIAGRAM, LOOP, SUBROUTINE and SYSTEM ANALYSIS.

PROGRESS CHART A graph or, more commonly, a *bar chart*, showing the time when the various operations required for the construction of the building should commence and finish. The actual times can be shown on the same chart to see how far the progress made conforms to the original intention. *See also* CRITICAL PATH METHOD.

PROJECTION A method of representing a three-dimensional object on a sheet of drawing paper. The most common form is the *orthographic* projection, which shows the object by means of three separate drawings: the *plan*, the *elevation* and the *side elevation*. In the *first-angle* projection (commonly used in Great Britain) each view is placed so that it represents the side of the object remote from it in the adjacent view. In the *third-angle* projection (commonly used in America and Australia) each view is placed so that it represents the side of the object near to it in the adjacent view. While the orthographic projection is true to scale in every respect, it fails to give *pictorial* representation of the object. In the *oblique parallel* projection the elevation is drawn as for the orthographic projection, and the plan and side elevation are then attached to the same picture at an angle of 45°. The object invariably looks too deep if drawn this way, and consequently an artificially fore-shortened scale is sometimes used along the 45° lines. In the *isometric* projection all three orthographic projections are drawn at an angle. Vertical lines remain

vertical, but all lines which are horizontal in the orthographic projection are drawn at 30° to the horizontal in the isometric projection. While the plan, elevation and side elevation are all given equal prominence, the object frequently appears distorted. However, all dimensions on the vertical and on the 30° axes are accurately to scale. The *axonometric* projection is used to overcome the pictorial limitation of the isometric. Vertical lines remain vertical; the other axes are drawn at angles other than 30°, and they consequently have different scales. If the two horizontal axes are drawn at the same angle, the projection is *dimetric*, if they are drawn at different angles it is *trimetric*. The latter gives a better pictorial representation, but all three axes have different scales. An even better picture is obtained by the use of a *perspective* projection, which is the way the eye sees an object. Lines which are parallel in plan are drawn to converge on a vanishing point, and lines which are parallel in the elevation and side elevation are also drawn to converge on their vanishing points. The complete, or *three-point* perspective is difficult to draw precisely, and it is sufficient to use a *two-point* perspective (in which the parallel lines of the elevation and the side elevation only converge on two vanishing points). Even a *one-point* perspective is often sufficient; in this only the parallel lines of the side-elevation converge on a vanishing point. Although perspectives could be used to scale dimensions with specially drawn scales, this is rarely required. Because of the complexity of the projection, perspectives are therefore frequently sketched without undue concern for dimensional accuracy.

PROOF STRESS The nominal stress (*i.e.* load per unit original cross-sectional area) which produces a specified permanent STRAIN. It is common practice to specify either a 0·1 or a 0·2 per cent proof stress, *i.e.* a stress which produces a permanent strain of 0·1 per cent (1×10^{-3}) or 0·2 per cent respectively. A proof stress is specified for metals which exhibit significant PLASTIC deformation, without showing a marked YIELD STRESS.

PROPELLER FAN *See* CENTRIFUGAL FAN.

PROPORTIONAL LIMIT The greatest stress which a material can sustain without departing from a proportional stress/strain relationship in accordance with HOOKE'S LAW.

PROPORTIONAL RULES BASED ON $\sqrt{2}$ Rules based on the diagonal of the square received some attention in the revivals of the nineteenth century. $\sqrt{2} = 1\cdot414$ was used for setting out in Gothic cathedrals, and Leonardo da Vinci mentioned in his Notebooks a circular construction which adds the diagonal to the side: $\theta = 1 + \sqrt{2} = 2\cdot414$. Either can be elaborated into a THEORY OF PROPORTIONS.

PROPORTIONING OF MIXES *See* MIX PROPORTIONS.

PROPORTIONS *See* THEORY OF PROPORTIONS.

PROPPED CANTILEVER A beam with one built-in and one simple support (Fig. 75). It is hyperstatic.

PROSCENIUM In the ancient theatre, the space between the 'scene' or background, and the orchestra, *i.e.* the actual stage. In the modern theatre, the portion of the stage in front of the curtain.

FIG. 75. Propped cantilever.

PROTECTIVE FINISH OF METALS *See* GALVANISING, SHERARDISING, CADMIUM PLATING, ANODISING, and ELECTROCHEMICAL CORROSION. *See also* PRIMER.

PROTON A positively charged sub-atomic particle of mass $1\cdot66 \times 10^{-24}$ grams. Its charge is equal and opposite to that of an ELECTRON. Its mass is 1840 times that of an electron.

PROTRACTOR An instrument, usually in the form of a graduated circle or semi-circle, for plotting and measuring angles.

PROVING RING A device for accurately measuring a load, or for calibrating a testing machine. It consists of a steel ring (the thickness depending on the range of loading) whose deflection under an axial compressive force is measured, *e.g.* with a DIAL GAUGE. The ring is calibrated in a standard testing machine.

PRUSSIAN BLUE Ferric ferrocyanide, $Fe_4.[Fe(CN)_6]_3$.

PRUSSIC ACID A solution in water of hydrogen cyanide (HCN) which is highly poisonous.

PS Polystyrene.

PSALI *Permanent supplementary artificial lighting in interiors.* The deliberate use of artificial lighting in daytime. Rooms can be made deeper if the back part is lit by luminaires, and the windows are required only to light part of the room. *PSALI* is not intended to cover windowless rooms, or the use of artificial light on very dull days, when the level of daylight is insufficient.

psf Pounds per square foot.

psi Pounds per square inch.

PSYCHROMETER Instrument for measuring relative humidity. The simplest type is the WET-AND-DRY BULB THERMOMETER. *See also* ASSMAN PSYCHROMETER, DEW-POINT HYGROMETER, HAIR HYGROMETER, HYGROGRAPH, THERMOHYGROGRAPH and WHIRLING PSYCHROMETER.

PSYCHROMETRIC CHART A graphical representation of certain thermodynamic relations. The dry-bulb temperature is the abisicissa and the water content of the atmosphere (in pounds of water per pound of dry air, or as vapour pressure in millibars) is the ordinate. The relative humidity is then plotted as a series of curves, and the wet-bulb temperature as a series of diagonal lines. The ENTHALPY is also plotted as a series of diagonal lines. The chart can therefore be used to determine the amount of heat or cooling power required to change air at one temperature and humidity to that at another temperature and humidity by AIR CONDITIONING. The psychrometric chart at standard atmospheric pressure is given in the ASHRAE Guide, and in the handbooks of air conditioning firms.

PSYCHROMETRY The study of the properties of a mixture of air and water vapour.

PTFE Polytetrafluorethylene.

PUDDLE A mixture of clay, water, and sometimes sand, worked while wet into a water-impervious layer. It is used as a cut-off wall to prevent the ingress or egress of water.

PUMICE A vesicular glass formed from the froth on the surface of gaseous LAVAS. It has a high silica content, and is thus classed as an acid rock. Its highly porous structure makes it suitable as a lightweight aggregate for concrete. The sharp edges of the gas vesicles enable pumice to be used as an abrasive.

PUMPED CONCRETE Concrete which is transported through a pipe or hose by a pump.

PUMPING The ejection of water or mud from the joints between paving slabs under load, due to the accumulation of water in the sub-base.

PUNCHING SHEAR The shear caused by the tendency of a column, column head, or column base to punch through a foundation slab, a FLAT SLAB or a FLAT PLATE (Fig. 76).

PUNKAH A large swinging fan, traditional to the hot–humid regions of Asia. It is usually made of cloth stretched on a rectangular frame, hinged at the top, suspended from the ceiling or the rafters, and worked by a cord. *See also* CEILING FAN.

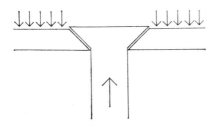

FIG. 76. Punching shear through a flat plate. In concrete structures failure occurs in DIAGONAL TENSION at 45°, and the failure surface is a frustrum of a cone.

PUNKAH LOUVRE A circular air ejector which can be rotated about a joint. It is widely used for ventilating ships. If the temperature is high it can be directed on the occupant of a cabin. At other times the air can be admitted without causing undue draughts.

PURLIN A horizontal beam in a roof, at right angles to the trusses or rafters. It carries the roofing material, if it is in sheet form, or the COMMON RAFTERS, if the roof consists of tiles, shingles or slates.

PUTLOG Cross piece in a scaffold. It normally rests on the LEDGERS and supports the SOFFIT boards or sheets.

PUTTY Compounds used for glazing windows. The most common is one made from powdered chalk and linseed oil.

PVA Polyvinyl acetate.

PVC Polyvinyl chloride.

PYCNOMETER An instrument for measuring density. The most common type is a jar or bottle of known volume. It is weighed with and without its contents.

PYLON Originally a gateway, particularly one in Egypt marked by truncated pyramidical towers. Later a tower or mast as such.

PYRAMID A solid bounded by plane surfaces, one being a polygon with any number of sides, and the others being triangles. The most elementary pyramid is the *tetrahedron*, which is a regular *polyhedron,* bounded by four equilateral triangles. The most common type of pyramid has a square base; the ancient Egyptians built huge stone monuments in this form. The top of the pyramid is called the *apex.*

PYROMETER Instrument for measuring temperatures. The term is used for instruments which measure high temperatures beyond the range of thermometry, and (less commonly) for instruments which measure temperature by electrical means.

PYROMETRIC CONE A small, pyramid-shaped cone whose composition is adjusted so that it melts at a definite temperature. The cones are made in series to cover a range of furnace temperatures, and the series constitutes a high-temperature thermometer. *See* SEGER CONE.

PYTHAGORAS' THEOREM 'The square on the longest side of a right-angled triangle equals the sum of the squares on the other two sides.' This theorem is credited to the Greek philosopher who lived in the sixth century BC.

Q

QUADRANGLE A rectangular space or court, the sides of which are occupied by a large building, such as a college.

QUADRANT An arc of a circle, forming one quarter of its circumference. It subtends an angle of 90° or $\frac{1}{2}\pi$ radians at the centre. Hence an instrument for measuring *altitudes* with a graduated quarter circle; this has now been superseded by the *sextant*, a reflecting instrument which measures angles up to 120°.

QUADRATIC EQUATION An algebraic equation which contains the square, but no higher powers, of an unknown quantity.

QUADRILATERAL Plane figure bounded by four straight lines. If two of the lines are parallel, it is called a *trapezium*. If two pairs of lines are parallel, it is called a *parallelogram*. A right-angled parallelogram is a *rectangle*. An equilateral parallelogram is a *rhombus*. A right-angled rhombus is a *square*.

QUALITY CONTROL Statistical control of the performance of products. *See also* FREQUENCY DISTRIBUTION CURVE and COEFFICIENT OF VARIATION.

QUANTITY SURVEYOR *See* BILL OF QUANTITIES.

QUANTUM The minimum, indivisible quantity of radiant energy.

QUARRY TILE A hard-burnt unglazed CLAY TILE.

QUARTER BEND A bend through an arc of 90°.

QUARTER CLOSURE *See* QUEEN CLOSER.

QUARTER ROUND A moulding which presents a profile of a quarter of a circle, inserted into corners.

QUARTER-WAVE PLATE Crystal plate inserted into a POLARISCOPE at 45°. The first plate breaks up the plane-polarised waves into two equal plane-polarised waves in directions at 45° to the axis of the polariser. On emerging from the quarter-wave filter, these two components have a relative path retardation of a quarter of the wavelength of the MONOCHROMATIC light. The second quarter-wave plate imparts a relative retardation by the same amount, but in the opposite sense. It is thus possible to show the ISOCHROMATICS without the ISOCLINICS (which are liable to cause confusion). Quarter-wave plates are usually cut from mica or quartz which has been split down to the required thickness.

QUARTZ A mineral consisting of crystalline silica (SiO_2). It is a major component of many igneous and sedimentary rocks. Although pure quartz is colourless and transparent (*rock-crystal*), it is commonly coloured by impurities. Many varieties are used as gemstones, *e.g.* AGATE, *amethyst, cairngorm,* ONYX and *tiger-eye.* SAND consists of small quartz grains with some impurities.

QUARTZITE A rock consisting of firmly compacted quartz grains.

QUATREFOIL *See* FOIL.

QUATTROCENTO Literally four hundred; a term used for the fifteenth century in the literature on Italian architecture, *i.e.* the RENAISSANCE. Similarly *trecento* is the fourteenth, *cinquecento* the sixteenth, and *seicento* the seventeenth century.

QUEEN CLOSER Also called *queen closure* or *quarter closure*, a brick cut in half along its length to keep the BOND correct at the corner of a brick wall. *See also* HALF BAT and KING CLOSURE.

QUEEN CLOSURE *See* QUEEN CLOSER.

QUEEN-POST ROOF TRUSS A traditional timber truss similar to the KING-POST truss, except that it has no central king post, but instead two queen posts on each side of the centre (Fig. 54b).

QUENCHING Rapid cooling of steel from an elevated temperature, normally by immersing it in oil or water. The usual effect of quenching is to produce a hard steel, by suppressing the formation of PERLITE from austenite, and forming MARTENSITE instead. Quenching is often followed by TEMPERING.

QUICK SET *See* FALSE SET.

QUICKLIME Calcium oxide (CaO). It is the raw material for HYDRATED LIME, which is used in lime mortar.

QUICKSAND A sand through which water moves *upwards* at sufficient speed to hold it in suspension. It has therefore negligible bearing capacity. The remedy is to reduce the flow of water. Although quicksand is due to water flow, the condition is more likely to occur with sand which is uniformly-grained and fine-grained.

QUICKSET LEVEL *See* DUMPY LEVEL.

QUICKSILVER Mercury.

QUOIN An outer corner of a wall. *Also spelled coin.*

R

RABBIT, or **RABBET** *See* REBATE.

RADBURN PLAN A concept for segregating pedestrian and vehicular traffic, advocated by *Lewis Mumford*, and first tried at Radburn, New Jersey, in the 1920s. It involves the construction of service roads at the rear of the houses, and pedestrian access in front, the two being separated by overpasses.

RADIAL SHRINKAGE The drying SHRINKAGE of timber at right angles to the growth rings. It is less than the *tangential shrinkage* which is normally considered; for many timbers the ratio is about 1:2.

RADIAN (rad) The angle between two radii of a circle, cut off on the circumference of an arc equal in length to the radius. 1 radian $=$ 57·295 78 degrees. 1 degree $=$ 0·017 453 3 radians. *See also* STERADIAN.

RADIANT ENERGY Energy consisting of electromagnetic waves, such as light or radiant heat.

RADIANT HEAT Heat transmitted to a body by electromagnetic waves, as distinct from heat transferred by CONDUCTION or CONVECTION. *See* INFRA-RED RADIATION.

RADIATION Energy transmitted by electromagnetic waves. It ranges in decreasing order of wavelengths from radio waves, heat rays, infra-red rays, visible light, ultra-violet rays, X-rays, gamma rays, to cosmic rays. *See also* STEFAN-BOLTZMANN LAW.

RADIOGRAPHY *See* X-RAYS.

RADIUS OF CURVATURE *See* NAVIER'S THEOREM.

RADIUS OF GYRATION A convenient short-hand notation for $(I/A)^{\frac{1}{2}}$, where I is the moment of inertia and A is the cross-sectional area. The radius of gyration is required for certain dynamic problems, and for computing the SLENDERNESS RATIO in buckling problems.

RAFT FOUNDATION A slab of concrete, usually reinforced, extending under the whole area of a building and projecting beyond the outside walls. It is used when loads are exceptionally heavy or the bearing capacity of the soil is low.

RAFTER A sloping timber extending from the wall plate to the ridge. One series of rafters suffices, if the roof covering consists of galvanised sheets, or some other sheet material. Tiles, shingles or slates are supported on *common rafters*, which are carried by the PURLINS. These in turn are

supported by *principal rafters*. The common rafters are parallel to the principal rafters, but more closely spaced and smaller in size since they carry less load.

RAG FELT *See* ROOFING FELT.

RAIA Royal Australian Institute of Architects, Canberra.

RAIC Royal Architectural Institute of Canada, Ottawa.

RAIN GAUGE An instrument for measuring rainfall. It usually consists of a funnel from which the water drips into a cylinder, graduated in inches or millimetres of rainfall per funnel area.

RAKE The slope, or angle of inclination to the vertical.

RAKED JOINT A mortar joint between bricks or blocks which has been scraped clean of mortar for about $\frac{3}{4}$ in (20 mm) back from the face. It may be left raked, or else it may subsequently be POINTED (Fig. 94).

RAKING SHORE *See* SHORE.

RAMMED EARTH CONSTRUCTION *See* PISÉ DE TERRE.

RANDOM ASHLAR *See* ASHLAR.

RANDOM SAMPLE *See* SAMPLE.

RANGE OF MEASURED DATA The difference between the largest and the smallest measurement.

RANKINE COLUMN FORMULA One of the oldest formulae for the practical design of SLENDER columns.

216

The EULER FORMULA gives the correct load P_e for very slender columns, but it is unsafe for the buckling of columns with intermediate slenderness ratios (which includes most of the practical problems in the design of buildings). Similarly the *short-column* load P_s is unsafe for slender columns. The Scottish engineering professor, W. J. Macquorn Rankine proposed in 1858 a gradual transition (Fig. 77) for the column load P

$$\frac{1}{P} = \frac{1}{P_s} + \frac{1}{P_e}$$

See also PERRY–ROBERTSON and SECANT FORMULA.

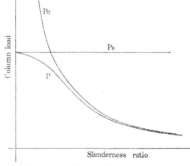

FIG. 77. Rankine column formula. The Rankine column load, P, becomes tangential to the short column load, P_s, at a slenderness ratio of zero, and to the Euler column load, P_e, at infinity.

RANKINE FAILURE CRITERION A FAILURE CRITERION which assumes that the material fails when the maximum direct stress reaches a limiting value. It gives good results for brittle materials, such as glass and concrete, failing in tension. *See also* COWAN CRITERION.

RANKINE THEORY A theory for the EARTH PRESSURE developed by a granular soil (loose sand or gravel) on a retaining wall, proposed by W. J. M. Rankine in 1857. It gives the magnitude of the *active earth pressure* as

$$\frac{1 - \sin \varphi}{1 + \sin \varphi} \, ph$$

where ρ is the soil density, φ its ANGLE OF INTERNAL FRICTION, and h is the depth of the soil below the surface. For most granular soils, φ is approximately 35°, and the active earth pressure is then $0.27 \, ph$. The magnitude of the *passive earth pressure* is

$$\frac{1 + \sin \varphi}{1 - \sin \varphi} \, ph = 3.7 \, ph$$

approximately. The theory does not apply to cohesive soils.

RAPID-HARDENING CEMENT *See* HIGH-EARLY-STRENGTH CEMENT.

RATCHET A wheel with saw-like teeth, into which a *pawl* (or single tooth) may catch for the purpose of preventing reversed motion.

RATIONAL NUMBER An INTEGER or a FRACTION. *See also* IRRATIONAL NUMBER.

REACTION The opposition to an action, *e.g.* to the downward pressure of a loaded beam, frame, etc. For static equilibrium the reaction must balance the action.

REACTION TURBINE A turbine in which the jets or nozzles are on the moving wheel, as opposed to an *impulse turbine*, which has fixed jets impinging on the moving wheel.

REACTIVE CONCRETE AGGREGATE Aggregate which is capable

217

of combining chemically with Portland cement, under normal conditions, and may thus cause harmful expansion. *See* ALKALI-AGGREGATE REACTION.

READY-MIXED CONCRETE Concrete mixed at a central plant and transported to the building site so that it arrives in a PLASTIC condition. Frequently some or all of the mixing water is added during transportation. Ready-mixed concrete eliminates storage of materials and mixing at the building site. It consequently saves space and makes for a cleaner building site.

REAL NUMBER A number which is not IMAGINARY or complex, *i.e.* a RATIONAL or an IRRATIONAL NUMBER.

REAMER A tool used for finishing holes which have previously been drilled.

REBATE A recessed timber edge, designed to receive a door, window sash, or some other piece. Also spelt *rabbet* and *rabbit*.

RECIPROCAL DIAGRAM A graphical solution for the design of isostatic trusses, proposed by the British physicist Clerk Maxwell in 1864. It consists of a series of *triangles* and *polygons of force*, drawn together so that all have at least one line in common (Fig. 78). We start by drawing the external forces (which are given) to scale. Starting at the left-hand support, we can now draw a line parallel to the tie a1 and another to the rafter c1. The intersection gives the point 1. We proceed until all points have been found, and then scale the length of the lines to obtain the forces in the members. *Bow's notation* is commonly used for this construction because of its convenience. It consists of numbering the spaces in and around the truss, instead of the joints. *See also* METHOD OF SECTIONS and RESOLUTION AT THE JOINTS.

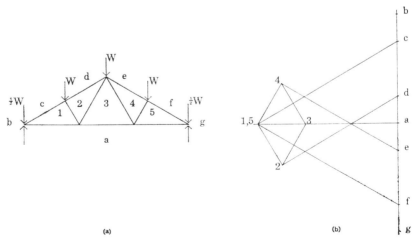

(a)   (b)

FIG. 78. Reciprocal diagram (b) for roof truss (a).

RECIPROCAL THEOREM A theorem which relates the deflection of a structure at two points, and the reciprocal (or corresponding) loads at the same points. It is the basis of several methods for the theoretical and indirect model (*e.g.* BEGGS' DEFORMETER) analysis of hyperstatic structures. The theorem was published by the English physicist Clerk Maxwell in 1864, and proved by the Italian engineer E. Betti in 1872.

RECIPROCATING ENGINE An engine consisting of a piston moving *backward and forward* in a cylinder. The reciprocating motion is generally converted into rotary motion by means of a connecting rod and a crankshaft or flywheel. A *rotary engine* produces rotary motion directly.

RECONDITIONING OF COLLAPSED TIMBER *See* COLLAPSE.

RECONSTITUTED STONE *See* CAST STONE.

RECRYSTALLISATION The formation of new crystals, *e.g.* ANNEALED crystals from previously strain-hardened crystals.

RECTANGLE *See* QUADRILATERAL.

RECTANGULAR CO-ORDINATES *See* CARTESIAN CO-ORDINATES.

RECTANGULAR FRAME *See* FRAME.

RED LEAD Red oxide of lead (Pb_3O_4). It is commonly used as a rust-inhibiting primer on steel. *See also* LITHARGE.

RED OXIDE Red IRON OXIDE, a pigment which does not inhibit corrosion. *See* RED LEAD.

REDUCING POWER The strength of a white pigment in reducing the colour of a coloured pigment. It is a measure of the paleness of tint produced, and is the opposite of STAINING POWER.

REDUCING VALVE A valve set to maintain a minimum pressure of steam in a heating system. If the pressure falls below the specified minimum, the valve opens automatically to feed water into the boiler.

REDUCTION A chemical process involving a decrease in the state of *oxidation*.

REDUCTION IN AREA The contraction in cross-sectional area which occurs when a ductile material is tested. It is one of the quantities normally measured during the test.

REDUNDANCY In structural engineering, a structural member or restraint in excess of those required for STATIC equilibrium, as defined by MÖBIUS' LAW. A structure with redundancies is HYPERSTATIC, one without redundancies is ISOSTATIC.

REDUNDANT FRAME *See* RIGID FRAME and TWO-PIN PORTAL.

RE-ENTRANT CORNER An *internal* angle or corner, the opposite of a *salient corner*.

REFECTORY A room for refreshments, particularly the dining hall in a college or monastery.

REFERENCE GRID A grid constituted of two mutually perpendicular sets of parallel reference lines. *See* MODULAR GRID.

REFLECTED GLARE GLARE produced by specular reflections of luminous objects, especially reflections appearing on or near the object viewed.

REFLECTION The change in direction which occurs when light or sound strikes a *mirror, i.e.* a surface capable of reflection without appreciable diffusion. The angle of reflection equals the angle of incidence. *See also* REFRACTION.

REFLECTIVE INSULATION A metal sheet which reflects infra-red radiation, and therefore reduces the amount of heat entering a building, particularly via the roof. It commonly takes the form of aluminium foil, 0·006 in (0·15 mm) or less in thickness, backed by kraft paper.

REFLECTOR *See* PARABOLIC REFLECTOR.

REFRACTION The change in direction which occurs when a ray of light passes from one medium to another with a different density. The ratio of the sine of the angle of incidence to the sine of the angle of refraction is the *refractive index. See also* REFLECTION.

REFRACTORY BRICK A brick made from a refractory material.

REFRACTORY MATERIAL One which can withstand a high temperature, and is thus suitable for lining a furnace. CHINA CLAY, ALUMINA, SILICA, DOLOMITE and *chromite* make suitable refractories.

REFRIGERATION CYCLE *See* ABSORPTION CYCLE and COMPRESSION CYCLE.

REFUSAL A term applied to the resistance offered by a FRICTION PILE to continued driving. It is often defined as the failure to penetrate more than $\frac{1}{2}$ in (13 mm) in five blows.

REGISTER An air grille in a ventilating or air conditioning duct or in a chimney, which is controlled by a manual damper.

REGNAULT HYGROMETER *See* DEW-POINT HYGROMETER.

REINFORCED BRICKWORK *or* **MASONRY** Brickwork or masonry, which has reinforcement embedded (usually in the joints) to increase the resistance to flexural tension.

REINFORCED CONCRETE Concrete which contains reinforcement, normally of steel, to improve its resistance to tension. It is designed

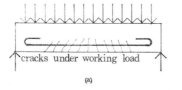

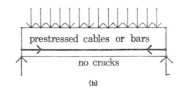

cracks under working load   (a)

prestressed cables or bars   no cracks   (b)

FIG. 79. Reinforced concrete cracks under working loads, prestressed concrete only, cracks under an overload: (a) reinforced concrete; (b) prestressed concrete.

on the assumption that the two materials act together in resisting the stresses in the composite material due to the loads. PRESTRESSED CONCRETE, in which the concrete is prestressed by steel before the loads are applied, is normally considered to be a separate form of construction (Fig. 79). *See also* BALANCED DESIGN, COMPRESSION REINFORCEMENT, T-BEAM and WHITNEY STRESS BLOCK.

REINFORCEMENT FOR CONCRETE *See* COLD-DRAWN WIRE, COLD-WORKING, DEFORMED BAR, DISTRIBUTION REINFORCEMENT, EXPANDED METAL, FOUR-WAY REINFORCEMENT, HELICAL REINFORCEMENT, HOOP REINFORCEMENT, LATERAL REINFORCEMENT, LONGITUDINAL REINFORCEMENT, MESH REINFORCEMENT, REINFORCED CONCRETE, SECONDARY REINFORCEMENT, TENDON and TWO-WAY REINFORCEMENT.

RELATIVE DENSITY Specific gravity.

RELATIVE HUMIDITY The ratio of the quantity of water vapour actually present in the air to that present at the same temperature in a water-saturated atmosphere. It is commonly expressed as a percentage. *See also* HUMIDITY and WET-BULB TEMPERATURE.

RELATIVE SETTLEMENT *See* DIFFERENTIAL SETTLEMENT.

RELAXATION METHOD A technique devised by *Sir Richard Southwell* in 1935 for the solution of mathematical equations by successive approximations. The MOMENT DISTRIBUTION METHOD developed independently by *Hardy Cross* is a special case of the relaxation method.

RELAXATION OF STEEL Decrease in the stress of steel due to CREEP in the steel under prolonged strain. The term is also applied to the LOSS OF PRESTRESS due to shrinkage and creep of the concrete, which results in a decreased steel strain.

RELAY A device in which a small electrical power is used to control a larger electrical power. Many relays employ electromagnets.

RELIEF The elevation or projection of a design from a plane surface in order to give a solid appearance, as opposed to INTAGLIO. Relief may be high, middle or low (BAS RELIEF).

REMOULDED CLAY Clay whose internal structure has been disturbed. Remoulding generally causes a loss of shear strength and an undesirable gain in compressibility. *See* SENSITIVITY RATIO.

RENAISSANCE The art, science and architecture resulting from the re-birth (*rinascimento*) of the classical tradition in Italy, particularly in Florence, in the fifteenth century. It influenced the whole of European, and through it, Colonial Architecture. In recent years the term Renaissance has had a more restricted meaning in relation to Italian architecture, comprising the period from about 1420 to 1500; the style of the sixteenth century is now called MANNERISM. From the structural point of view, the most remarkable achievement of the Renaissance was the Dome of Florence Cathedral, erected by Filippo Brunelleschi between 1420 and 1436. It is ribbed, DOUBLE-WALLED SHELL with a span of 137 ft 9 in (42 m), bigger than those of St Peter's in

Rome and St Paul's in London, built subsequently.

RENDERING The application of mortar or plaster by means of a float or trowel.

REPEATED LOADING *See* FATIGUE.

REPOSE *See* ANGLE OF INTERNAL FRICTION.

REPRESENTATIVE SAMPLE *See* SAMPLE.

RESILIENCE The ability of a material to absorb and return strain energy without permanent deformation.

RESIN (*a*) The produce from the secretion of certain plants and trees, consisting of highly POLYMERISED substances. They are insoluble in water, but soluble in certain organic solvents, and fusible. Most resins are hard and brittle. (*b*) Synthetic imitation of a natural resin. The term 'synthetic resin' is often used as a synonym for a THERMOSETTING plastic. *See also* ROSIN.

RESISTANCE *See* MOMENT OF RESISTANCE.

RESISTANCE STRAIN GAUGE *See* ELECTRICAL RESISTANCE STRAIN GAUGE.

RESISTANCE WELDING Welding two parts by holding them in tight contact with electrodes through which a heavy alternating current flows momentarily, causing them to fuse together.

RESOLUTION AT THE JOINTS A method for the design of isostatic trusses, proposed by the Russian engineer D. J. Jourawski in 1850. Since each joint (as well as the whole truss) is in equilibrium, it is possible to resolve horizontally and vertically to produce 2 *j* equations for *j* joints. According to MÖBIUS' LAW, the truss has 2 *j* − 3 members, in which we must determine the forces. The method thus produces three check equations. *See also* TENSION CO-EFFICIENTS, RECIPROCAL DIAGRAM and METHOD OF SECTIONS.

RESOLUTION OF A FORCE Taking the *components* of a force, usually horizontally and vertically.

RESONANCE The phenomenon of minimum acoustical impedance as the frequency is varied. The maximum velocity of motion occurs at this *resonant frequency*. The same principles apply to other mechanical vibrations. *See also* DECAY FACTOR and FORCED VIBRATION.

RESONANT ABSORBER *See* HELMHOLTZ ABSORBER.

RESONATOR *See* HELMHOLTZ ABSORBER.

RESULTANT The vector sum of two or more forces.

RETAINING WALL A wall, usually battered, which retains a weight of earth or water. It may rely for its stability on the mass of the masonry or concrete (*gravity retaining wall*), or on the cantilever strength of the wall (which may be of steel or timber, but is usually of reinforced concrete). In a *cantilever retaining* *wall* the weight of soil lying above the heel is utilised to improve the stability of the wall (Figs. 10, 21 and 80).

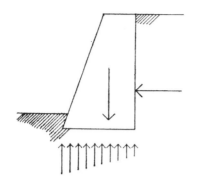

Fig. 80.   Gravity retaining wall.

RETARDED HEMIHYDRATE PLASTER   *See* PLASTER OF PARIS.

RETARDER   A substance which slows down a chemical reaction, as opposed to an ACCELERATOR. In concrete, an additive which delays the setting of the cement, and thus allows more time for placing the concrete.

RETICULATE   To divide so as to form a network.

RETICULE   *See* COLLIMATION LINE.

RETURN   (*a*) A return circuit, pipe, or duct. (*b*) A change in the direction of a wall or other building component, usually at right angles.

REVEAL   (*a*) The part of the jamb between the frame and the outside wall, which is revealed, inasmuch as it is not covered by the frame. (*b*) The visible side of an open fireplace. (*c*) The entire jamb or vertical face of an opening.

REVERBERATION CHAMBER
A chamber for acoustic research in which all surfaces are reflecting (and not absorbing) and non-parallel. For accurate work, the chamber should have a large volume. The ISO recommends that the volume should be between 180 and 200 m^3 for the measurement of SOUND ABSORPTION coefficients. (*See also* IMPEDANCE TUBE.)   For   SOUND-INSULATION measurements a transmitting and a receiving chamber are required, between which the test sample is placed. Reverberation chambers for insulation measurements are commonly built in three units: two rooms placed above one another for floor measurements, and two adjacent rooms for wall measurements. *See also* ANECHOIC CHAMBER.

REVERBERATION TIME   The period of time required for sound at a certain frequency to decay, after its source has been silenced, by 60 decibels, *i.e.* to 1×10^{-6} of its original loudness. Rooms which have a large volume, built of materials which have a low absorption, tend to have a long reverberation time. For design purposes, the reverberation time (T) is usually calculated by the *Sabine formula*

$$T = 0.049 \frac{V}{A} \text{ seconds}$$

where V is the volume of the room in cubic feet, and A is the total absorption in square feet SABINS. In metric units

$$T = 0.16 \frac{V}{A} \text{ seconds}$$

where V and A are in cubic metres and square metres respectively.

REVERSED DOOR A door opening in the direction opposite to that considered regular, *e.g.* a cupboard door opening inward.

REVETMENT (*a*) A protective covering to a soil or soft rock surface, to protect it against scouring by water or other effects of weathering. It may consist of grass or other plants, bundles of brushwood, asphalt, stone slabs, or concrete. (*b*) A gravity RETAINING WALL.

REVOLUTION *See* SURFACE OF REVOLUTION.

REVOLVING SHELF A shelf designed to give access to inner corners where two cupboards meet, by rotating it on a vertical pivot. Also called a *lazy Susan*.

REYNOLDS' CRITICAL VELOCITY The velocity at which flow changes from streamline to turbulent.

REYNOLDS' NUMBER A non-dimensional ratio used for assessing the similarity of motion in viscous fluids, named after the British hydraulic engineer, who proposed it in 1883.

$$\text{Reynolds' number} = \frac{Lv}{v}$$

where L is any typical length, v is the velocity of the fluid, and v is its kinematic viscosity. In model analysis, in order to satisfy the PI-THEOREM, the Reynolds' number must be the same for the structure and the model.

In practice it is often difficult to satisfy this condition, and some distortion may have to be accepted.

RH Relative humidity.

RHEOLOGICAL MODEL A physical model used to illustrate a mathematical theory in the field of RHEOLOGY (Fig. 81). Rheological models are mainly composed of springs (for ELASTIC deformation), dashpots (for viscous or CREEP deformation), and heavy weights resting on a rough surface which offers frictional resistance (for PLASTIC deformation). *See also* HOOKE, KELVIN, MAXWELL, NEWTON, SAINT VENANT BODY and STRESS–STRAIN DIAGRAM OF CONCRETE.

RHEOLOGY The science of the flow of materials. In building science, rheological studies are valuable for determining the flow of liquid concrete, paint or mastics; and for the CREEP deformation of timber and concrete.

RHEOSTAT A resistance which can be adjusted to vary the total resistance in the electrical circuit.

RHOMBUS *See* QUADRILATERAL.

RIB (*a*) A small beam projecting from a concrete slab. (*b*) The vertical member of a T-BEAM. Also called WEB.

RIBA Royal Institute of British Architects, London.

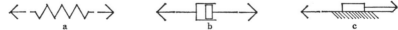

FIG. 81. Rheological models for elastic deformation (a), viscous deformation (b) and plastic deformation (c).

RIBBED SHELL A shell stiffened with ribs. It seems likely that the ribs of Gothic vaults served as a constructional aid, rather than as structural supports for the shells. Although most early reinforced concrete shells had ribs (partly due to the fear that thin shells might buckle), they are now rarely used, because they greatly add to the cost of the formwork. *See* VIOLLET-LE-DUC'S DICTIONARY.

RIBBED SLAB A panel composed of a thin slab reinforced by ribs in one or two directions (usually at right angles to one another). The latter is also called *waffle slab* because of its appearance. It is commonly used with mushroom columns in a FLAT SLAB, and the slab is then made solid in the area surrounding the column for extra shear resistance. Also called *concrete joist construction.*

RIBBON DEVELOPMENT The uncontrolled construction of houses along main roads leading out of towns.

RICH CONCRETE A concrete with a high cement content, as opposed to a *lean concrete.*

RICHTER SCALE Scale devised by C. F. Richter in California for determining the magnitude of an earthquake. For this purpose 'standard earthquake' is defined as one providing a maximum trace amplitude of 0·001 mm on a certain type of seismograph at a 'standard distance' of 100 km. The magnitude on the Richter scale was originally defined as the logarithm (to the base 10) of the ratio of the amplitude of any earthquake at the 'standard distance' to that of the 'standard earthquake'. To make the scale independent of the type of recording instrument it was later defined by the equation

$$\log E = A + B \log M$$

where M is the magnitude on the Richter scale, E is the energy in ergs, and A and B are constants. The smallest earthquakes recorded by very sensitive instruments at short distances have an energy of 10^5 ergs, and the largest an energy of 10^{25} ergs. The variation on the Richter scale is from 0 for the smallest earthquake to 8·6 for the largest recorded. *See also* MERCALLI SCALE.

RICS Royal Institute of Chartered Surveyors, London.

RIDGE A horizontal line caused by the junction of two sloping roof surfaces (Fig. 45).

RIDGE BOARD A horizontal board at the apex of the roof, set on edge. The rafters are nailed to it.

RIGHT ANGLE An angle of 90°.

RIGHT ASCENSION *See* AZIMUTH.

RIGHT-HAND Supposedly a fitting convenient to a right-handed person. A *right-hand screw* is one which tightens clockwise. A *right-hand stair* is one which has its handrail on the right going up. In Great Britain a right-hand door is one whose hinges are on the right, when it swings *towards* the hand opening it. However, in the USA a *right-hand door* is one whose hinges are on the right, when it swings *away* from the hand opening it. A *right-hand lock* is one which fits a right-hand door, according to local convention.

RIGID FOAM *See* EXPANDED PLASTIC.

RIGID FRAME A frame with RIGID JOINTS, which are capable of transmitting moments, as opposed to a PIN-JOINTED structure. *See also* Fig. 73 and PORTAL.

RIGID JOINT A joint which is capable of transmitting the full moment at the end of a member to the other members framing into the joint. If no strain energy is dissipated in deforming the joint, the angle between the members remains unaltered; *e.g.* if the angle between a beam and a column was 90° before loading, it remains 90° under load even though the remainder of the frame is deformed. The fully rigid joint is an ideal which cannot be precisely achieved; however the assumption of full joint rigidity is frequently made in reinforced concrete and in welded steel frames. *See also* SEMI-RIGID JOINT and PIN JOINT.

RIGIDITY *See* MODULUS OF RIGIDITY.

RILEM Réunion Internationale des Laboratoires d'Essais et des Recherches sur les Matériaux et les Constructions, Paris.

RIM LOCK A lock which is screwed to the face of the door, as opposed to a MORTISE lock.

RING TENSION *See* HOOP TENSION.

RISE (*a*) The vertical height of an arch, roof truss, or rigid frame. The vertical height of a cable structure is its *sag*. (*b*) The vertical distance from the top of one tread to the top of the next tread of a staircase. *See also* GOING. (*c*) The total vertical distance which a stair rises from floor to floor.

RISER A vertical board under the TREAD of a step in a staircase.

RITTER'S METHOD *See* METHOD OF SECTIONS.

RIVET A headed shank for making a permanent joint between two pieces of metal. It is inserted into holes in the two pieces, and closed by forming a head on the projecting shank. In steel structures the head must be formed while the rivet is red-hot, and because of the inconvenience of handling the red-hot rivets, riveting is being gradually replaced by welding and by HIGH-TENSILE BOLTING. The heads of aluminium rivets can be formed cold, provided the diameter of the rivets is not too large, and EXPLOSIVE RIVETS can be used in inaccessible locations.

RMS Root mean square.

ROCK *See* SEDIMENTARY, IGNEOUS *and* METAMORPHIC ROCK.

ROCK FLOUR Silt-sized crushed rock.

ROCK WOOL *See* MINERAL WOOL.

ROCKWELL HARDNESS TEST A method for determining hardness by indenting the test piece with a diamond cone or a hard steel ball, and measuring the depth of penetration. A smaller load is used for the *Rockwell superficial hardness* test.

ROCOCO The name given to the last phase of Baroque architecture.

ROLL ROOFING *See* BUILT-UP ROOFING.

ROLLED STEEL JOIST (RSJ) A rolled steel section of I-shape. It is the most common type for small and medium-sized steel beams and columns.

ROLLED STEEL SECTION Any hot-rolled steel section, including joists, angles, channels and rails.

ROLLING FRICTION *See* FRICTION.

ROMAN CEMENT *See* NATURAL CEMENT.

ROMAN CONCRETE The concrete used in Ancient Rome. The art of making concrete was lost after the fall of the Roman Empire, to be rediscovered only in the late eighteenth century in England. The predominant type consisted of lime, sand and broken brick (*opus caementitium*). Some interaction took place between the brick (which contains alumina) and the lime, but the concrete was not HYDRAULIC. For special work the Romans used a hydraulic concrete made with POZZOLANA.

ROMAN ORDERS *See* ORDERS.

ROMANESQUE ARCHITECTURE The style current in Europe from about the ninth century until the advent of Gothic. It includes both the *Anglo-Saxon* and the NORMAN architecture of England.

RÖNTGEN RAYS *See* X-RAYS.

ROOF *See* GABLE ROOF and MANSARD ROOF. Also ROOF TRUSS, CRUCKS, RAFTER, DOME, VAULT, SHELL and SUSPENSION ROOF.

ROOF TRUSS *See* BELFAST, FINK, HOWE, PRATT and WARREN TRUSS; KING-POST, QUEEN-POST and HAMMER-BEAM ROOF; also KNEE BRACE and TRUSSED RAFTER.

ROOFING FELT Waterproof felt (rag felt soaked in asphalt, bitumen or tar) used in BUILT-UP ROOFING (also called *composition roofing* or *roll roofing*). Asbestos felt, consisting mainly of asbestos fibres, is also used.

ROOT MEAN SQUARE The square root of the mean value of the squares of individual test results or statistical data.

ROSE WINDOW A circular window with patterned tracery arranged like the spokes of a wheel. It was widely used in Gothic architecture.

ROSETTE *See* STRAIN ROSETTE.

ROSIN A RESIN obtained as a residue from the distillation of turpentine. It is used as a varnish, as a drier in paint, and as a soldering flux. Also called *colophony*.

ROSTRUM Literally the prow of a ship. The platform from which the tribunes addressed the people in the *Forum Romanum* was decorated with the prows of ships taken in war; hence a speaker's platform. *See also* BEMA.

ROTARY ENGINE *See* RECIPROCATING ENGINE.

ROTARY VENEER *See* WOOD VENEER.

ROTATIONAL SURFACE *See* SURFACE OF REVOLUTION.

ROTUNDA A circular building, or a circular room within a building, especially one roofed with a dome.

ROUGHCAST Plaster mixed with small stones or shells, used on the outside of buildings.

ROWLOCK *See* BRICK ON EDGE.

RPM Revolutions per minute.

RSJ Rolled steel joist.

RUBBED FINISH A finish obtained by using a surface abrasive to remove irregularities.

RUBBER, RUBBER-BASE PAINT *See* LATEX.

RUBBER MOUNTING A vibration-isolation mounting for a machine which is liable to transmit vibration to its supports.

RUBBLE Rough stones of irregular shape and size. They may result from quarrying, from the demolition of old buildings, or (more rarely) from the natural disintegration of large pieces of rock.

RULED SURFACE A SURFACE on which it is possible to draw straight lines. Consequently such a surface can be generated by a straight line. Ruled surfaces are important for shell construction, because the concrete formwork can be easily constructed from straight pieces of timber. A surface is *singly ruled* if at every point only a single straight line can be drawn. It is *doubly ruled* if at every point two straight lines can be drawn. *Cylindrical shells* (Figs. 24 and 68) and *conoids* (Figs. 68 and 96) are singly ruled. HYPERBOLIC PARABOLOIDS (Fig. 53) and *hyperboloids of revolution* (Fig. 95) are doubly ruled.

RUST Hydrated iron oxide ($2Fe_2O_3.3H_2O$) formed on unprotected iron by exposure to air and moisture.

RUSTIC BRICK A brick whose surface has been roughened by impressing it with a pattern, or by coating it with sand. Also called *texture brick*.

RUSTICATED COLUMN A circular column whose shaft is interrupted by square blocks.

RUSTICATION (*a*) Construction of natural stonework with recessed joints. It was popularised by the Renaissance architecture of Florence in the fifteenth century. Rusticated stonework may be *smooth* or *diamond pointed* (both with a dressed surface), *cyclopean* (rock-faced, with a finish as quarried, or an artful imitation thereof), or VERMICULATED. (*b*) Simulation of rusticated stonework is *stucco*. This was probably introduced by Andrea PALLADIO. (*c*) Simulation of rusticated stonework in concrete, by moulding a groove in the concrete surface with a *rustication strip*.

RUSTICATION STRIP A strip of timber fixed to concrete formwork to produce a 'rusticated' surface, *i.e.* a series of grooves imitating the rustication of natural-stone blocks.

S

S Chemical symbol for sulphur.

SAA Standards Association of Australia, Sydney.

SABIN Unit of sound absorption, equal to one square foot or one square metre of open window. It is named after W. C. Sabine, Professor of Natural Philosophy at Harvard University, who founded the science of architectural acoustics at the end of the nineteenth century.

SABINE FORMULA *See* REVERBERATION TIME.

SABS South African Bureau of Standards, Pretoria.

SACK *See* BAG.

SACRED CUT A THEORY OF PROPORTIONS, published by T. Brunés (*Ancient Geometry*, Copenhagen, 1967). It is based on a circle, whose radius equals the diagonal of a square, which in turn is produced by quartering a square measuring 10 by 10 units (the decimal system, obtained by counting on one's fingers). This gives $10/\sqrt{2} = 7 \cdot 07$.

SACRIFICIAL PROTECTION The property of certain metals, for example zinc used in GALVANISING, of protecting steel although the coating may not cover the entire surface. Any small uncovered areas are protected from progressive attack by the products of corrosion.

SADDLE SURFACE A doubly curved surface of negative GAUSSIAN CURVATURE. If a DOME may be said to correspond to the top of a mountain, then a saddle corresponds to a mountain pass. The term is synonymous with *anticlastic surface*.

SAFETY FACTOR *See* FACTOR OF SAFETY.

SAFETY GLASS (*a*) Glass containing thin wire reinforcement. (*b*) Glass laminated with transparent plastic; this prevents splinters flying if the glass is broken. (*c*) Glass, toughened by heat treatment, which breaks into small fragments without splintering.

SAG *See* RISE.

SAGGING MOMENT A positive bending moment, such as occurs in simply supported or continuous beams near mid-span (Fig. 66); it causes a 'sagging' deformation. *See also* HOGGING MOMENT.

SAINT-VENANT BODY In rheological terminology, a PLASTIC MATERIAL. The rheological model of this body, named after the French nineteenth century mathematician, is a heavy weight, resting on a rough surface. The weight does not move, when pulled, until the force is sufficient to overcome the frictional resistance. Thereafter it moves with a constant force. This is in accordance with the definition of plastic deformation (Fig. 81).

SAINT-VENANT'S PRINCIPLE 'Forces applied at one part of an elastic structure will induce stresses which, except in the region close to that part, will depend almost entirely upon their resultant action, and very little on their distribution.' The principle was originally proposed by the French mathematician in 1864, and rephrased in the above form by Sir Richard Southwell in 1923.

SAL AMMONIAC Ammonium chloride (NH_4Cl).

SALIENT CORNER *See* RE-ENTRANT CORNER.

SALT A substance obtained by the union of an acid and a base radical, or by displacing the hydrogen of an ACID by a metal. In *double salts* the replacement is by two metals. Most salts are crystalline. *Common salt* is sodium chloride (NaCl) which crystallises in the cubic system.

SALT GLAZE Glaze formed on stoneware by shovelling salt into the hot kiln.

SAMPLE A group of members of large POPULATIONS, used to give information as to the larger quantity. A *random sample* is selected without bias, so that each member has an equal chance of inclusion. A *representative sample* is selected to be representative of the whole population.

SAMPLE, BOREHOLE *See* BORE-HOLE SAMPLE.

SAND Naturally occurring deposits of COHESIONLESS sediment, ranging from 10 mm ($\frac{3}{8}$ in) to 0·1 mm (0·004 in), and resulting from the disintegration of rock. Most sands are white or yellow and consist of SILICA; however, *black sands* occur naturally near volcanic rock deposits, and *coral sands* occur near coral reefs. Sand is used as FINE AGGREGATE

for concrete, in cement and in lime MORTAR, and in CALCIUM-SILICATE BRICKS.

SAND BLASTING Abrading a surface, such as concrete, by a stream of sand ejected from a nozzle by compressed air. It may be used merely to clean up construction joints, or it may be carried deeper to EXPOSE the aggregate, or to produce a sculpture. *See also* SANDING.

SAND-FACED BRICK A facing brick coated with sand to give it a RUSTIC finish.

SAND–LIME BRICK *See* CALCIUM-SILICATE BRICK.

SANDHEAP ANALOGY An ANA-LOGUE for the plastic torsional resistance of a section, proposed by A. Nadai, which is an extension of the MEMBRANE ANALOGY. If the cross-section subject to torsion is cut from a piece of cardboard or sheet metal, suspended in a horizontal position, and covered with sand flowing from an orifice above until it spills over the edges, the natural slope of the sand represents the constant plastic shear stress. The volume of the sand, which can be readily ascertained by weighing, gives a measure of the section's torsional resistance moment (Fig. 82).

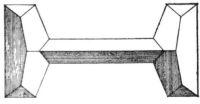

FIG. 82. Sandheap analogy for the torsional strength of an I-section of the type used in prestressed concrete.

SANDING The operation of finishing surfaces, particularly those of wood, with sandpaper or some other abrasive. It may be done by hand, or with a machine employing a belt or a revolving disc, faced with sandpaper. *See also* SAND BLASTING.

SANDSTONE Sedimentary rock containing a large proportion of rounded silica grains, generally ranging from 1 mm to 0·1 mm in diameter. The sand is cemented into a solid mass by a matrix which may be composed of silica (*siliceous sandstone*), of lime (*calcareous sandstone*), or of iron ore (*ferruginous sandstone*). The finer-grained sandstones are easily carved if the matrix is sufficiently soft (*see* FREESTONE). Sedimentary rocks composed of larger sand particles are called *gritstones*. *See also* QUARTZITE.

SANDWICH CONSTRUCTION Composite construction with a light, insulating core, which would have inadequate strength without outer layers of higher density and greater strength.

SANT'ELIA, ANTONIO *See* FUTURISM.

SANTORIN EARTH A natural pozzolana found on a Greek island of that name (also called Thera). It has been used since Antiquity.

SAP The watery fluid circulating in trees, which is necessary for their growth.

SAPWOOD The outer layers of the wood of a tree in which food materials are conveyed and stored during the life of the tree. They are usually of lighter colour.

SARKING A layer of boards or bituminous felt, or both, laid as an undercovering for tiles and other roofing. It may serve as thermal insulation, or to prevent ingress of water in roofs laid to a low pitch.

SASH A frame into which window panes are set. It normally implies a sliding frame in a *sash window*, but the term is also used for *casement windows* and fixed windows.

SASH BARS The strips of wood which separate the panes of glass in a window sash composed of several panes. Also called *muntins*.

SASH WINDOW A window contained in a cased frame which slides, as opposed to a CASEMENT WINDOW. The normal sash window (also called a *vertical sash* or *balanced sash*) slides up and down, and is balanced by sash cords passing over sash pulleys and balanced by counterweights. It was introduced into England in the seventeenth century, and was a prominent feature of residential building from the Restoration to the Georgian period, and subsequently. The less commonly used *sliding sash window* moves horizontally. Also called a *double-hung window*.

SASHLESS WINDOW Window composed of panes of glass which slide along parallel tracks in the window frame towards each other to leave openings at the sides.

SATURATED AIR Air which contains as much water vapour as it is capable of absorbing, *i.e.* air at the *dew point*. If the moisture content is increased, CONDENSATION or fogging occurs. The amount of water vapour, which air can absorb, decreases with temperature, so that air which is partly saturated may become fully saturated as it is cooled.

SATURATED SOLUTION A solution containing the maximum amount of a particular substance which it can dissolve at that particular temperature.

SATURATED VAPOUR PRESSURE *See* VAPOUR PRESSURE.

SAW ARBOR The spindle on which a circular saw is mounted.

SAW DOCTOR A craftsman who sharpens, sets and maintains the saws in a sawmill. *See* SWAGE.

SAWDUST CONCRETE Concrete whose aggregate consists mainly of sawdust, *i.e.* the waste product of timber. It has a low strength, but it can be *nailed*.

SAWTOOTH ROOF *See* NORTH-LIGHT ROOF.

Sb Chemical symbol for *antimony* (stibium).

SCAFFOLD A temporary structure of steel, timber or aluminium, to support men and materials during construction, or for shoring concrete during hardening.

SCAGLIOLA An imitation of ornamental marble, made from cement or lime, gypsum, marble chips and colouring matter. It was used in Ancient Rome, and became particularly popular in the seventeenth and eighteenth centuries.

SCALAR *See* VECTOR.

SCALE *See* MILLSCALE *and* OCTAVE.

SCALE DRAWING A drawing which shows all parts of the objects illustrated in the same proportion of their true size. Some pictorial PROJECTIONS are not scale drawings, and sketches are not normally drawn to scale.

SCANTLING A piece of timber of comparatively small dimensions.

SCARF JOINT An END JOINT between two pieces of timber, tapered to form sloping surfaces which match. It may be glued or bolted. A *stepped* or *hooked* scarf joint is one in which the jointing plane is discontinuous, and both pieces are machined to form matching steps or matching hooks. This facilitates alignment of the two ends.

SCHMIDT HAMMER An instrument used for the NON-DESTRUCTIVE TESTING of hardened concrete. It is based on the proposition that the rebound of the steel hammer from the concrete surface is proportional to the compressive strength of the concrete.

SCHWEDLER DOME A braced dome (devised by a German engineer of that name in the late nineteenth century) consisting of hoops and meridianal bars, connected together to form a series of trapezia, lined up along horizontal polygonal rings. To stiffen the structure, each trapezium is divided into two triangles by a diagonal; however, symmetrical loading introduces no stresses in these diagonals, if the dome is pin-jointed.

SCIENCE OF ARCHITECTURE Those parts of architecture amenable to scientific analysis. Today this is often held to include almost the whole of the practice of architecture, from structural design to business management. One aspect invariably excluded is architectural aesthetics. It is proper to note that *scientia* in Renaissance texts means the THEORY OF PROPORTIONS, the only part of architectural practice amenable to mathematical treatment at that time. The term *ars* generally referred to traditional craft practices. Thus the meaning of the terms *science* and *art of architecture* has become practically the reverse of the Renaissance terminology.

SCLEROMETER An instrument for the determination of hardness by means of a scratch with a diamond pyramid.

SCOTCH BOND *See* BOND (*a*).

SCRAFFETO *See* GRAFFITO.

SCRATCH-COAT The first coat of stucco or plaster applied to a surface in three-coat work.

SCREED (*a*) A heavy rule used for forming a concrete surface to the desired level or shape. Also called *screed board, tamper* or *strikeoff*. (*b*) The operation of striking off concrete lying above the desired level or shape. (*c*) A layer of concrete or mortar laid to finish a floor surface and hide the construction joints. This is sometimes called *jointless flooring*. (*d*) A bed of mortar laid as a base for ceramic or glass tiles. (*e*) A layer of concrete placed on a 'flat' roof to provide the correct gradient for drainage.

SCREEN *See* MESH.

SCREEN ANALYSIS *See* PARTICLE SIZE ANALYSIS.

SCREENINGS The rejects from screening a granular material. They may consist of either oversize or undersize particles.

SCREW ANCHOR A shell which expands and wedges itself into a hole drilled for it when a screw is inserted into it.

SCREW DISLOCATION *See* DISLOCATION IN A CRYSTAL.

SCREW JACK A lifting device actuated by means of a square-threaded screw. Its lifting capacity is more limited than that of a *hydraulic jack*.

SCREW NAIL A fastening device used where the holding power required exceeds that of a nail. It is intended to be driven with a hammer, but removed with a screw driver.

SCREW PILE A pile with a spiral-bladed shoe, which is twisted into the ground. Because of the low torsional strength of reinforced concrete, the pile is rarely precast; instead the driving shaft is usually withdrawn, and the pile cast in the BORED hole.

Se Chemical symbol for *selenium*.

SEAL *See* WATER SEAL.

SEALING COMPOUND *See* JOINT SEALANT *and* MEMBRANE CURING.

SEASONING *See* KILN *and* NATURAL SEASONING.

SEASONING CHECK Separation of wood extending longitudinally, formed during drying. It is commonly caused by the immediate effect of a dry wind or hot sun on freshly sawn timber. It usually extends only a few inches in length, whereas a longitudinal SHAKE may be several feet long.

sec Secant of an angle; sec θ = $1/\cos \theta$.

SECANT (*a*) A straight line which intersects a curve. *See also* TANGENT. (*b*) A CIRCULAR FUNCTION.

SECANT COLUMN FORMULA A formula for the design of columns, which is more sophisticated than the RANKINE FORMULA; it contains a secant term. The formula deals with the buckling of eccentrically loaded columns, but it can be used for concentrically used columns, treating the unavoidable imperfections in long columns as equivalent to an initial slight eccentricity of loading, defined by the empirical constant in the secant formula. It is widely used for the design of high-strength steel and aluminium structures. *See also* PERRY–ROBERTSON FORMULA.

SECANT MODULUS OF ELAS-TICITY Many materials (*e.g.* concrete) do not strictly conform to HOOKE'S LAW because of deviations caused by INELASTIC behaviour. If the deviation is significant, it becomes necessary to define the MODULUS OF ELASTICITY as the tangent or secant to the stress–strain curve. In Fig. 83 E_s is the *secant modulus*, and E_t the *tangent modulus* at a specified stress *f*. E_{ti} is the *initial tangent modulus*. *See also* Fig. 91.

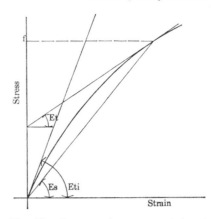

FIG. 83. Secant and tangent modulus of elasticity. E_{ti} = initial tangent modulus; E_t = tangent modulus at stress *f*; E_s = secant modulus at stress *f*.

SECOND MOMENT OF AREA The sum of the products obtained by multiplying each element of an area, d*A*, by the square of its distance from the reference axis, drawn through the CENTROID of the section. It is therefore

$$\sum y^2\,\mathrm{d}A \quad \text{or} \quad \int y^2\,\mathrm{d}A$$

The second moment of area is a geometric property essential to the solution of bending problems. It is frequently misnamed the MOMENT OF INERTIA.

SECONDARY BEAM A beam carried by the main or primary beams, and transmitting its load to them.

SECONDARY COLOUR A colour obtained by mixing two or more PRIMARY COLOURS.

SECONDARY REINFORCEMENT Reinforcement which is subsidiary to the main reinforcement, such as

DISTRIBUTION BARS in slabs, STIRRUPS in beams, and LATERAL REINFORCEMENT in columns.

SECONDARY STRESSES Stresses which are of secondary importance, and do not determine the main dimensions of a structural member. The member may have to be checked for secondary stress after it has been dimensioned for the primary stresses.

SECRET NAILING Driving nails (*e.g.* sideways into a joint) so that they cannot be seen. Also called *blind nailing* or *concealed nailing*.

SECTION MODULUS A convenient short-hand notation for the ratio I/y, where I is the SECOND MOMENT OF AREA, and y is the extreme FIBRE DISTANCE. The section modulus is used extensively in the design of beams (*see* NAVIER'S THEOREM) and section moduli for standard sections in steel, aluminium, timber and concrete are available in *section tables*. *See also* PLASTIC MODULUS.

SECTION PROPERTIES The geometric properties normally listed in *section tables*, *i.e.* the cross-sectional area, the second moment of area and the modulus of the section.

SECTION TABLES *See* SECTION PROPERTIES.

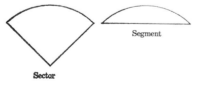

Sector

Segment

FIG. 84. Sector and segment of a circle.

SECTOR OF A CIRCLE A figure bounded by two radii and an *arc* (Fig. 84).

SEDIMENTARY ROCK A rock produced as a sediment on the floors of oceans, lakes or rivers, or on land. It may consist of fragments of pre-existing rock, or the hard parts of organisms. *See* LIMESTONE, SANDSTONE and SHALE. *See also* IGNEOUS and METAMORPHIC ROCK.

SEDIMENTATION *See* STOKES' LAW.

SEEPAGE *See* PERMEABILITY.

SEGER CONE A type of PYROMETRIC CONE, made of a mixture of clay and salt of known melting point.

SEGMENT OF A CIRCLE A figure bounded by a straight line (a SECANT) and an ARC (Fig. 84).

SEICENTO *See* QUATTROCENTO.

SEISMIC LOADING *See* EARTHQUAKE LOADING.

SEISMOGRAPH An instrument for recording the magnitude and frequency of an earthquake. *See also* RICHTER SCALE.

SELENIUM A non-metallic element used in photoelectric cells and in rectifiers. Its chemical properties are somewhat similar to those of sulphur. Its chemical symbol is Se, its atomic number is 34, its atomic weight is 78·96, and its specific gravity is 4·45. It has a melting point of 220°C, and its valency is 2, 4 or 6. There are several allotropic forms. The two most important are red selenium, and grey (or 'metallic') selenium which is a conductor of electricity when illuminated.

SELF-CLOSING DOOR *See* FIRE DOOR.

SELF-STRESSING CONCRETE *See* EXPANSIVE CEMENT.

SELF-SUPPORTING WALL A wall which carries its own weight, but no superimposed load.

SEMI-CIRCULAR ARCH *See* CIRCULAR ARCH.

SEMICONDUCTOR A material intermediate between an insulator and a conductor. Its conductivity can generally be controlled.

SEMI-FLEXIBLE JOINT *See* SEMI-RIGID JOINT.

SEMI-LOGARITHMIC GRAPH PAPER (SEMI-LOG PAPER) *See* LOGARITHMIC SCALE.

SEMI-RIGID JOINT A joint which is designed to permit some rotation, either in steel or in reinforced concrete construction. It is intermediate between a RIGID JOINT and a PIN JOINT. Also called a *semi-flexible joint* or a *partially fixed joint*.

SENSIBLE HEAT The heat absorbed or emitted by a fluid or solid when temperature changes, without a change of state, as distinct from the LATENT HEAT.

SENSIBLE HEAT FACTOR The ratio of sensible heat to total heat (sensible heat plus latent heat).

SENSITIVITY RATIO The ratio of the unconfined compressive

strength of clay in its undisturbed state and after REMOULDING. For some clays this is only slightly above unity, but for clays sensitive to remoulding the ratio may be as high as eight.

SEPTIC TANK A tank for the purification of domestic sewage, used in districts not served by sewer pipes. Disintegration of organic matter is affected by natural bacterial action, and the effluent is discharged into the ground.

SERVICE CORE A vertical unit in a high-rise building, which contains the vertical runs of most of the mechanical and electrical services, particularly the elevators (lifts). It is frequently the first part of the building to be erected, and used for vertical transportation during construction. The service core usually contributes substantially to the building's capacity to resist horizontal forces, such as wind and earthquake loads.

SESQUI One-and-a-half times, *e.g.* Fe_2O_3 (ferric oxide) is also called iron sesqui-oxide.

SET Strain remaining after removal of stress. *See also* INELASTIC and SETTING.

SETBACK The withdrawal of the face of a building to a line some distance from the boundary of the property, or from the street. A building may be set back at the level of the ground floor, or merely at the upper floors. The amount of setback required frequently depends on the height of the building, and it affects the amount of daylight received.

SETTING OF CONCRETE The initial stage in the chemical reaction between cement and water when the concrete stiffens and loses the fluidity necessary to fill the formwork. It gains significant strength only during the next, or HARDENING, stage. *See also* INITIAL SET.

SETTLEMENT *See* CONSOLIDATION and DIFFERENTIAL SETTLEMENT.

SEXTANT *See* QUADRANT.

sg Specific gravity.

SGRAFFITO *See* GRAFFITO.

SHAFT The trunk of a classical column between the base and the capital.

SHAKE A partial or complete separating between adjoining layers of wood, due initially to causes other than drying. A *falling shake* is one caused by the felling of the tree. A *water shake, ring shake* or *cup shake* is one occurring between two adjacent growth rings. A *heart shake* or *star shake* is one extending from the pith of the tree, and existing in the log before conversion. A *wind shake* is one caused by wind action on the growing tree. A *transverse shake* runs across the fibres, and a *longitudinal shake* parallel to them. *See also* SEASONING CHECK.

SHALE A laminated and fissile sedimentary rock consisting primarily of clay and silt particles.

SHALLOW DOME *See* HOOP FORCE.

SHAPE FACTOR *See* PLASTIC MODULUS.

SHAPING MACHINE *See* MILLING MACHINE.

SHEAR BOX A laboratory SHEAR TEST which determines the strength of soil by applying a shear force directly to a soil sample, contained in a box split horizontally (Fig. 85).

SHEAR FORCE The resultant of all the vertical forces acting at any section of a beam (or slab) on one side of the section (Figs. 27, 76 and 86). This force tends to shear or cut through the beam. The shear force is the vertical resultant of the STATICAL equilibrium equations at the section, while the BENDING MOMENT is the moment resultant (and generally the more important consideration).

SHEAR LEGS A pair of poles lashed together at the top, with a pulley hung from the lashing. They are used for lifting moderately heavy loads.

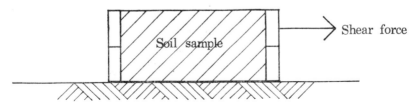

FIG. 85. Shear box for testing soil sample.

237

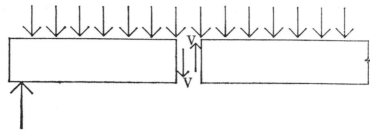

FIG. 86. Shear force *V* at an imaginary cut in a beam.

SHEAR MODULUS OF ELAS-TICITY Modulus of rigidity.

SHEAR REINFORCEMENT Reinforcement designed to resist the shear or diagonal tensile stresses in concrete. Shear reinforcement is generally required only when shear stresses are in excess of the permissible.

SHEAR STRAIN *See* STRAIN.

SHEAR STRENGTH *See* COULOMB'S EQUATION.

SHEAR STRESS *See* STRESS.

SHEAR TESTS FOR SOIL *See* SHEAR BOX and TRIAXIAL COMPRESSION TEST (which are laboratory tests), UNCONFINED COMPRESSION TEST (a laboratory or site test) and VANE TEST (a site test).

SHEAR WALL A wall which resists shear forces in its own plane due to wind, earthquake forces, explosions, etc.

SHEATH An enclosure for post-tensioned tendons. The tendon is placed in the concrete enclosed in the sheath, and bond is prevented until after the tendon has been pre-stressed. It is common practice to

bond tendons after stressing by grout injection to protect them from corrosion, and to increase the ultimate strength by establishing bond.

SHEAVE A grooved wheel for a pulley.

SHED ROOF A roof having one slope only, with one set of rafters which fall from a higher to a lower wall. Also called *lean-to roof*.

SHEET Material produced in thin layers. Sheet metal is thinner than metal *plate*, but thicker than metal *foil*. Metal *strip* is narrower than sheet metal.

SHEET GLASS Glass of the type used in windows. It is thinner than PLATE GLASS. It is specified either by thickness, or by weight, in ounces per square foot. (18-ounce glass is $\frac{1}{12}$ in thick; 32-ounce glass is $\frac{5}{32}$ in thick.)

SHEET PILE A PILE in the form of a plank, driven in close contact with others to provide a tight wall to resist the lateral pressure of water, adjacent earth, or other materials. It may be made interlocking if made of metal, or tongued and grooved if made of timber or concrete. *See also* WELLPOINT DEWATERING.

SHELF LIFE Maximum interval during which a perishable material, *e.g.* an adhesive, may be stored, and remain in usable condition. *See also* POT LIFE.

SHELL CONSTRUCTION Construction using thin curved slabs. *See also* MEMBRANE THEORY, BEAM THEORY, GAUSSIAN CURVATURE, SURFACE OF REVOLUTION, SURFACE OF TRANSLATION, DEVELOPABLE SURFACE, RULED SURFACE, DOUBLE-WALLED SHELL, RIBBED SHELL, EDGE BEAM, BARREL VAULT, NORTHLIGHT SHELL, DOME, SHALLOW DOME, SADDLE, HYPERBOLIC PARABOLOID, UMBRELLA SHELL and FOLDED PLATE ROOF.

SHELL ROOF *See* SHELL CONSTRUCTION.

SHELLAC An incrustation, formed on certain tropical trees by an insect, which is a natural resin. The purified form is *lac*. It is soluble in alcohol, and is used in spirit varnishes and in *French polish*.

SHERARDISING A method of applying a protective zinc coating to steel. The steel parts are packed in boxes filled with sand mixed with zinc, and heated to a temperature below the melting point of zinc. *See also* GALVANISING.

SHF Sensible heat factor.

SHIELDED-ARC WELDING *See* ARGON ARC WELDING and HELI-ARC WELDING.

SHIM A thin piece, usually tapered, which is driven into a joint to level or plumb a structural member.

SHINGLE (*a*) Thin piece of wood, or other material, used for covering sloping roofs. (*b*) Rounded stone of variable size and shape, but coarser than SAND.

SHOE *See* PILE SHOE.

SHOOTING CONCRETE Placing of SHOTCRETE.

SHOP WELDING, SHOP RIVETING Welding or riveting carried out on the workshop, as opposed to work carried out on the SITE.

SHORE A temporary support, usually sloping (*raking shore*), but occasionally horizontal (*flying shore*), or vertical (*dead shore*).

SHORING Posts or props of timber or other material, used in compression as temporary support for excavations, formwork, or propping of unsafe structures.

SHORT CIRCUIT An accidental connection, of zero or low resistance, joining two sides of an electrical circuit.

SHORT-COLUMN FORMULA The strength of a column, which has no tendency whatsoever towards BUCKLING (Fig. 77), is

$$P = fA$$

where A is the cross-sectional area, and f is the failing stress, *i.e.* the yield stress of steel, or the crushing strength of concrete. Note that, unlike the EULER FORMULA, the SCF is independent of the modulus of elasticity.

SHORT CYLINDRICAL SHELL *See* BEAM THEORY OF SHELLS.

SHORT TON   *See* TON.

SHOTBLASTING Cleaning a steel surface by projecting steel shot against it with compressed air or with a centrifugal steel impeller.

SHOTCRETE Cement mortar or concrete placed under pressure through the nozzle of a CEMENT GUN. Also known as *gunite*, or as *pneumatically applied mortar*.

SHRINKAGE Contraction due to moisture movement. The two building materials most affected are timber and concrete. The maximum fractional shrinkage of timber is about 100×10^{-3} parallel to the growth rings, 50×10^{-3} at right angles to the rings, and 1×10^{-3} along its length. Timber is SEASONED before use to reduce shrinkage after the timber has been fixed in the building (*see also* RADIAL SHRINKAGE). The shrinkage of concrete is less, about 3×10^{-4}; but it cannot be reduced by storage before use where concrete is cast in place together with (non-shrinking) reinforcement.

SHRINKAGE-COMPENSATING CEMENT *See* EXPANSIVE CEMENT.

SHRINKAGE CRACKING Cracking caused by SHRINKAGE when contraction is resisted by restraints.

SHRINKAGE JOINT A CONTRACTION JOINT.

SHRINKAGE LIMIT The limiting water content for a clay soil, below which a reduction in water content causes no further decrease in volume. It is often accompanied by a change in colour, and marks the limit

between the plastic and the solid state.

SHRINKAGE LOSS LOSS OF PRESTRESS caused by the shrinkage of the concrete.

SHRINKAGE REINFORCEMENT Secondary reinforcement designed to resist shrinkage stresses in the concrete. *See also* DISTRIBUTION REINFORCEMENT.

SHRINKAGE STRESSES Stresses caused when the shrinkage of concrete (or timber) is resisted by restraints or non-shrinking primary metal reinforcement.

SHUTTERING Formwork.

SHUTTERS Protective covering for the outside of windows. Shutters usually consist of vertically hinged wooden frames, and are closed at night, taking the place of curtains. *Louvred* shutters are used for ventilation, in conjunction with inward-opening *casement* windows. *Steel shutters* are used in fortifications, or in ordinary buildings if rioting is a frequent occurrence.

Si Chemical symbol for *silicon*.

SI UNITS The units used in the *Système International d'Unités*. This is a modified version of the metric system recommended for international usage, and adopted by the British, Australian, New Zealand and South African governments. It differs from the CGS UNITS: (i) by using millimetres and metres in preference to centimetres; (ii) by using the NEWTON in place of the kilogram as the unit of force and weight; and (iii) by using the JOULE as the sole unit of measuring work and energy, replacing the conventional erg, calorie, watt-hour, and kilogram-force metre. *See also* Appendix G.

SIDE-HUNG WINDOW *See* CASE-MENT WINDOW.

SIDE-SWAY Sideways movement of a frame, or of a member of a frame, due to wind or other lateral loads, due to the arrangement of the vertical loads, or due to plastic collapse (Fig. 57). *See* CARRY-OVER MOMENT.

SIDING Wall cladding for small frame building, other than masonry or brick. The term is particularly used in America, and includes CLAPBOARD, metal, asbestos cement and asphalt. *See also* WOOD SIDING.

SIEMENS MARTIN PROCESS *See* OPEN HEARTH PROCESS.

SIENNA *See* IRON OXIDE.

SIEVE ANALYSIS *See* PARTICLE SIZE ANALYSIS.

SIGHT RAIL A horizontal board set at some specified height above, say, the INVERT level of a drain. With a vertical pole it is possible to set out the drain between two sight rails accurately without surveying instruments.

SIL Speech interference level.

SILICA *Silicon dioxide*, SiO_2. About 60 per cent of the earth's crust consists of silica, and it is the chief constituent of sand and clay. Its crystalline form is QUARTZ.

SILICA BRICK *See* CALCIUM SILICATE BRICK.

SILICA GEL A colloidal form of silica made by treating sodium silicate with hydrochloric or acetic acid. It has a high capacity for adsorbing water vapour, and it is used as a drying agent in instrument cases, etc.

SILICON A non-metallic element, the second most abundant in the earth's crust (next to oxygen). It is an amorphous brown powder, or a grey crystalline substance; however, it commonly occurs as SILICA, or in the form of silicates which are constituents of most rocks, clays and soils. It is a major constituent of Portland cement, and an alloying element for steel and aluminium. Silicon's chemical symbol is Si, its atomic number is 14, its valency is 4, its atomic weight is 28·06, its specific gravity is 2·4 and its melting point is 1420°C.

SILICON CARBIDE *See* CARBORUNDUM.

SILICONE A heat-stable compound in which silicon atoms are linked up with oxygen atoms, the remaining valencies of the silicon atoms being saturated with hydrogen or organic radicals. There are many different silicones. All are chemically inert, and they are used as sealants, insulators and lubricators, *e.g.* in water-resistant films, heat resistant paints, synthetic rubbers, or resins for electrical insulation.

SILL (*a*) The lowest horizontal member of a frame for a house or other structure. (*b*) The horizontal member below a door or window opening.

SILT Natural deposit resulting from the disintegration of rock, whose particles are intermediate in size between sand and clay, ranging from about 2 to 50 micrometres in diameter.

SILVER A metal of characteristic 'silvery' colour. Its chemical symbol is Ag, its atomic number is 47, its atomic weight is 107·88, and its specific gravity is 10·50. It has a valency of 1, and a melting point of 960°C. *Sterling silver*, which was the legal standard of British coinage from the sixteenth century to 1920, is 92·5 per cent silver and 7·5 per cent copper.

SILVER SOLDER *See* SOLDERING.

SILVER STEEL Bright drawn steel containing about $1–1\frac{1}{4}$ per cent of carbon.

SIMILARITY *See* DIMENSIONAL ANALYSIS.

SIMPLE BEAM *or* **SIMPLY SUPPORTED BEAM** A beam without restraint or CONTINUITY at the supports. *See also* BUILT-IN BEAM.

SIMPLE FRAME An isostatic frame.

SIMPSON'S RULE A rule for the evaluation of an irregular area, or for graphical integration. Let the area be divided into an *even* number n of parallel strips of width x. The lengths of the boundary ordinates, or separating strips, are measured, and

these are:

$$y_0, y_1, y_2, \ldots, y_{n-1}, y_n$$

The area of the figure is then

$$\tfrac{1}{3}x[y_0 + y_n + 2(y_2 + y_4 + \cdots \\ \cdots + y_{n-2}) \\ + 4(y_1 + y_3 + \cdots + y_{n-1})]$$

The PRISMOIDAL rule is an extension of Simpson's rules to solid geometry, and it is used for measuring the volume of an excavation. Let the excavation be divided into a series of vertical strips of width x, and let the areas at each section be A_0, A_1, $A_2, \ldots, A_{n-1}$ and A_n, where n is an even number. Then the volume follows from the same formula, substituting A for y. Simpson's rule is more accurate than the TRAPEZOIDAL RULE.

SIMULTANEOUS EQUATIONS Several equations written in terms of the same unknowns. Each equation can solve only one unknown quantity. If there are three unknowns, for example, three equations are required.

sin The sine of an angle (*see* CIRCULAR FUNCTIONS).

SINE THEOREM A trigonometric theorem relating the sine of an angle to the length of the side opposite (*see* Fig. 87).

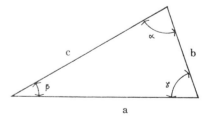

FIG. 87. Sine theorem, $(a/\sin \alpha) = (b/\sin \beta) = (c/\sin \gamma)$.

SINE WAVE A simple harmonic wave represented by the general equation

$$y = a \sin 2\pi \left(\frac{t}{T} - \frac{x}{\lambda} \right)$$

where x and y are the Cartesian co-ordinates of the curve, and t is the time. T (the period), λ (the wavelength), and a are constants. A *cosine wave* has the same shape, displaced by half a wavelength. *See also* FOURIER SERIES, HARMONIC and SOUND WAVE.

SINGAPORE INDEX *See* EQUATORIAL COMFORT INDEX.

SINGLE-ACTING *See* DOUBLE-ACTING.

SINGLE-PITCH ROOF A roof which slopes only in one direction, such as a *shed roof* or a *lean-to roof*.

SINGLE-SIZED AGGREGATE Aggregate in which most of the particles lie between narrow limits of size. It is usually produced by removing larger and smaller particles by sieving. *See* PEA GRAVEL.

SINGLY CURVED SURFACE A surface with zero GAUSSIAN CURVATURE, as distinct from a doubly curved surface. It is DEVELOPABLE and RULED.

SINGLY RULED *See* RULED SURFACE.

sinh HYPERBOLIC sine.

SIPHON A closed pipe which rises partly above the hydraulic gradient of the pipe. Provided the siphon has been PRIMED, and the pipe rises nowhere above the head due to

atmospheric pressure (approximately 30 ft or 10 m), it conveys water. The term *inverted siphon* is often used for a sagging pipe, even though this presents no problems of siphonage.

SITE WELDING, SITE BOLTING, SITE RIVETING Welding, bolting or riveting carried out on the site, as opposed to work carried out in the workshop. Also called *field welding*, etc.

SIZE A thin, pasty substance used as a sealer, binder or filler. It generally consists of a diluted glue, oil or resin.

SIZE ANALYSIS *See* PARTICLE SIZE ANALYSIS.

SKELETON CONSTRUCTION Construction in which the loads are transmitted to the ground by a FRAME, as opposed to construction with load-bearing walls.

SKEW Oblique, at an angle to the main direction.

SKIRTING A finishing board which covers the joint between the wall and the floor of a room. Also called *baseboard*.

SKY COMPONENT (OF THE DAYLIGHT FACTOR) The ratio of the part of daylight illumination received directly from a sky of assumed or known luminance distribution, to the illumination on a horizontal plane due to an unobstructed hemisphere of this sky. A special case of this component is obtained when the sky is of uniform luminance and the window apertures are unglazed; it is called the *sky factor*. This is defined as the part of the daylight illumination which would be received directly through unglazed openings from a sky of uniform luminance, to the illumination on a horizontal plane due to an unobstructed hemisphere of this sky; direct sunlight is excluded. The term has special legal significance in England.

SKY FACTOR *See* SKY COMPONENT.

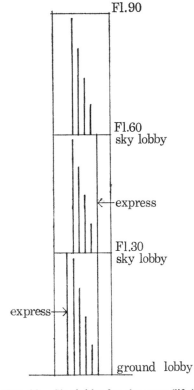

FIG. 88. Sky lobby for elevators (lifts). The 31st to the 59th floors are served by elevators from a lobby on the 30th floor, which in turn is served by an express elevator from the ground lobby. The same procedure is repeated between the 61st and 89th floors.

SKY LOBBY An elevator (lift) lobby at an upper floor. A 90-storey building, for example, could be divided into three sections of 30 storeys, each with its separate elevator system and elevator lobby. The two sky lobbies are served by express elevators. By stopping the upper elevator units at the sky lobbies, an appreciable amount of space is saved (Fig. 88).

SKY-LUMINANCE DISTRIBUTION *See* MOON-SPENCER SKY.

SKYLIGHT A window placed in a flat or sloping roof.

SKYSCRAPER A term originally coined for the 10-storey Montauk Building in CHICAGO, built in 1882 with load-bearing walls. Buildings of this height were originally made possible by the development of the passenger elevator, but soon rose much higher with the development of steel FRAME CONSTRUCTION.

SLAB In concrete construction, a flat surface which forms the floor or roof. Slabs may be directly supported on the ground (on GRADE); spanning *one way* between supporting beams, with only DISTRIBUTION STEEL at right angles; spanning *two-way* between supporting beams in both directions and with main reinforcement in both directions (*see* Fig. 58); spanning two-way and directly supported on the columns (Fig. 39) without beams (FLAT PLATE); or supported on enlarged column heads (FLAT SLAB). When the slab spans both ways, it is frequently divided into MIDDLE STRIPS and COLUMN STRIPS in each direction (*see* Fig. 31). *See also* BEAM-AND-SLAB FLOOR.

SLACK The scheduling flexibility available for an activity in a PERT network. It is equivalent to the total FLOAT in a CPM network.

SLAG WOOL *See* MINERAL WOOL.

SLAKED LIME *See* HYDRATED LIME.

SLATE A fine-grained metamorphic rock formed from clay, silt, shale or volcanic ash by high pressure. This gives the slate a cleavage across the original bedding planes. The material can be split into thin slabs. The terms *princess, duchess, countess,* etc., used in conjunction with slate for covering roofs refer to the size of the pieces, not to their quality.

SLENDERNESS RATIO The ratio of the EFFECTIVE LENGTH OF A COLUMN to its least RADIUS OF GYRATION, *i.e.* the radius of gyration in the direction of the smaller second moment of area. The longer the column, and the thinner it is, the more likely is it to BUCKLE.

SLICED VENEER *See* WOOD VENEER.

SLIDE RULE A device for performing mechanically simple mathematical processes. The common slide rule has two scales, which may be slid past one another to perform addition or subtraction of the *scales.* Both scales are LOGARITHMIC, so that the slide rule actually performs multiplication and division. The slide rule is, in effect, a simple ANALOGUE COMPUTER.

SLIDING FORMWORK *See* CLIMBING FORMWORK.

SLIDING SASH WINDOW *See* SASH WINDOW.

SLIP CIRCLE An assumed line of shear failure of a *clay* slope, which produces a rotational or cylindrical slide. The resistance of the slope to failure then equals the shear strength of the clay, multiplied by the surface of failure (the product of the length of the circular arc and the length of the slope).

SLIP LINES *See* LÜDERS' LINES.

SLIPFORM *See* CLIMBING FORMWORK.

SLOPE The inclination of a surface, particularly to the horizontal. The slope of structural members due to elastic deformation is usually expressed in RADIANS.

SLOPE DEFLECTION A technique devised independently by G. A. Maney and A. Bendixen in 1914 for the solution of the bending moments in rigid frames and continuous beams by a series of simultaneous equations. Since all joints are assumed RIGID, the change in slope and the deflection is the same for all members framing into any one joint, and these equalities supply sufficient equations to solve the redundancies of the HYPERSTATIC STRUCTURE.

SLOPE OF TIMBER GRAIN The angle between the axis of a piece of timber and the general direction of the grain. *See also* STRAIGHT GRAIN.

SLOTTED ANGLE Steel angle pre-punched with slotted holes. It is used for shelving and other utility structures, particularly those of a temporary nature.

SLOW-BURNING CONSTRUC-TION (*a*) Construction with materials treated to make them more FIRE-RESISTING. (*b*) Construction with heavy timber sections which are protected by the layer of charcoal formed during initial combustion. Such frames are more fire-resisting than unprotected steel frames. *See also* FIRE PROTECTION OF STEEL FRAMES.

SLOW-BURNING INSULATION Insulating material which chars or burns without a flame or blaze. Some plastic materials used for thermal insulation are highly COMBUSTIBLE.

SLUMP TEST A method of measuring the WORKABILITY of freshly mixed concrete. The concrete is placed in a mould, which consists of a truncated cone, 12 in (30 cm) high, with a base diameter of 8 in (20 cm), and a top diameter of 4 in (10 cm). The mould is then lifted, and the subsidence measured to the nearest $\frac{1}{2}$ in. *See also* BALL TEST and COMPACTING FACTOR TEST.

SLURRY A fluid mixture of cement and water, or of sand, cement and water.

SMALL CIRCLE *See* GEODETIC LINE.

SMOKE TUNNEL A WIND-TUNNEL in which a smoke generator is used to indicate the movement of the air.

Sn Chemical symbol for *tin* (stannum).

SNAPHEAD The conventional, ap-proximately hemispherical, head used on steel rivets.

SNOW GUARD A board which prevents snow from sliding off a sloping roof.

SNOW LOAD The superimposed load assumed to result from severe snow falls in any particular region. Snow loads range from zero in most parts of Australia to 60 lb/ft^2 (3 kN/m^2) in Northern Canada.

SOAPSTONE *See* TALC.

SOCKET *See* SPIGOT.

SODIUM A very reactive ALKALI metal. Its chemical symbol is Na, its atomic number is 11, its atomic weight is 22·997, and its specific gravity is 0·98. It has a melting point of 97·7°C, and a valency of 1.

SODIUM SILICATE *See* WATER-GLASS.

SOFFIT The underside of any horizontal member of a structure, *e.g.* a beam or a slab.

SOFT SOLDER *See* SOLDER.

SOFT WATER *See* HARD WATER.

SOFTBOARD A low-density fibre-board, as opposed to a *hardboard*.

SOFTWARE The means whereby the operation of a digital computer is controlled, *i.e.* the PROGRAMS which cause it to carry out calculations or other operations. *See also* HARDWARE.

SOFTWOOD *See* HARDWOOD.

SOIL AUGER *See* AUGER.

SOIL CEMENT A mixture of Portland cement and locally available soil. It serves as a soil stabiliser.

SOIL CLASSIFICATION *See* PARTICLE SIZE ANALYSIS and ATTERBERG LIMITS.

SOIL DRAIN Drain which carries sewage or trade effluent to the sewer, as opposed to a STORM DRAIN.

SOIL MECHANICS A term coined in 1936 by Karl Terzaghi, who systematised the subject, to embrace all aspects of the scientific study of soils as engineering materials. *See* ATTERBERG LIMITS, CONSOLIDATION, COULOMB'S EQUATION, DIFFERENTIAL SETTLEMENT, EARTH PRESSURE, PARTICLE SIZE ANALYSIS, RANKINE THEORY, SHEAR TESTS and SLIP CIRCLE.

SOIL PROFILE A vertical section showing the variation of the soil below the surface of a site.

SOIL SAMPLE *See* BOREHOLE SAMPLE.

SOLAR HEAT The heat produced by solar radiation. It ranges from about 445 BThU/ft^2 per hour (1400 W/m^2) at noon at the outer edge of the earth's atmosphere when it is closest to the sun to about 415 (1300) when it is furthest from the sun. At the earth's surface the maximum solar heat gain through ordinary window glass ranges up to 250 BThU/ ft^2 of sash area per hour (800 W/m^2) in the tropics, reducing to 220 (700) at a latitude of 50°.

SOLAR ORIENTATION The position of the building in relation to the north (or south), with particular reference to the amount of sunshine falling on the walls and windows, and the penetration of the sun through the windows into the building. *See also* SUNSHADING.

SOLARIUM A room, terrace or balcony, generally with some glass walls, exposed to the rays of the sun. The term is particularly used for hospitals and sanatoria; for private houses *sun room* is more common.

SOLARSCOPE A device for studying, with the aid of MODELS, sunlight penetration and the shadows to be cast by and on buildings. There are two types. In one the model is placed on a platform, and a lamp at the end of a long arm is moved to imitate the position of the sun at various times of the day and the year; the latitude can also be altered. This apparatus is simple to use and direct-reading, but the location of the 'sun' at the end of an arm conflicts with the requirement of 'infinite' distance. The other type consists of a platform which can be rotated in altitude and AZIMUTH, which must be calculated for the sun's position. The sun is represented by a horizontal light at the far end of a long room. This is more accurate, but the interpretation requires some computation. For either type the platform can be made transparent, to obtain a view of internal shadows. Also called a *heliodon*.

SOLDERING A process for joining two pieces of metal by means of *solder*, *i.e.* an alloy which has a lower melting point than the pieces to be joined. For satisfactory jointing, the surfaces to be joined must be kept free from oxide films, and this is accomplished by using a *flux* which melts at a lower temperature than the solder. For the lower-strength solders, known as *soft solders, zinc-chloride* is a suitable flux. For the higher-strength solders (*hard solder* or *silver solder*) BORAX is a suitable flux. *Plumber's solder,* commonly used in building, consists of an alloy of lead and tin. 'Coarse' solder (consisting of three parts lead to 1 part tin) melts at 250°C; 'fine' solder (consisting of equal parts of lead and tin) melts at 188°C. *See also* BRAZING.

SOLENOID A multi-turn coil of wire wound in uniform layers on a cylindrical former. When carrying a direct current it behaves like a bar magnet. It is commonly used in RELAYS, switches, CIRCUIT BREAKERS, and brakes.

SOLID DOOR A flush door with a solid core, as opposed to a HOLLOW-CORE DOOR.

SOLID SOLUTION *See* PHASE DIAGRAM.

SOLID-WEB JOIST A conventional joist with a solid web formed by a plate or a rolled section, as opposed to an OPEN-WEB JOIST.

SOLIDS, GEOMETRICAL *See* POLYHEDRON.

SOLIDUS LINE The line separating the solid phase from the liquid + solid phase. In a PHASE DIAGRAM it shows the variation of the composition of an alloy with temperature when melting is complete (Fig. 71).

SOLSTICE The 21st June and 22nd December, when the sun passes through the first point of Cancer and the first point of Capricorn respectively, and attains its maximum distance from the celestial equator. Consequently these are the longest and shortest days of the year. *See also* EQUINOX.

SOMMER (Also spelt *summer*) *See* BREAST SUMMER.

SONE A unit of loudness. It is directly proportional to loudness, whereas the PHON has a logarithmic scale. 1 sone = 40 phons. A noise twice as loud is 2 sones or 50 phons; a noise half as loud is $\frac{1}{2}$ sone or 30 phons. The relation is, however, not quite exact at higher sound levels, because the two scales were determined in different laboratories.

SOREL'S CEMENT *See* OXY-CHLORIDE CEMENT.

SOUND *See also under* ACOUSTIC and NOISE.

SOUND ABSORPTION The property of a material which reduces REVERBERATION *within* a room, as distinct from SOUND INSULATION. It has little effect on sound transmission through a wall or floor. *See also* ACOUSTIC BOARD, ACOUSTIC PLASTER, HELMHOLTZ ABSORBER, IMPEDANCE TUBE, REVERBERATION CHAMBER and SUSPENDED ABSORBER.

SOUND ABSORPTION COEFFICIENT The ratio of the sound energy absorbed by a surface to the energy incident upon the surface. Its value ranges from about 0·01 for a polished marble to 1·0 for the absorbing fibreglass wedges used in ANECHOIC CHAMBERS. *See also* NOISE REDUCTION COEFFICIENT.

SOUND, AIRBORNE *See* AIR-BORNE SOUND.

SOUND ATTENUATION Reduction in sound intensity. *See also* DECIBEL.

SOUND FREQUENCY ANA-LYSER An instrument used in conjunction with a *sound level meter* to determine the distribution of sound frequencies in a mixed noise or sound. (*See* WHITE NOISE.) An *octave-band analyser* covers a constant-percentage bandwidth of 2, *i.e.* the upper cut-off frequency is the OCTAVE of the lower cut-off frequency. A *half-octave-band analyser* has a frequency band of $\sqrt{2}$.

SOUND, IMPACT *See* IMPACT SOUND.

SOUND INSULATION Reduction of sound passing *through* a wall, as opposed to SOUND ABSORPTION. Insulation against AIRBORNE SOUND requires MASS; the more massive the wall or floor, the better the insulation; thus a thick layer of concrete is excellent insulation against airborne sound. *See also* REVERBERATION CHAMBER. Insulation against IMPACT SOUND is best achieved by means of a FLOATING FLOOR.

SOUND KNOT *See* KNOT.

SOUND LEVEL METER An instrument for measuring the level of a sound, irrespective of frequency. It is usually calibrated in DECIBELS. The distribution of the sound over the sound spectrum is determined with a SOUND FREQUENCY ANALYSER.

SOUND LEVEL RECORDER An automatically recording sound level meter.

SOUND MIRROR *See* ECHO and WHISPERING GALLERY.

SOUND PERFUME *See* BACK-GROUND NOISE.

SOUND PRESSURE *See* DECIBEL and EFFECTIVE SOUND PRESSURE.

SOUND REDUCTION FACTOR The reduction in the intensity of sound produced, for example, by a partition, expressed in decibels; it may vary with the wavelength of the sound. Its reciprocal is the *acoustical transmission factor*.

SOUND SPECTRUM ANALYSER A SOUND FREQUENCY ANALYSER.

SOUND TRANSMISSION *See* SOUND INSULATION, SOUND REDUCTION FACTOR and TRANSMISSION LOSS.

SOUND WAVES The pressure waves within the audible spectrum. The complex waveforms can be broken down into SINE WAVES of various frequencies.

SOUNDING BOARD A reflecting surface placed above a rostrum or pulpit to direct the sound towards the audience. *See also* ACOUSTICAL CLOUD.

SOUNDNESS Freedom of a metal casting or of concrete from cracks, flaws and fissures; this is sometimes checked by listening to the sound which the casting makes when struck. The term is also used to denote freedom from excessive volume change, and from deterioration due to exposure to the weather.

SOUNDPROOFING *See* SOUND INSULATION, DISCONTINUOUS CONSTRUCTION and RUBBER MOUNTING.

SOUTHLIGHT ROOF *See* NORTHLIGHT ROOF and Fig. 68.

SOUTHWELL'S METHOD *See* TENSION COEFFICIENT and RELAXATION METHOD.

SOW *See* PIG.

sp gr Specific gravity.

SPACE FRAME A FRAME which can only be solved by considering its behaviour in space, *i.e.* in two mutually perpendicular planes at the same time. Space frames may be ISOSTATIC or HYPERSTATIC (*see* MÖBIUS' LAW). All frames exist in space, but most can be resolved into a series of plane frames, which makes their design and construction much simpler than that of true space frames. *See also* GEODESIC DOME, LATTICE STRUCTURE and TENSION COEFFICIENT.

SPACE LATTICE (*a*) A synonym for a triangulated SPACE FRAME. (*b*) *See* BODY-CENTRED CUBIC LATTICE, FACE-CENTRED CUBIC LATTICE and CLOSE-PACKED HEXAGONAL LATTICE.

SPACE-TIME The four-dimensional continuum used in the theory of relativity. Some mid-twentieth century artists have been looking to contemporary science for inspiration, and the notion of four-dimensional space thus entered into art and architectural criticism. While it can be argued that design was conceived two-dimensionally before the discovery of *perspective* in the Renaissance, the concept of time as the fourth dimension of architectural design has no scientific basis.

SPAN The distance between the supports of a structure. The world's longest span at the time of writing is 4260 ft (1300 m) in the Verrazano Narrows Bridge, New York (completed in 1964). *See also* EFFECTIVE SPAN.

SPANDREL The part of the wall between the head of a window and the *sill* of the window above it. The term is also used as a synonym for *spandrel beam*, which is a beam placed within a spandrel, or a structural beam on the edge of a building frame. *Spandrel* also denotes the triangular infilling under the outer string of a stair, and the triangular infilling above the extrados of an arch, between the abutment and the crown.

SPATTERDASH A rich mixture of cement and coarse sand thrown hard onto a brick or concrete wall, to form a thin, coarse-textured, continuous coat.

SPECIFIC GRAVITY The ratio of the mass of a given volume of a substance to the mass of an equal volume of water (at 4°C, when water has its minimum volume). Since water weighs 1 gram per cubic centimetre, the density of water in metric units is numerically equal to its specific gravity, which is a dimensionless ratio. In FPS units, the specific gravity must be multiplied by the density of water which is $62 \cdot 4$ lb/ft^3.

SPECIFIC HEAT The ratio of the quantity of heat required to raise a substance through a given temperature range, to that required to raise the same mass of water through the same temperature range.

SPECIFIC SURFACE The total surface area of the particles contained in a unit weight or absolute unit volume of a material. The smaller the particle size, or the finer the powder, the greater the specific surface. *See also* FINENESS MODULUS, PARTICLE SIZE ANALYSIS and STOKES' LAW.

SPECTRUM *See* VISIBLE SPECTRUM.

SPEECH SPECTRUM The frequency range of speech, particularly that part which matters for intelligibility.

SPEED *See* VELOCITY.

SPELTER An alloy containing about 99 per cent zinc.

SPHERE *See* SURFACE OF REVOLUTION.

SPHERICAL DOME *See* DOME.

SPHERICAL REFLECTOR *See* PARABOLIC REFLECTOR.

SPHERICAL TRIGONOMETRY *See* TRIGONOMETRY.

SPIGOT The plain end of a length of pipe which is fitted into an enlarged *socket* or *bell* at the beginning of the next pipe. The *spigot-and-socket joint* (also called *bell-and-spigot* joint) is made tight by CAULKING.

SPIRAL REINFORCEMENT *See* HELICAL REINFORCEMENT.

SPIRAL STAIR *See* HELICAL STAIR.

SPIRE A tall, needle-like pyramid, rising from a tower or roof, usually above a church. The base of the pyramid is frequently a regular OCTAGON.

SPIRIT LEVEL An instrument for testing horizontal or vertical alignment. It consists of a wooden straightedge incorporating a slightly curved glass tube filled partially with alcohol. The horizontal position is indicated by the central location of the air bubble. The term is also applied if the tube is filled with a liquid other than alcohol.

SPIRIT STAIN A dye dissolved in METHYLATED SPIRITS, usually with SHELLAC or some other resin as a binder. It is used for darkening a wood surface, but emphasises the grain of the timber less than a *water stain* (in which water is the solvent for the dye). *See also* OIL STAIN.

SPIRITS OF ALUM Sulphuric acid.

SPIRITS OF SALT Hydrochloric acid.

SPIRITS OF SULPHUR Sulphureous acid.

SPIRITS OF VITRIOL Sulphuric acid.

SPL Sound pressure level, usually in accordance with the decibel scale.

SPLAY An inclined surface, *i.e.* a large BEVEL or CHAMFER, running across the full width of the surface.

SPLICE A joining of two structural pieces. The joint is generally designed to be as strong, or stronger, than the pieces to be joined.

SPLICE OF REINFORCING BARS Transfer of force from one bar to another. This may be achieved by welding or by mechanical connection. However, the normal procedure is to overlap the bars in the concrete, usually without touching; the force is transferred by bond between the concrete and the steel.

SPLIT-LEVEL HOUSE A house in which one or more rooms are above the level of the ground floor (which commonly contains the kitchen and dining area), but below the level of the floor above (on which the bedrooms are normally placed). The split-level floor usually consists of the main living room, and its ceiling is higher than that of the other rooms. Against the advantage of extra height must be set the inconvenience caused by the steps between the various levels.

SPLIT-RING CONNECTOR A TIMBER CONNECTOR inserted in a ring-shaped groove pre-cut in both pieces of timber, made with a hole saw.

SPLITTING TENSILE TEST Determination of tensile strength of concrete by testing a cylinder (*see* CYLINDER STRENGTH) on its side in compression. The cylinder splits across the vertical diameter. Also known as *Brazilian test* and *diametral compression test*.

SPOIL Material excavated which is in excess of the fill required.

SPONTANEOUS COMBUSTION An instantaneous bursting into flames

of a mixture of substances, due to the evolution of heat through chemical action between them.

SPOT WELDING Joining two or more overlapping pieces by local fusion of small areas or spots.

SPRAYED ASBESTOS Asbestos blown on to a surface with a spray gun. It provides thermal insulation which is rot-proof, vermin-proof and incombustible. It also serves as lightweight FIRE-PROTECTION OF STEEL STRUCTURES, since 1 in of it is as effective as 2 in of concrete. Its weight is approximately 7 per cent of that of concrete. Now illegal in New York and some other cities for health reasons; mineral wool fibre is used instead.

SPRAYED CONCRETE *See* SHOT-CRETE.

SPRING WASHER A steel ring cut once and bent into a shallow helix. Used as a washer, it prevents the nut from unscrewing.

SPRING WOOD *See* EARLY WOOD.

SPRINGINGS *or* SPRING The level at which an arch springs from its supports. *See also* CROWN.

SPRINKLER SYSTEM A system which sprinkles a fire with water as soon as it breaks out, and thus extinguishes it, or controls it until the fire brigade arrives. It usually consists of pipes installed in or below the ceiling throughout the building. Branches projecting from these pipes are sealed by *sprinkler heads* which open at a predetermined temperature, usually 155°F (68°C). The sprinkler head may be sealed by a metal plug, which melts at the predetermined temperature; or by a plastic plug which contains liquid bursting it at the predetermined temperature. It is important that this temperature be set low enough to control a fire in its early stages, but also high enough to ensure that the system does not go off on a hot day without a fire, since the damage done by the water is often more than the damage done by a small fire.

SQUARE (a) A surface measure used in some parts of America, England and Australia, equal to 100 square feet. (b) *See* QUADRI-LATERAL. (c) An L-shaped tool for setting out right angles.

SQUARE DOME A spherical dome on a square base (Fig. 89). In a reinforced concrete shell dome it is only necessary to provide ties or frames which offer the necessary restraints (b). In a masonry dome either squinches or pendentives are needed to provide the transition from the square to the sphere. *Squinches* are short arches constructed across the angle of the square in tiers. *Pendentives* are corbelled structures in the form of spherical triangles, and they are prominent features on many *Byzantine* domes (a). In *Moslem* domes the pendentives or squinches are sometimes exaggerated to suggest the form of *stalactites* in a limestone cave.

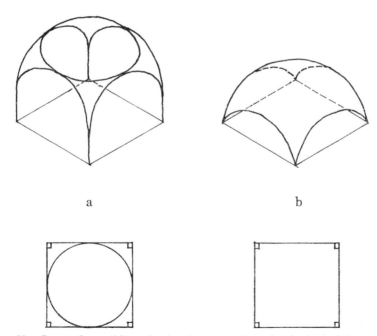

a   b

FIG. 89.   Square dome: (a) pendentives for supporting a hemi-spherical dome on a square base; (b) spherical dome squared by cuts which are stiffened by transverse frames.

SQUARE MATRIX A matrix in which the number of rows equals the number of columns.

SQUARED LOG *See* balk.

SQUARED STONE *See* ashlar.

SQUINCH *See* square dome.

STABILITY *See* breakdown, buckle, buttress, depolymerisation, differential settlement, dimensional stability, retaining wall and slip circle.

STABILISED SOIL Soil which has been treated with a binder to reduce its movement. Suitable binders are Portland cement, waste oil, bitumen, resin or a more stable soil, but low cost is a prime consideration.

STADIUM Originally a course for foot-racing, one stadium ($\frac{1}{8}$ Roman mile) in length. Hence a place for athletic exercises, together with the tiers of seats for spectators, and appropriate service rooms. An *indoor stadium* also has a roof, and this usually presents major structural problems because of the span required.

STAIN *See* oil stain and spirit stain. *See also* paint.

STAINED GLASS A decorative panel or window composed of pieces of coloured glass, joined by means of lead beads or concrete. The glass colouring is not a stain, but is fired into the glass.

STAINING POWER The amount of colour given to a white pigment by a given amount of coloured pigment. It is the opposite of reducing power.

STAINLESS STEEL Steel which is highly resistant to atmospheric corrosion, and attack by organic and dilute mineral acids. It is used for cutlery, facing panels for curtain walls, etc. There are many different alloys, but all contain between 8 and 30 per cent chromium, in addition to a smaller amount of other elements, especially nickel.

STAIR One *step* in a flight of stairs; also the entire flight. *See also* baluster, riser and tread.

STAIR WELL A space around which a staircase is disposed.

STALACTITE WORK *See* square dome.

STANCHION (*a*) A column, particularly of structural steel. (*b*) An upright bar placed intermediate between the mullions to strengthen a leaded light.

STANDARD DEVIATION The square root of the *variance*, *i.e.* the root of the average of the squares of the *deviations* of a number of observations from their mean value. It is a measure of the spread of the observations, and it is necessary first to square the variances, and then take the root of their mean, because otherwise the positive and negative variances would largely cancel out. By squaring the deviations, all values become positive. *See also* coefficient of variation.

STANDARD FIRE TEST *See* fire resistance grading.

STANDARD OVERCAST SKY
See MOON-SPENCER SKY.

STANDARD (PETERSBURG or PETROGRAD) *See* BOARD FOOT.

STANDARD SECTION A metal section which has been standardised. In the case of hot-rolled sections, this is much cheaper than a section specially made to order; however, the difference is less marked for cold-rolled and extruded sections.

STANDARD TEMPERATURE AND PRESSURE (STP) Temperature of 0°C and pressure of 1 atmosphere.

STANDARD WIRE GAUGE (SWG) An arbitrary series of numbers used for standardising the diameter of wires. There are several such standards, some limited to specific materials. The most commonly used are the British (or Imperial) Standard Wire Gauge and the American Brown and Sharpe Wire Gauge. The abbreviation SWG without prefix means the former.

STANDARDISED NORMAL VARIATE *See* GAUSSIAN CURVE.

STANDARDS INSTITUTES *See* AFNOR, ASTM, BSI, CSA, DIN, ISI, ISO, SAA, SABS and USASI.

STANDING-WAVE TUBE An IMPEDANCE TUBE.

STAPLE A loop of bent wire, sharpened to two points, to be used as a fastener.

STARTER A device for starting an electric motor and accelerating it to its normal speed.

STARVED JOINTS Glued joints which do not contain enough adhesive, due to the use of insufficient adhesive, excessive pressure, or adhesive of inadequate viscosity. Also called *hungry joints*.

STATIC FRICTION *See* FRICTION.

STATIC LOAD A load which is not DYNAMIC, *i.e.* a normal dead or superimposed load.

STATIC MOMENT A term used for the first MOMENT OF AN AREA.

STATIC PENETRATION TEST A PENETRATION TEST in which the testing device is pushed into the soil by a measurable force, as opposed to a *dynamic* test which employs a specified number of blows with a standard hammer.

STATICALLY DETERMINATE *See* ISOSTATIC STRUCTURE.

STATICALLY INDETERMINATE *See* HYPERSTATIC STRUCTURE.

STATICS The branch of the science of mechanics which deals with forces in EQUILIBRIUM; as opposed to DYNAMICS. The condition of static equilibrium is generally expressed in terms of three equations:

$$\Sigma H = 0$$
$$\Sigma V = 0$$
$$\Sigma M = 0$$

which means that the forces are zero in the *horizontal* and *vertical* directions, and the *moments* about any one chosen point are zero. If a structure can be solved with these equations alone, it is called ISOSTATIC. If there are more unknown structural restraints than can be solved by these equations, it is called HYPERSTATIC.

STATISTICS Numerical data systematically collected; the science of collecting numerical data. *See also* GAUSSIAN CURVE, LAW OF LARGE NUMBERS, POPULATION, PROBABILITY, SAMPLE and STANDARD DEVIATION.

STAUNCHION *See* STANCHION.

STEADY FLOW *See* STREAMLINE FLOW.

STEAM CURING Accelerating the CURING of precast concrete by exposing to steam in an oven at ordinary pressure, or at high pressure in an AUTOCLAVE.

STEATITE *See* TALC.

STEEL A malleable alloy of iron with a carbon content between 0·1 and 1·7 per cent. Iron with a lower carbon content is classified as WROUGHT IRON and iron with a higher carbon content as CAST IRON. ALLOY STEEL contains other elements in addition to carbon. Prior to the invention of the BESSEMER PROCESS, steel could only be produced at great expense (*see* DAMASCENE STEEL). *See also* Fig. 71, ALPHA IRON, ANNEALING, AUSTENITE, CEMENTITE, FERRITE, GAMMA IRON, HIGH-SPEED STEEL, HIGH-STRENGTH STEEL, MARTENSITE, NORMALISING, OPEN-HEARTH PROCESS, PERLITE, QUENCHING, REINFORCEMENT FOR CONCRETE, STAINLESS STEEL and TEMPERING.

STEEL FRAME *See* FRAME CONSTRUCTION.

STEEL PLATE, SHEET *See* SHEET.

STEEL REINFORCEMENT *See* REINFORCEMENT.

STEELYARD An instrument for weighing, which consists of a lever with unequal arms, with a single weight moving along a graduated scale. It is the type of balance used in Ancient Rome.

STEFAN–BOLTZMANN LAW The total RADIATION from a black body is proportional to the fourth power of the absolute temperature, or sT^4, where T is the temperature in degrees Kelvin, and s is a constant which equals $5·735 \times 10^{-5}$ erg/cm^2 deg^4 sec ($5·735 \times 10^{-8}$ W/m^2 deg^4).

STEP *See* STAIR.

STERADIAN (sr) Unit of solid angle. The solid angle which, having its vertex in the centre of a sphere, cuts off an area of the surface of the sphere equal to that of a square having sides of length equal to the radius of the sphere. It is used in lighting calculations, *see* LUMEN.

STEREOCHEMISTRY The study of the spatial arrangements of atoms in complex molecules.

STEREOGRAM A drawing or photograph which can be viewed three-dimensionally. One common method is to superimpose one print of a STEREOSCOPIC pair in one colour on the other in a different colour, and viewing the composite picture through two appropriately coloured glasses.

STEREOGRAPHY The science of perspective PROJECTION.

STEREOMETRY Solid geometry.

STEREOPHONIC SOUND Sound produced by several loudspeakers to give the impression of auditory perspective.

STEREOSCOPE An instrument for viewing a stereoscopic pair of photographs three-dimensionally. It consists of two lenses set at the correct distance apart to correspond with the separation of the *camera* lenses.

STEREOSCOPIC CAMERA A camera designed to give two displaced images (called a *stereoscopic pair*) by means of two matched lenses and shutters, so that the pair when viewed by both human eyes in a *stereoscope* gives a three-dimensional view of the object photographed.

STEREOSCOPIC VISION The three-dimensional vision seen by the two human eyes due to their being set a small distance apart.

STEREOTOMY The science of making sections of solid bodies.

STEREOTYPE The printing process in which a solid plate of type-metal, cast from a papier-mâché mould taken from the surface of a *forme* of type, is used for printing, instead of the forme itself. Hence something which is repeated constantly without change, like so many printed sheets run off the same stereotype.

STEVINUS *See* PARALLELOGRAM OF FORCES.

STIFF FRAME *See* RIGID FRAME.

STIFF-JOINTED FRAME *See* RIGID FRAME.

STIFFENER A small member added to a thin section to prevent BUCKLING; *e.g.* an angle welded or riveted to the web of a deep steel or aluminium girder, which strengthens the web plate against buckling in diagonal compression due to the shear force.

STIFFNESS Resistance to deformation. In rigid frames the flexural stiffness, which determines the MOMENT DISTRIBUTION between members, is defined as EI/L, where E is Young's modulus, I is the second moment of area, and L is the effective length.

STIFFNESS METHOD A matrix solution for HYPERSTATIC STRUCTURES, intended for evaluation by electronic digital computer, in which the equations are framed in terms of the unknown joint displacements.

STIJL, DE *See* DE STIJL.

STILE LIBERTY The Italian version of ART NOUVEAU.

STILL A vessel for the distillation of liquids.

STILT HOUSE A house built on stilts, to allow passage of air under the floor, and to provide shaded storage or living space. It is a traditional form of construction in many hot-humid countries.

STIRRUP In concrete construction, reinforcement to resist shear. It is normally a bar of U-shape, properly anchored to the longitudinal steel, and placed perpendicular to it.

STOKES' LAW A formula developed by G. G. Stokes in 1851 for the velocity of sedimentation (or settlement) of spherical particles in a liquid. From the observed terminal velocity the particle diameter, and thus the SPECIFIC SURFACE, may be derived:

$$d = \left[\left(\frac{18\mu z}{(\gamma_s - \gamma_l)}\, t\right)\right]^{\frac{1}{2}}$$

where d is the particle diameter, z is the depth to which it has settled, t is the time taken for it to settle, μ is the viscosity of the liquid (for water at 20°C, 1 centipoise, or 1 mN . s/m²) and γ_s and γ_l are the specific gravity of the solids (generally assumed to be 2·7) and of the liquid (1·0 for water). The sample is taken with a pipette at a depth z after a time t, and the amount of solid is measured. *See* PARTICLE-SIZE ANALYSIS.

STONE *See* ROCK *and* COARSE AGGREGATE.

STOPPING Filling cracks and nail holes with putty before painting.

STORE *See* COMPUTER STORE.

STOREY *See* FLOOR.

STORM DOOR An extra door for protection against bad weather.

STORM DRAIN Drain which carries rain water, as opposed to a SOIL DRAIN.

STORY *See* FLOOR.

STP Standard temperature and pressure.

STRAIGHT ARCH A FLAT ARCH.

STRAIGHT GRAIN Timber grain which is straight and in line with the

axis of the piece, not *sloping*. *See also* FIGURE.

STRAIGHT-LINE THEORY A theory based on a linear relationship; specifically, a structural theory based on HOOKE'S LAW.

STRAIGHTEDGE A long piece of wood or metal whose edges are true and parallel. It is used for setting out, and for testing the accuracy of straight lines in buildings.

STRAIN Change in the dimensions or shape of a body per unit length or angle. A *shear strain* is a distortion caused by shear STRESSES. A *direct strain* is an elongation or shortening caused by tensile or compressive stresses respectively. Strains may be ELASTIC or INELASTIC.

STRAIN ENERGY Mechanical energy stored up in a stressed material. Strain energy methods are normally part of the elastic theory, and it is therefore implied that the strain is within the elastic limit. The elastic strain energy is the potential energy of deformation. It equals the work done by the external forces in producing the strains, and it is recoverable. For convenience, strain energy is usually divided into strain energy due to direct forces, shear forces, bending moments and twisting moments, and each part is treated separately. In many problems the strain energy due to bending is much greater than the other components which may then be neglected without serious loss of accuracy.

STRAIN ENERGY METHOD The oldest method for the solution of HYPERSTATIC STRUCTURES, published by the Italian engineer A. Castigliano in 1870. It is based on his *second theorem* which states that 'the partial differential coefficient of the total strain energy of a frame with respect to the load in a redundant member is equal to the initial lack of fit of that member.' If the frame has been force-fitted or prestressed, the initial lack of fit must be introduced into the equations; otherwise the partial differential coefficient is nil. One equation may be set up for each REDUNDANCY of the frame. The strain energy method is still the most convenient method for solving hyperstatic arches, bow girders, and other curved members. However, for linear hyperstatic frames, other methods are more suitable.

STRAIN GAUGE An instrument for measuring strain, also called an *extensometer* or *tensometer*. *See* ELECTRICAL RESISTANCE STRAIN GAUGE, HUGGENBERGER TENSOMETER, DEMEC STRAIN GAUGE, ACOUSTIC STRAIN GAUGE and CAPACITANCE STRAIN GAUGE. *See also* BRITTLE COATING, PHOTOELASTICITY, MICRO-GRID and MOIRÉ FRINGES.

STRAIN HARDENING *See* COLD WORKING.

STRAIN ROSETTE Device measuring strain at one point in three directions. The unknown quantities are the magnitude of the two (mutually perpendicular) PRINCIPAL STRESSES and their direction. If all three are unknown, then three measurements are required at each point. Strain rosettes are usually arranged along 60° (equilateral triangle) or two at right angles, and the third at 45°.

STRAND Wires twisted around a centre wire or core; a TENDON made in the form of a strand.

STREAMLINE FLOW Fluid flow which is continuous, *steady* and *laminar*, as in a viscous fluid. The upper limit is REYNOLDS' CRITICAL VELOCITY, above which the flow becomes *turbulent, i.e.* unsteady and eddying.

STRENGTH *See* LIMITING STRENGTH OF MATERIAL, STRESS and TESTING MACHINE.

STRENGTH OF MATERIALS A conventional sub-division of the theory of structures, which deals with the calculation of stresses and strains due to tension, compression, shear, torsion and flexure, and any combination thereof. The *theory of structures* proper is then considered to cover the stresses and strains in structural members when they are combined into trusses, frames, etc. *See also* MATERIALS SCIENCE.

STRESS Internal force per unit area, considering an infinitesimally small part of a body. When the forces are tangential to the plane they are called *shear stresses*; when they are perpendicular to the plane, they are called *direct stresses*. Direct stresses may be *compressive* or *tensile*, depending on whether they act towards or away from the plane of separation. The deformation caused by the stress is called *strain*.

STRESS ANALYSIS *See* EXPERIMENTAL STRESS ANALYSIS and PRINCIPAL STRESS.

STRESS CIRCLE *See* MOHR CIRCLE.

STRESS CONCENTRATION A local high stress, or crowding of the ISOSTATIC LINES, caused by a sudden change in section, such as occurs at a NOTCH, the base of a screw thread, or a hole. Stress concentrations are particularly serious in brittle materials where they may lead to premature failure. In plastic materials local plastic yielding reduces the high stresses at the point of concentration, and raises them over a wider zone (Fig. 90). They are shown up particularly well by PHOTO-ELASTICITY.

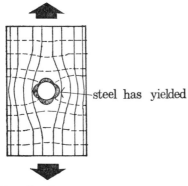
steel has yielded

FIG. 90. Stress concentration around a hole, relieved by localised plastic yielding of the material.

STRESS CORROSION Corrosion of a metal, accelerated by its being highly stressed.

STRESS DIAGRAM *See* STRESS–STRAIN DIAGRAM.

STRESS-GRADING OF TIMBER Grading timber mechanically into several categories of strength. The most common machine is based on an empirical relation between the strength and the deflection of timber. Each piece of timber is deflected at

several points along its length, and the deflection category marked by means of a spot of dye. The timber is then classified by its colour markings.

STRESS RELAXATION *See* RELAXATION OF STEEL.

STRESS RELIEVING Heating of a metal or alloy, followed by slow cooling, to relieve internal stresses built up by hot or cold working. *See* ANNEALING, NORMALISING and TEMPERING.

STRESS–STRAIN DIAGRAM The diagram obtained by plotting the stresses in a test specimen against the strains. It is used to assess the structural suitability of materials, since it shows the strength of the material, its elastic and inelastic deformation, and its ductility or brittleness.

STRESS–STRAIN DIAGRAM OF CONCRETE The relation between the stress and strain in concrete up to failure (Fig. 91). The slight curvature of the diagram at low loads is due to the combined effect of elastic deformation and creep. At about 60 per cent of the maximum stress, MICRO-CRACKS begin to form, and produce pseudo-plastic deformation. This can be represented on a rheological model (Fig. 92) by substituting for the elastic spring of a visco-elastic KELVIN SOLID a series of brittle springs in parallel which break progressively with increasing strain; the maximum compression of the springs before brittle fracture is shown by the arrows. Although the pseudo-plastic behaviour of concrete due to micro-crack formation differs from the truly plastic behaviour of mild steel, it can be utilised in

260

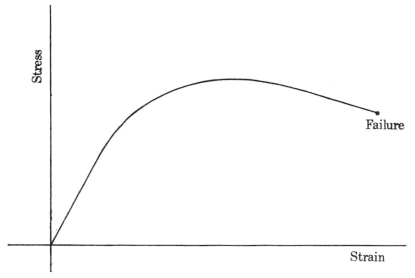

FIG. 91. Stress–strain diagram of concrete tested to failure.

assessing the ultimate strength, as in the WHITNEY STRESS BLOCK.

STRESS–STRAIN DIAGRAM OF MILD STEEL The relation between stress and strain in mild steel up to failure (Fig. 93).

STRESS TRAJECTORY *See* ISO-STATIC LINE.

STRESSCOAT *See* BRITTLE COATING.

STRESSED-SKIN CONSTRUCTION A form of construction in which the outer skin acts with the framework to contribute to the membrane and flexural strength of the unit, instead of being merely a cladding which protects the inside from the weather. The term was

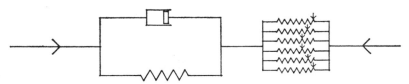

FIG. 92. Rheological model for deformation of concrete up to failure. The left-hand unit, which consists of a spring and a dashpot in parallel, models the creep deformation. The right-hand unit consists of a large number of parallel brittle springs which fracture when they are compressed to the arrow; it produces elastic deformation at low loads, the pseudo-plastic deformation due to the fractured springs (representing the micro-cracks) at high loads.

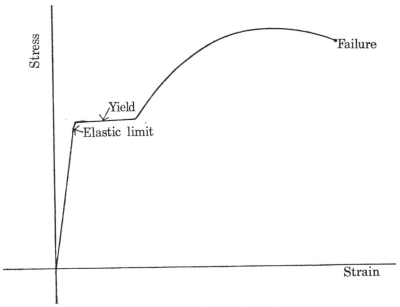

FIG. 93.   Stress–strain diagram for mild steel.

originally applied to aircraft frames, and later to prefabricated houses in LIGHT-GAUGE construction. *See also* GEODETIC CONSTRUCTION.

STRETCHER A brick, block or stone laid with its length parallel to the wall. Usually stretchers are interspersed with HEADERS to achieve a proper bond. *See* BOND (*a*).

STRIKE OFF *See* SCREED.

STRING COURSE A continuous projecting horizontal band set in the surface of a wall, sometimes moulded. Its function is partly decorative and partly to throw the water off the facade.

STRING POLYGON *See* LINK POLYGON.

STRINGER (*a*) A horizontal piece of steel or timber, connecting uprights in a framework and supporting the floor (or the rails of a railroad bridge). (*b*) The inclined member which supports the treads and risers of a stair.

STRIP *See* SHEET.

STRIP FOOTING A footing for a wall, or a joint footing for a line of columns.

STROBOSCOPE Instrument for the inspection of objects rotating at high speed. It can be timed to light up the object only when it is in the same position, so that it appears stationary.

STRUCK JOINT A mortar joint formed with a recess at the bottom of the joint, by pressing the trowel in at the lower edge. This work can be done as the wall goes up, and it is therefore more durable than POINTING. A *struck joint* is suitable only for interior work because it would collect water at the lower edge. For exterior work, a *weather-struck joint*, or *weather joint* is produced by pressing the trowel in at the upper edge, so that the recess is formed at the top of the joint, and the water is thrown off the joint (Fig. 94).

STRUCTURAL ANALYSIS AND DESIGN Structural analysis consists of determining the load-bearing capacity of a structure whose dimensions are known or have been assumed. Structural design consists of determining the structural sizes required in the members of a structure for a given system of loads. The two processes can be treated separately in ISOSTATIC structures, since the forces and moments can be determined from the loads and the general geometry of the structure without knowing the cross-sectional dimensions of the members. In the design of HYPERSTATIC structures it is generally necessary to assume the sizes of the members before any structural calculations can be made, and the processes of design and analysis cannot easily be separated.

STRUCTURAL MODEL *See* MODEL ANALYSIS.

STRUCTURE-BORNE SOUND WAVES *See* AIRBORNE SOUND and DISCONTINUOUS CONSTRUCTION.

STRUT A compression member, the opposite of a *tie*.

STUB MORTISE A MORTISE which does not pass entirely through a timber.

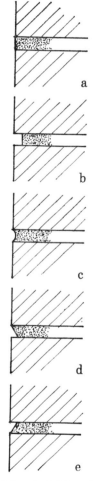

FIG. 94. Joints in brickwork: (a) flush joint; (b) raked joint; (c) tooled joint; (d) struck joint; (e) weather-struck joint, or weathered joint.

STUB TENON A TENON cut to fit into a STUB MORTISE, or a short tenon used at the lower end of a post to prevent it from slipping out of position.

STUCCO Smooth external plastering, as opposed to *rough cast*. The term originally had a wider meaning, and in historical books it is used also for interior and decorated work.

STUD (*a*) An upright timber. (*b*) A threaded rod or bolt without a head. It may be fixed to a steel frame by resistance *welding*, or a pointed stud may be shot with a *stud gun* into timber, masonry or concrete.

STUD WALL A timber-framed wall. The studs, or vertical members, are usually spaced at 12–14 in (0·3–0·4 m) centres.

STUFFINESS A feeling of being in a close or ill-ventilated room. Thomas Bedford carried out experiments in the 1930s which suggest that the sensation was more likely to occur in warm rooms, in the absence of ventilation touching the thermal receptors in the skin. By contrast the sensation of *freshness* is associated with cooler conditions, and with sensations recognisable as those of touch, caused by air movement.

STYRENE–BUTADIENE COATING A lacquer-type paint, which can be used on masonry and wood.

SUBJECTIVE BRIGHTNESS *See* LUMINOSITY.

SUBLIMATE Solid obtained by the direct condensation of vapour without passing through the liquid state. This is possible only for materials whose melting point and boiling point are very close together.

SUBROUTINE A part of a digital computer PROGRAM defining a particular operation which is executed at several different stages during the complete program. A *library subroutine* is a standard subroutine, such as the inversion of a matrix, usually provided by the computer manufacturer, which is available to the programmer, to avoid the need for re-programming standard operations.

SUBSIDENCE Settlement caused by mining operations.

SUBSONIC *See* MACH NUMBER.

SUB-STATION *See* TRANSFORMER.

SUBSTRUCTURE *See* SUPER-STRUCTURE.

SUBTOPIA A term coined by Ian Nairn in 1955 to denote 'the world of universal low-density mess' extending from the suburbs to the country, 'a foreground of casual and unconsidered equipment, litter and lettered admonition. . . . Reduction of density where it should be increased, reduction of vitality by false genteelism. . . '.

SUCTION *See* WIND PRESSURE.

SUCTION RATE The amount of water absorbed by a brick in one minute. Also called *absorption rate*.

SULFUR *See* SULPHUR.

SULPHUR A non-metallic element of yellow colour. Its chemical symbol is S, its atomic number is 16, its atomic weight is 32·06, and its valency is 2, 4 or 6. α-sulphur crystallises in rhombic form, has a lemon-yellow colour, melts at 112·8°C and has a specific gravity of 2·07. β-sulphur crystallises in monoclinic form, has a deeper yellow colour, melts at 119·0°C and has a specific gravity of 1·96. Also spelt *sulfur*.

SULPHURIC ANHYDRIDE Sulphur trioxide (SO_3). It combines with water to form *sulphuric acid* (H_2SO_4).

SULLIVAN, LOUIS H. *See* CHICAGO SCHOOL and FUNCTIONALISM.

SUMMER (also spelt *sommer*) *See* BREASTSUMMER.

SUMMER COMFORT ZONE *See* COMFORT ZONE.

SUMMER WOOD *See* EARLY WOOD.

SUN ROOM *See* SOLARIUM.

SUNLIGHT PENETRATION *See* SOLARSCOPE *and* WALDRAM DIAGRAM.

SUNSHADING Controlling the entry of the sun into a building by means of LOUVRES, projecting EAVES, projecting balconies or vertical slats. Sunshading devices are prominent features on the facade of the building, and their visual aspect must therefore be considered as well as their technical efficiency. *See also* BRISE SOLEIL *and* SOLAR ORIENTATION.

SUPER COOLING Cooling a liquid below its normal freezing point. In the case of a mixture, cooling below the LIQUIDUS in the phase diagram.

SUPERIMPOSED LOAD The load superimposed on the DEAD LOAD of the building. The term is generally synonymous with LIVE LOAD, although a distinction is sometimes made between the superimposed dead load caused by movable partitions, etc., and the live load caused by people.

SUPERPOSITION 'If a material is elastic and obeys HOOKE'S LAW, the relation between load and deformation is linear, and the effect of the various loads can therefore be computed separately and subsequently added, or superimposed on one another.' The *principle of superposition* greatly simplifies the design of elastic structures. It cannot, however, be used where the load-deformation relation is non-linear, as in some elastic problems relating to suspension structures or to buckling.

SUPERSONIC *See* MACH NUMBER and ULTRASONIC.

SUPERSTRUCTURE The structure above the main supporting level, as opposed to the foundation or *substructure*.

SUPPLEMENTARY ANGLE *See* COMPLEMENTARY ANGLE.

SUPPORT MOMENT The negative bending moment at a fixed-ended or continuous support.

SURCHARGED WALL A retaining wall which has an embankment on its top.

SURFACE *See also* RULED SURFACE, DEVELOPABLE SURFACE, MINIMAL SURFACE and GAUSSIAN CURVATURE.

SURFACE FINISH OF CON-CRETE *See* BÉTON BRUT, EXPOSED AGGREGATE and FORMWORK.

SURFACE OF REVOLUTION A surface generated by rotation. For example, a *sphere* is generated by the rotation of a circle, a *paraboloid* by a parabola, and an *ellipsoid* by an ellipse. A *hyperboloid of revolution* is generated by a straight line at an angle to a vertical axis (Fig. 95). *See also* TORUS and SURFACE OF TRANS-LATION.

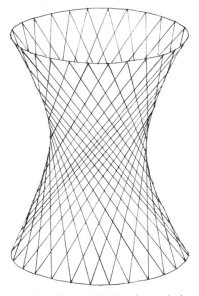

FIG. 95. Hyperboloid of revolution. This is both a surface of revolution and a ruled surface.

SURFACE OF TRANSLATION A surface generated by the motion of a plane curve parallel to itself over another curve. A *cylinder* (Fig. 24) is generated by a curve, such as a circle, moving along a straight line. A HYPERBOLIC PARABOLOID (Fig. 53) is generated by an inclined straight line moving along two other inclined straight lines, or alternatively by a convex parabola moving along a concave parabola; it is an *anticlastic* surface. An *elliptical paraboloid* is generated by a convex parabola moving over another convex parabola, and it is consequently a *synclastic* surface. A *parabolic conoid* is generated by a straight line moving over a flat parabola at one end, and a more strongly curved one at the other (Figs. 68 and 96). Other conoids are similarly generated over different curves. *See also* SURFACE OF REVOLUTION and RULED SURFACE.

FIG. 96.　Parabolic conoid.

SURFACE SEALER *See* PAINT.

SURFACE SPREAD OF FLAME *See* FLAME SPREAD.

SURFACE TENSION A property possessed by liquid surfaces whereby they appear to be covered with a thin elastic membrane in a state of tension. It is due to the unbalanced molecular cohesive forces near the surface. *See* CAPILLARY ACTION.

SURFACE WATER The run-off after rain, as opposed to soil or waste water. The two are often drained separately, and the surface water discharged at a suitable place in the open to save sewer capacity.

SURFACE WATERPROOFER
Waterproofing concrete and other materials by painting a liquid on the surface, as opposed to an INTEGRAL WATERPROOFER. The liquid may be colourless, or a pigmented paint. Many surface waterproofers contain silicone or epoxy resins.

SURVEYOR'S LEVEL (*frequently abbreviated to level*) A telescope mounted on a tripod through adjustable screws which very precisely determines differences in level over appreciable distances. The telescope is set horizontally with the aid of a SPIRIT LEVEL attached to it, and sights are then taken on a graduated staff, placed on the ground and held vertically at the various points whose levels are required. The differences between the graduations of the staff, as seen through the telescope, equal the differences in level. The same result can be obtained with a THEODOLITE, but the surveyor's level is a cheaper instrument, and easier to use if all readings are to be level. *See also* DUMPY LEVEL and CLINO-METER.

SUSPEN-ARCH A composite structure consisting of arches and suspension cables, proposed by the Central American engineer Paul Chelazzi in the 1950s.

SUSPENDED ABSORBER A prefabricated space absorber suspended within a room to improve its acoustic performance. It may be hung from the ceiling, from the structural system, or from a secondary suspension system, such as stretched wires. *See also* ACOUSTICAL CLOUD, whose purpose is to *reflect*.

SUSPENDED CEILING A FALSE CEILING suspended from the floor above.

SUSPENDED FORMWORK Formwork which is suspended from the supports for the concrete floor to be cast, and not propped from below.

SUSPENDED SPAN A short span freely supported from the ends of cantilevers, as in a GERBER BEAM.

SUSPENSION CABLE A cable hanging freely. If carrying mainly its own weight, it assumes the shape of a *catenary*. If carrying mainly a load uniformly distributed in plan it assumes the shape of a *parabola*.

SUSPENSION ROOF A roof supported by suspension cables (Fig. 97). It may take the form of a dished dome, *i.e.* a series of cables hanging from an outer compression ring and terminating at a central tension ring, or a dished cylindrical shell. Although the cable is the most efficient structural type for long spans, the suspension roof has found only limited application in architecture. One reason is the fact that very large SPANS, of the magnitude used in bridges, are not required for buildings. Another disability is the flexibility of the structure, and its lack of insulation.

SWAGE The spread of a tooth on each side of a timber saw to provide a clearance for the blade. *See* SAW DOCTOR.

SWAY *See* SIDE-SWAY.

SWAY BRACE A diagonal member, or a pair of diagonals, designed

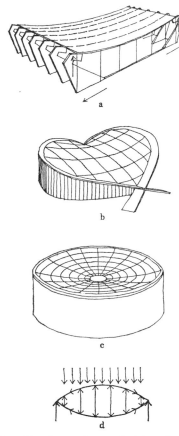

FIG. 97. Suspension roofs. (a) roof with parallel cables, anchored to banked seats; (b) roof with parallel cables supported by crossed arches; (c) radial cables forming a dished dome; (d) radial cables, doubled for aerodynamic stability, forming a dome (bicycle wheel roof).

to resist wind or other horizontal forces acting on a light structural frame. *See also* SHEAR WALL.

SWG   Standard wire gauge.

SYNCLASTIC   *See* GAUSSIAN CURVATURE.

SYNTHETIC RESIN   *See* RESIN.

SYNTHETIC STONE   Cast stone.

SYSTEM ANALYSIS   The definition and interpretation of problems for computer-aided solution. A system analyst is a person who prepares a FLOW DIAGRAM showing the interrelation of man and computer in the solution of the problem. In a large organisation this is a separate function. The computer programmer interprets the system flow diagram by preparing a detailed flow diagram, and coding the problem. The computer operator is the man who actually handles the HARDWARE.

SYSTEM BUILDING   INDUSTRIALISED BUILDING in accordance with a *closed* system.

SYSTEMATIC ERROR   *See* COMPENSATING ERROR.

T

TACK WELD   A weld not designed to carry a load, but to make a non-structural connection.

TALC   Acid metasilicate of magnesium ($H_2Mg_3Si_4O_{12}$). It is a soft mineral, with a soapy or greasy feel, which can be used for intricate carvings. Also called *soapstone* or *steatite*. Finely ground talc is called *French chalk*.

TAMPER   *See* SCREED.

tan The tangent of an angle. *See* CIRCULAR FUNCTIONS.

T & G JOINT Tongue and groove joint.

TANGENT (*a*) A straight line which just touches a curve at a single point, but does not intersect it. *See also* SECANT. (*b*) A CIRCULAR FUNCTION.

TANGENT MODULUS OF ELASTICITY *See* SECANT MODULUS and Fig. 83.

TANGENTIAL SHRINKAGE *See* RADIAL SHRINKAGE.

TANGENTIAL STRESS A *shear stress*.

TANK *See* HOOP TENSION.

TAPE PUNCH *See* CARD PUNCH.

TAPE READER A device which reads the holes in computer tape, and translates them into alphanumeric characters on a typewriter, *i.e.* normal typescript. A *card reader* does the same with computer cards.

TAPER The gradual reduction in size, *e.g.* the narrowing of a column towards the top or the bottom.

TAPERED WASHER A bevelled washer for use with, for example, a standard rolled steel joist which has non-parallel flanges.

TAPESTRY Textile fabric decorated with ornamental designs or pictorial subjects, particularly one used for hanging on a wall.

TAR A bituminous substance obtained from the destructive distillation of coal. It has a lower melting point than ASPHALT.

TARE The weight of a vehicle, as distinct from its load.

TARMACADAM *See* MACADAM.

TARPAULIN A covering of canvas impregnated with tar or paint, used to protect materials or unfinished work against rain.

TATAMI The Japanese mat, which provides the MODULE for the design of traditional Japanese houses. Each room is an exact multiple of the size of the mat, which measures approximately 6 ft by 3 ft (approximately 2 m by 1 m).

T-BEAM (*a*) In metal, a section shaped like the letter T. It is an I-BEAM with only one flange. (*b*) In concrete, the T-beam is commonly formed by the RIB and the portion of the floor slab above it. It is part of a MONOLITHIC beam-and-slab floor, not a separate T-shaped beam. *See* EFFECTIVE FLANGE WIDTH and L-BEAM.

TEFLON *See* POLYTETRAFLUORETHYLENE.

TELAMONES *See* CARYATIDS.

TEMPERA PAINTING A mural painting technique widely used in the Middle Ages and the Renaissance, which produces transparent colours on GESSO. The powdered pigment is bound with egg or gum arabic, and thinned with water. *See also* DISTEMPER.

TEMPERATURE *See* °C and °F.

TEMPERATURE GRADIENT
The change in temperature per unit length, *e.g.* through a wall which is warm on one side and cold on the other.

TEMPERATURE MOVEMENT
Thermal expansion and contraction.

TEMPERATURE REINFORCE-MENT *See* DISTRIBUTION REINFORCEMENT.

TEMPERATURE STRESS A stress caused by a change in temperature. All materials expand with rising temperature and contract with falling temperature. Stresses are caused only if the movement is restrained.

TEMPERED HARDBOARD HARDBOARD which has been treated during manufacture to improve water resistance and strength. Its density usually exceeds 60 lb/ft^3 (960 kg/m^3).

TEMPERING Heating hardened steel to a few hundred degrees and cooling it slowly to reduce the brittleness induced by the MARTENSITE. The steel loses some hardness and strength in the process. Many bright steels acquire a characteristic colour on tempering, caused by an oxide film, which can be used to determine the temperature to which they have been heated. For plain carbon tool steels, heated for a normal period, these are: straw 225°C, yellow–brown 255°C, red–brown 265°C, purple 275°C, violet 285°C, dark blue 295°C, and light blue 310°C. At higher temperatures the skin turns grey. *See also* ANNEALING.

TEMPLATE or **TEMPLET** A sheet or light frame of wood or metal, used for marking out work to be done, or as a guide for a cutting tool.

TENACITY A term generally synonymous with the ultimate tensile strength.

TENDON A bar, wire, strand, or cable of high-tensile steel, used to impart PRESTRESS to concrete when the element is tensioned.

TENEMENT A rented apartment. The term now usually implies one in a building in the poorer part of the city, and one which is overcrowded and lacks adequate sanitation.

TENON *See* MORTISE.

TENSION A direct pull in line with the axis of the body, and therefore the direct opposite of *compression*.

TENSION COEFFICIENT A notation introduced by the British mathematician Sir Richard Southwell in 1920 to simplify Jourawski's method of RESOLUTION AT THE JOINTS. It is defined as the force in the member of a truss, divided by its length. The sines and cosines resulting from resolution of the forces thus become Cartesian co-ordinates, and the simultaneous equations of equilibrium are readily written in matrix form and evaluated by computer. The method is particularly useful for the solution of isostatic SPACE FRAMES, for which the other methods available for plane frames are not practicable.

TENSION RING *See* HOOP TENSION.

TENSION STRUCTURE *See* PNEUMATIC ROOF and SUSPENSION ROOF.

TENSION WOOD Abnormal wood which may be formed on the upper sides of branches in inclined stems of hardwood trees.

TENSOMETER *See* STRAIN GAUGE.

TENSOTAST *See* DEMEC STRAIN GAUGE.

TERA *See* BILLION.

TERMINAL VELOCITY *See* STOKES' LAW.

TERMITE An insect which shuns light, and is highly destructive to seasoned timber, especially soft wood. Australian cypress pine and several eucalypts are naturally resistant, but most timbers require protection by TERMITE SHIELDS or IMPREGNATION. Termites are killed by arsenic or creosote. Also called *white ant*.

TERMITE SHIELD A protective shield placed between the foundation piers and a timber floor, around pipes, etc. It usually consists of galvanised iron, bent down at the edges. Termites cannot stand daylight and they can only get past the inedible iron by building an earthlike shelter tube; this shows up on the metal cap. Moreover, termites are very reluctant to pass over the downward bend.

TERRA COTTA Burnt clay units for ornamental work. Their colour varies from yellow to reddish brown. They are very durable, even unglazed. The glazed terra cotta is called FAIENCE. *See also* CERAMIC VENEER.

TERRACE Originally a raised level earth surface for walking on sloping ground, sometimes provided with a balustrade. Hence an enclosed level platform in front of a house; a gallery or a balcony attached to a house; a row of houses on a raised platform; and any row of houses of uniform style.

TERRAZZO Marble-aggregate concrete that is either cast-in-place as a TOPPING or precast. It is subsequently ground smooth for decorative surfacing on floors or walls.

TESSERA *See* MOSAIC.

TEST CYLINDER, CUBE *See* CYLINDER STRENGTH.

TESTING MACHINE A machine used for loading test pieces, usually to destruction, to determine their *deformation* and *strength*. The most common machines load the specimen either in tension or compression, but flexural, FATIGUE and IMPACT tests are also common. A simple testing machine can be made by using a hydraulic jack in conjunction with a PROVING RING. Most precision testing machines also use hydraulic jacks for loading, but screws, with or without levers, or dead weights may be employed for smaller loads. Testing machines with a capacity of more than 500 tons are rare.

TETRAHEDRON *See* POLYHEDRON.

TEXTURE BRICK *See* RUSTIC BRICK.

TEXTURE PAINT A paint which can be manipulated after application to give a textured finish.

tg Continental abbreviation for tangent (TAN).

THATCH A roof covering of reed, straw, or rushes. It has a high insulating value, but burns very easily. The fire risk can be somewhat reduced by soaking the thatch in a fire-resisting solution before laying.

THEATRE A place for viewing (particularly dramatic plays), either in the open air or in a building specially constructed for the purpose. *See also* AMPHITHEATRE.

THEATRE DIMMERS Variable resistances installed in lighting circuits, which control the amount of light produced by varying the voltage.

THEODOLITE A surveyor's instrument for measuring horizontal and vertical angles. It has a telescope rotating on a horizontal (*trunnion*) axis, to which is attached the vertical measuring circle. The trunnion axis is carried on a forked standard fixed to the upper plate, whose rotation relative to the lower plate measures the horizontal angles. The lower plate is fixed through levelling screws to a tripod. Modern theodolites almost invariably are capable of transiting about the trunnion axis, and in America the instrument is commonly called a TRANSIT. *See also* SURVEYOR'S LEVEL.

THEOREM OF THREE MOMENTS A theorem, derived by the French mathematician B. P. E. Clapeyron in 1857, for calculating the redundant support moments in CONTINUOUS beams (*see* Figs. 18 and 66). Once these REDUNDANCIES have been determined, the problem becomes isostatic.

THEORIES OF FAILURE *See* FAILURE CRITERIA.

THEORY OF PROPORTIONS A theory which is intended to enable a designer to produce beautiful proportions by mathematical rules. Several proportional theories have been produced, mostly in the nineteenth century, by comparison with measured drawings of classical masterpieces. A theory which fitted a sufficient number of acknowledged works of art was thought to prove that the rule had been used by the original designer. A new work conforming to the rule would then be more likely to have beautiful proportions than one which ignored the rule. However, the same works of art have been used to prove quite different mathematical rules, which suggests that none of the theories are on a firm scientific foundation. Psychological tests have not supported any of the proportional theories. Most observers cannot distinguish between designs based on the GOLDEN SECTION (1:1·618), HARMONIC PROPORTIONS (1:1·667), or some simple arbitrary rule, such as 1:1·5. At the same time, experimental psychology has provided no alternative rules. *See also* PROPORTIONAL RULES BASED ON $\sqrt{2}$, SACRED CUT, and SCIENCE OF ARCHITECTURE.

THEORY OF STRUCTURES *See* STRENGTH OF MATERIALS.

THERMAL CAPACITY The capacity for storing heat or cold. In cold climates thick walls with a high thermal capacity are useful to conserve fuel. In hot-dry climates thick walls are also used to keep the house cooler during the hottest part of the day. In hot–humid climates, ventilation is more important, and light construction with a low thermal capacity is traditional.

THERMAL COMFORT *See* COMFORT ZONE.

THERMAL CONDUCTANCE The amount of heat which passes through a unit area of a building material, of the thickness normally used in the building, due to a difference of temperature of 1° between the two faces. It may be expressed in British or metric units. The thermal conductance is related to the THERMAL CONDUCTIVITY (*K-value*) which is heat passing through a unit thickness. It is also related to the THERMAL TRANSMITTANCE (*U-value*), which is based on temperatures measured in the air beyond the roof or wall, instead of temperatures measured on the surface of the building material.

THERMAL CONDUCTION The process of heat transfer through a material medium in which heat is transmitted from particle to particle, not as in convection by movement of particles, nor by radiation.

THERMAL CONDUCTIVITY Rate of transfer of heat along a body by conduction, measured by the amount of heat per unit surface area flowing per hour through a unit thickness for a temperature difference of 1°. It is frequently called the *K-value*. For most building materials the K-value is approximately proportional to their densities. *See also* THERMAL CONDUCTANCE.

THERMAL CONDUCTOR A material which readily transmits heat by conduction.

THERMAL EXPANSION Increase in the length of members of a building due to an increase in temperature. Unless provision is made for *expansion joints*, thermal movement is likely to produce cracking. The *coefficient* of thermal expansion is the expansion of the material per unit length per degree temperature change; it has different values for Fahrenheit and Celsius units, but is otherwise the same for metric and British units.

THERMAL INSULATION *See* INSULATION, INSULATING BOARD, and THERMAL RESISTANCE.

THERMAL MOVEMENT *See* THERMAL EXPANSION.

THERMAL RADIATION *See* RADIANT HEAT.

THERMAL RESISTANCE The resistance to the passage of heat provided by the roof, wall or floor of a building. It is the *reciprocal* of the THERMAL TRANSMITTANCE or U-value.

THERMAL STRESS Stress produced by thermal movement which is resisted by the building. If the thermal stresses are higher than the capacity of the materials to resist them, expansion or contraction joints are required. Thermal stresses are particularly important in brittle materials, such as concrete and brick, because of their tendency to crack at comparatively small tensile stresses.

THERMAL TRANSMITTANCE The amount of heat transmitted through a roof, wall or floor due to a temperature difference in the *air* on both sides. It is commonly known as the U-value, and expressed as the amount of heat per unit surface area transmitted per hour for a temperature difference of 1°. The U-values of various forms of construction have been determined experimentally. Some countries, particularly those with a cold climate, lay down maximum permissible U-values. Also called *air-to-air heat-transmission co-efficient*. See also THERMAL CONDUCTANCE.

THERMAL UNIT   *See* BThU.

THERMIC BORING A method of boring holes into concrete by means of a high temperature, produced by burning a steel lance packed with steel wool, which is ignited and kept burning by a gas such as an oxyacetylene mixture.

THERMIONICS Literally the science dealing with the emission of electrons from hot bodies. It generally concentrates on the subsequent behaviour and control of such electrons, particularly *in vacuo*. A *thermionic vacuum tube* can be used for the rectification, amplification or detection of electric currents.

THERMISTOR A temperature-sensitive resistance element of metallic oxide, whose electrical resistance decreases with increase in temperature.

THERMOCOUPLE A thermometer consisting of a pair of electric wires so joined as to produce a *thermo-electric effect*. When the ends of two dissimilar metals are joined, an electric current is produced by a change in temperature which is proportional to the temperature difference between the hot and the cold junctions. Thermocouples are remote-reading thermometers, but for room temperatures they are not as accurate as the conventional mercury-in-glass type. A cheap combination of wires consists of copper and CONSTANTAN. Precious metals, such as platinum and rhodium are required for high temperatures.

THERMODYNAMICS The study of the relation between heat and energy. The two often-quoted laws of thermodynamics are: *First law:* 'Heat and mechanical energy are mutually convertible; there is a constant relation between the amount of heat lost and energy gained, or vice versa, which is called the MECHANICAL EQUIVALENT OF HEAT'. *Second law:* 'Heat can never pass spontaneously from a colder to a hotter body; consequently the ENTROPY of the universe tends to a maximum'.

THERMO-ELECTRIC EFFECT *See* THERMOCOUPLE.

THERMOHYGROGRAPH A clock-driven recording instrument which records both the dry-bulb and the wet-bulb temperature, or the dry-bulb temperature and the relative humidity.

THERMOPLASTIC Becoming soft when heated and hard when cooled. *See also* THERMOSETTING.

THERMOSETTING Becoming rigid on heating due to chemical reaction, usually between a resin and a HARDENER. Thermosetting resins cannot normally be softened, and they do not soften significantly on heating. *See also* THERMOPLASTIC and COLD-SETTING RESIN.

THERMOSTAT A device for maintaining a constant temperature. It is commonly used in conjunction with heating and air conditioning plants. *See also* BI-METALLIC STRIP.

THICK SHELL *See* MEMBRANE THEORY.

THIN SHELL *See* MEMBRANE THEORY.

THIN-WALLED SECTION *See* LOCAL BUCKLING.

THINNER Any volatile liquid which lowers the viscosity of a paint or varnish, and thus makes it flow more easily. It must be compatible with the medium of the paint. The most common thinners are TURPENTINE and WHITE SPIRIT.

THIRD-ANGLE PROJECTION *See* PROJECTION.

THIXOTROPIC Stiffening when left standing for a short period, and acquiring a lower viscosity on mechanical agitation. The process is reversible, and is characteristic of certain COLLOIDAL gels.

THREE-DIMENSIONAL PHOTO-ELASTICITY A technique for solving space problems, as opposed to plane photoelasticity. In 1935 *Solakian* discovered that for certain PHOTOELASTIC MATERIALS the stress patterns could be frozen above room temperature. The model can therefore be stressed in an oven, and cooled while still under load. It is then cut into parallel slices which are examined in a POLARISCOPE at room temperature. The three principal stresses are obtained by composition from the plane slices. Also called the *frozen-stress method.*

THREE-DIMENSIONAL SOUND *See* STEREOPHONIC SOUND.

THREE-DIMENSIONAL VISION *See* STEREOSCOPIC VISION.

THREE-HINGED FRAME *See* PORTAL.

THREE-PHASE An alternating-current system in which the currents, flowing in three independent circuits, are displaced in phase by 120 electrical degrees.

THREE-PIN FRAME *See* PORTAL and Fig. 73.

THREE-PRONG PLUG An electric plug which has two prongs connecting to the main circuit, and one to the earth.

THRESHOLD OF AUDIBILITY *See* HEARING THRESHOLD and PHON.

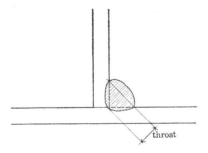

FIG. 98.   Throat of a fillet weld.

THROAT The minimum thickness of a fillet weld; it is the dimension which determines the strength (Fig. 98).

THRUST A pushing force exerted by one part of a structure on an adjoining part. The term is more particularly used for horizontal or inclined forces. *See* ARCH, BUTTRESS, RETAINING WALL and VIOLLET-LE-DUC'S DICTIONARY.

THRUST, LINE OF *See* LINE OF THRUST.

TIE (*a*) A tension member, the opposite of a *strut*. (*b*) In reinforced concrete columns, the lateral or HOOP REINFORCEMENT. (*c*) A WALL TIE.

TIE-BEAM The horizontal, lowest member of a roof truss.

TIED COLUMN A reinforced concrete column laterally reinforced with *ties* or HOOPS.

TIGHT-FITTING BOLT A BRIGHT BOLT.

TILE A thin slab used for covering a wall or floor. It may be made of unglazed or glazed ceramics (*see* CERAMIC MOSAIC and QUARRY TILE), of natural stone, concrete, asbestos cement, or various plastics (notably PVC).

TILT-UP CONSTRUCTION A method of precast concrete construction in which members are cast horizontally in a location adjacent to their final position, and tilted into place after removal of the moulds.

TILTING MIXER A small BATCH MIXER for concrete or mortar, which discharges its contents by tilting the entire drum.

TIMBER *See* LUMBER.

TIMBER CONNECTOR A device for connecting pieces of timber, other than a nail or a screw. The term includes SPLIT-RING CONNECTORS, shear-plate connectors, and toothed plate connectors, but not GANG-NAILS and NAIL PLATES.

TIMBER FRAME *See* BALLOON FRAME, BRICK VENEER, HALF-TIMBERED and SKELETON CONSTRUCTION.

TIMBRE The quality of a sound which distinguishes one instrument from another, and one voice from another. It derives from the particular combination and relative strength of the HARMONIC OVERTONES. *Tone colour* is the effect produced by a combination of timbres, *i.e.* of instruments and/or voices in a particular musical composition.

TIME, CRASHED or NORMAL *See* CRASH TIME.

TIME-SHARING A method of operating a large digital computer, whereby the CENTRAL PROCESSOR UNIT is combined with numerous input–output devices to make the optimal use of the central unit. The input–output devices may be widely separated, and connected to the central unit by telephone. If the capacity of the central unit is exceeded, some of the input–output devices are kept waiting for a short time.

TIME SWITCH A switch controlled by an electrical clock, which opens and closes a circuit at a predetermined time.

TIN A white metallic element, once widely used for table ware and other utensils (*pewter*), and one of the constituents of BRONZE; now mainly used

276

as a protective coating for steel in *tin-plate*. Its chemical symbol is Sn, its atomic number is 50, its atomic weight is 118·7, its specific gravity is 7·3, its melting point is 232°C, and its valency is 2 or 4.

TIN ROOF Literally a roof covered with *tin plated* steel sheet. In practice it usually means a roof covered with GALVANISED sheets (which are zinc plated).

TINTS Coloured pigments softened by white.

TL Transmission loss.

TNO Nederlands Centrale Organisatie voor Toegepast Natuurwetenschappelijk Onderzoek, the Dutch Government Organisation for Applied Scientific Research, which has several divisions interested in Building Research.

TOE *See* CANTILEVER RETAINING WALL.

TOLERANCE The permitted variation from a given dimension. It is of particular importance when components are factory produced by different manufacturers, since compliance with the specified tolerance is essential if the parts are to be fitted without cutting or filling gaps on the site. A tolerance may be negative (as for a partition to fit between two existing walls) or positive (as for a door frame to fit a given door), or both positive and negative (if there is no definite restriction either way). The *limits of size* are the two extreme sizes between which the actual size must lie, and the difference between them is the tolerance. *See also* MODULE.

TOLERANCE OF HIGH TEMPERATURE *See* COMFORT ZONE.

TOLERANCE OF NOISE *See* DAMAGE RISK CRITERION and BACKGROUND NOISE.

TON 1 long ton = 2240 lb; 1 short ton = 2000 lb. 1 metric ton (tonne) = 1000 kg = 0·984 long tons = 1·102 short tons. In the USA 'ton' usually means a short ton, but in most other English-speaking countries it commonly means a long ton. The metric ton is usually spelled 'tonne' in English, to distinguish it from the other two.

TON OF REFRIGERATION The cooling effect obtained when 1 (short) ton of ice at 32°F melts to water at 32°F in 24 hours. It is the unit heat flow rate used in designing air conditioning plants. 1 ton of refrigeration = 12 000 BThU per hour = 3517 watts (or joules per second).

TONE *See* HARMONIC.

TONE COLOUR *See* TIMBRE.

TONGUE AND GROOVE JOINT (T & G JOINT) A joint in timber, and also in precast concrete piles, with projecting and grooved edges which provide a sliding fit. Floor boards, in particular, are frequently made with T & G joints, to allow the timber to move with change in moisture content while maintaining a satisfactory joint (Fig. 99).

TOP-HAT SECTION A light-gauge metal section shaped ⎍.

TOP-HAT STRUCTURE A tall building frame with a stiffened upper floor. This reduces the cantilever

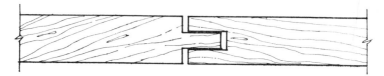

FIG. 99.   Tongue and groove joint.

deflection of the frame under lateral loading, and increases its strength, with consequent saving of structural material.

TOP-HUNG WINDOW   *See* CASE-MENT WINDOW.

TOPPING   A layer of high-quality concrete placed to form a floor surface on a concrete base. *See also* GRANO-LITHIC CONCRETE and TERRAZZO.

TOPSOIL   The layer of soil which by its HUMUS content supports vegetation. It is valuable for agriculture and gardening, but must usually be removed before the foundation of a building is put down.

TORQUE   *See* TWISTING MOMENT.

TORR   Unit used in vacuum technology, equal to 1 mm of mercury.

TORSION BUCKLING   Buckling of a column through rotation due to inadequate elastic stability.

TORSIONAL MOMENT   *See* TWISTING MOMENT.

TORSIONAL RIGIDITY   A measure of the stiffness of a member in resisting torsion. It is usually taken as the product GJ, where G is the MODULUS OF RIGIDITY and J is the POLAR MOMENT OF INERTIA.

TORUS   (*a*) A surface or solid generated by the revolution of a circle or other conic section about any axis, *e.g.* a solid ring of circular or elliptical section. (*b*) A large convex moulding of approximately semi-circular section, used especially at the base of a column.

TOUGHENED GLASS   *See* SAFETY GLASS.

TOUGHNESS   The ability to resist fracture by shock or impact.

TOWER CRANE   A crane with a swinging horizontal jib on top of a tower, whose height can be raised as the height of the building increases. *See also* DERRICK and GANTRY CRANE.

TRABEATED   In traditional construction, spanning with stone lintels (*i.e.* beams), as opposed to ARCUATED construction.

TRACERY   Delicate interlacing of lines or ribs, as for example in the upper part of a Gothic window.

TRADITIONAL CONSTRUCTION   *See* VERNACULAR CONSTRUCTION.

TRAJECTORY, STRESS   *See* ISO-STATIC LINE.

TRAMMEL   A beam compass used for scribing unusually large circular arcs.

TRANSDUCER A device for transforming mechanical vibrations or sound waves into electrical or magnetic energy (or vice versa).

TRANSEPT The transverse part of a cruciform church, as opposed to the NAVE.

TRANSFER BOND The bond stress resulting from the transfer of stress from a pre-tensioned tendon to the concrete. The diameter of the tendon is greatly reduced by the (tensile) prestress. Outside the concrete member, where the steel stress drops to nil, the full diameter of the wire is recovered *after transfer*, and it is gradually reduced as the tendon enters the concrete member. This sets up a high contact pressure between the concrete and steel, which in turn generates high frictional forces. These provide the necessary bond to anchor the end of a pre-tensioned tendon.

TRANSFER LENGTH The distance at the end of a pre-tensioned tendon necessary to develop the full tensile stress in the tendon. It is built up by TRANSFER BOND from the end. Also called *transmission length*.

TRANSFER OF PRESTRESS The process of transferring the anchorage of the prestress in the TENDONS from the POST-TENSIONING jacks or from a PRE-TENSIONING bed to the concrete member.

TRANSFORMATION TEMPERA-TURE Temperature at which one PHASE of an alloy system changes to another.

TRANSFORMED SECTION A hypothetical section of one material which has the same elastic properties as a composite section of two materials. It is a device for simplifying calculations for composite materials.

TRANSFORMER An electrical apparatus for converting energy from one voltage to another, either up or down. In a large building electrical power is received at a high voltage, and in a *sub-station* this is transformed to the standard voltage used in the building (normally 240 volts in Europe and Australia, 120 volts in America).

TRANSISTOR An electrical device which utilises SEMICONDUCTORS for purposes previously performed by electron discharge tubes, such as rectifying or amplifying an electrical current.

TRANSIT (*a*) The apparent passage of a heavenly body across the meridian of a place, due to the earth's daily revolution, *i.e.* the moment when it reaches its culmination or highest point. The sun's transit is *apparent noon*. (*b*) A TRANSIT THEODO-LITE.

TRANSIT THEODOLITE (*frequently abbreviated to transit*) A THEODOLITE which can be completely rotated about its horizontal axis. Virtually every modern instrument is designed to be able to do so.

TRANSITION TEMPERATURE OF STEEL The temperature at which the failure of a given type of steel changes from ductile to BRITTLE.

TRANSLATION Linear movement of a point in space without *rotation*.

TRANSLATION, SURFACE OF *See* SURFACE OF TRANSLATION.

TRANSLUCENT CONCRETE *See* BÉTON TRANSLUCIDE and GLASS-CONCRETE CONSTRUCTION.

TRANSLUCENT GLASS Glass which has been patterned so that it is *not* transparent.

TRANSMISSION *See* ABSORPTION (*b*).

TRANSMISSION FACTOR In lighting, the ratio of the LUMINOUS FLUX transmitted by a body to that which it received. The logarithm to the base 10 of the reciprocal of the transmission factor is called the *optical density*.

TRANSMISSION LENGTH *See* TRANSFER LENGTH.

TRANSMISSION LOSS Loss of sound pressure through acoustic barriers and mufflers. It is measured in DECIBELS.

TRANSMITTANCE, THERMAL *See* THERMAL TRANSMITTANCE.

TRANSOM *See* MULLION.

TRANSPARENCY *See* OPACITY.

TRANSVERSE FRAME *See* EDGE BEAM.

TRANSVERSE LOADING Loading perpendicular to a structural member, *e.g.* vertical loading on a horizontal beam.

TRANSVERSE REINFORCE-MENT *See* LATERAL REINFORCE-MENT.

TRAP A bend or dip in a SOIL DRAIN, so arranged that it is always full of water and provides a *water seal*, which prevents odours from entering the building.

TRAP DOOR A door, flush with the surface, in a floor, roof or ceiling, or in the stage of a theatre.

TRAPEZIUM A QUADRILATERAL with two parallel sides. If the lengths of the two parallel sides are y_1 and y_2, and the height of the trapezium, at right angles to them, is x, then the area of the trapezium is $\frac{1}{2}x(y_1 + y_2)$. This is the basis of the TRAPEZOIDAL RULE.

TRAPEZOIDAL RULE A rule for the evaluation of an irregular area, or for graphical integration. Let the area be divided into a number of parallel strips of width x. The lengths of the boundary ordinates, or separating strips, are measured and these are $y_0, y_1, \ldots y_{n-1}$, and y_n. The area of the figures is then

$$x(\tfrac{1}{2}y_0 + y_1 + y_2 + \cdots + y_{n-1} + \tfrac{1}{2}y_n)$$

This calculation is quicker, but less accurate, than SIMPSON'S RULE.

TRASS A natural POZZOLANA of volcanic origin found in Germany near the River Rhine.

TRAVELLER GANTRY A stationary gantry which carries a traveller, *i.e.* a hoist moving on rails across the top.

TRAVELLING GANTRY A gantry built on wheels so that it can travel.

TREAD The level part of a step in a staircase. Also the horizontal distance between one RISER and the next, exclusive of the nosing.

TRECENTO   *See* QUATTROCENTO.

TREFOIL   *See* FOIL.

TREMIE   A hopper with a pipe at the bottom, used for placing concrete under water.

TRIANGLE   *See* POLYGON.

TRIANGLE OF FORCES   *See* PARALLELOGRAM OF FORCES.

TRIANGULATION   A method used in the design of plane and space frames to ensure that they are isostatic. According to MÖBIUS' LAW, a truss whose members are all arranged in the form of adjacent triangles, is statically determinate.

TRIAXIAL COMPRESSION TEST A test on a sample (normally soil) contained in a rubber bag surrounded by liquid which exerts lateral pressure in two perpendicular directions. The vertical pressure (in the third direction) is applied by a piston. Unlike the UNCONFINED COMPRESSION TEST, it can be used on cohesionless soils. The triaxial pressures can be adjusted to simulate those in a foundation. Lateral pressure increases the load-bearing capacity of soil. *See* COULOMB'S EQUATION.

TRIBOPHYSICS   The physics of friction.

TRICALCIUM SILICATE   One of the principal components of PORTLAND CEMENT. Its chemical composition is $3CaO.SiO_2$, or C_3S in the notation used by cement chemists. It is the main constituent of the component named *Alite* by Tornebohm in 1897, before the chemical composition of cement had been properly established.

TRICLINIUM   A couch running around three sides of a table, on which ancient Romans reclined for meals. Hence a Roman dining room.

TRIFORIUM   The space formed between the sloping roof over the aisle of a large Gothic church and the vaulting over the aisle. It is usually an arcaded passage facing on the nave, just below the CLEARSTOREY windows.

TRIGONOMETRY   The mathematics of CIRCULAR FUNCTIONS. *Plane trigonometry* deals with triangles drawn on a plane surface. *Spherical trigonometry* deals with triangles drawn on the surface of a sphere.

TRILLION   *See* BILLION.

TRIM   (*a*) The edging of an opening in a colour or material different from that of the wall surface. (*b*) A generic term for architraves, skirtings, etc. which cover open joints. (*c*) A generic term for all visible interior finishing work, including hinges and locks.

TRIMETRIC PROJECTION   *See* PROJECTION.

TRIP COIL   A SOLENOID-operated circuit breaker.

TRIPOD   A three-legged support.

TRUE VOLUME OF A POROUS MATERIAL   The volume excluding both the open and the closed pores. *See also* BULK VOLUME.

TRUNNION AXIS The horizontal axis of rotation of a THEODOLITE.

TRUSS *See* ROOF TRUSS.

TRUSSED BEAM or PURLIN A beam or purlin stiffened with a tie rod.

TRUSSED RAFTER A triangulated rafter in a roof truss. The Fink truss is a large-span example (Fig. 38).

TRY SQUARE A gauge consisting of two pieces of metal accurately set at right angles, used for laying out work, and for testing finished work for squareness.

TUBULAR SCAFFOLDING Scaffolding built up from galvanised steel or aluminium tubes with clamps. The tubes are usually of 2 in (5 cm) external diameter.

TUCK POINTING An obsolete method of emphasising the joints in brick and natural stone by grooving the mortar to form a *tuck*, which is then filled with mortar or putty of a distinctive colour, usually white or black.

TUDOR ARCHITECTURE The architectural style of England during the reign of the Tudors, particularly Henry VIII and Elizabeth I. It marks the transition from late GOTHIC to the RENAISSANCE. Windows with mullions and transoms, and HALF-TIMBERING surviving from the Gothic period, were mixed with the fashionable Italian Renaissance forms.

TUMBLER SWITCH A switch operated by pushing a short lever up or down.

TUNG OIL An oil obtained from the seeds of *Aleurites cordata,* used in the manufacture of paints, varnishes and enamels. It has excellent water-resistance. Since the trees were found mainly in China and Japan, it is also called *China wood oil.*

TUNGSTEN A metallic element used as an alloy for hard steels, and as a filament in electric lamps. Its chemical symbol is W, its atomic number is 74, its atomic weight is 184, its specific gravity is 19·3, and its melting point is 3300°C.

TURBIDIMETER A device for the PARTICLE-SIZE analysis of finely divided material. Successive measurements are taken of the turbidity of a suspension of the fluid. *See* STOKES' LAW.

TURBINE A rotating prime mover, as distinct from a reciprocating engine.

TURBULENT FLOW *See* STREAMLINE FLOW.

TURNBUCKLE A coupling between the ends of two rods, one having a left-hand and the other a right-hand thread. Hence rotation of the buckle adjusts the tension in the rods. A simpler type has only one right-hand thread, and a swivel at the other end.

TURNED BOLT *See* BRIGHT BOLT.

TURNING Making an object on a lathe.

TURPENTINE A thinner obtained by distilling the sap of certain pine trees. *See also* WHITE SPIRIT.

TUSCAN ORDER One of the five Roman ORDERS. It is distinguished by the plain column, and the absence of decorative detail.

TWISTING MOMENT The moment of all the forces acting on a member about its polar axis, *i.e.* the moment normal to the section. The BENDING MOMENT is the moment about an axis in the plane of the section. The twisting moment is also called *torsional moment* or *torque*. *See also* MEMBRANE ANALOGY and SANDHEAP ANALOGY.

TWO-HINGE FRAME *See* PORTAL.

TWO-PART ADHESIVE A synthetic glue supplied in two parts, a powdered resin and an ACCELERATOR, which are mixed only just before use.

TWO-PIN FRAME *See* PORTAL and Fig. 73.

TWO-WAY REINFORCEMENT IN CONCRETE SLABS Reinforcement arranged in bands or bars at right angles to one another. This is the normal arrangement in two-way slabs, flat slabs and flat plates, *four-way reinforcement* now being obsolete.

TWO-WAY SLAB A slab spanning between beams in two directions, as distinct from a one-way slab, a flat SLAB or a FLAT PLATE (*see* Fig. 58).

U

U-BOLT A steel bar bent into a U-shape, and fitted with screw threads and nuts on each end. Also called a *clip*.

UF Urea formaldehyde.

UIA Union Internationale des Architects, Paris.

ULTIMATE STRENGTH The highest load which a test piece can sustain before breaking. It is not necessarily the basis for the ULTIMATE STRENGTH THEORY, *e.g.* the yield or proof stress of steel is used in limit design, not its ultimate strength.

ULTIMATE STRENGTH THEORY A theory for the strength of structures based on the plastic, or assumedly plastic, behaviour of engineering materials. The term is applied both to LIMIT DESIGN (which is a true ultimate strength theory assuming the formation of PLASTIC HINGES) and to conventional elastic design with PLASTIC STRESS DISTRIBUTION at the sections of maximum bending moment. (*See also* STRESS–STRAIN DIAGRAM OF MILD STEEL and OF CONCRETE, and Figs. 57 and 103.)

ULTRAMARINE A characteristic blue pigment, made by grinding LAPIS LAZULI. It is very expensive and most ultramarine sold today is an artificial pigment imitating the colour.

ULTRASONIC WAVE Mechanical vibration in a solid, liquid or gas which has a frequency higher than that of audible sound, and a speed equal to that of sound in the same medium. Waves of air pressure of high frequency, beyond the upper limit of the human audibility (about 18 000 Hz) can be generated with ultrasonic sirens or by electrical means (*see also* PIEZO-ELECTRIC EFFECT). Practical frequencies are generally above 1 000 000 Hz. Ultrasonic waves are reflected and refracted at the boundaries of a solid, and they can therefore be used for the NON-DESTRUCTIVE detection of cracks and flaws in concrete structures, in metal castings etc., using suitable equipment for recording the reflections received from ultrasonic impulses. By the same process, ultrasonic waves can be used to determine the thickness of a piece of material when only one surface is accessible.

ULTRA-VIOLET RADIATION
Electromagnetic radiation with wavelengths shorter than 3900 Å (390 nm), *i.e.* beyond the violet end of visible light. It forms part of the radiation received from the sun, and it has a destructive effect on some materials, including a number of plastics. Hence ultra-violet radiation cycles are included in some WEATHER-OMETERS.

UMBER *See* IRON OXIDE.

UMBRELLA SHELL A shell roof formed by four HYPAR shells (Fig.

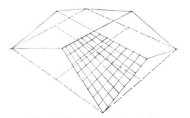

FIG. 100.   Umbrella shell.

100) or by other suitable arrangements of hypar shells.

UNCONFINED COMPRESSION TEST A compressive test on a sample of a material without lateral restraint, which is normal practice for structural materials. The term is mainly used for a soil test, to distinguish it from a TRIAXIAL COMPRESSION TEST. The test is of little use for COHESIONLESS soils, because their compressive strength without lateral confinement is too low for accurate measurement. However, it is a useful test for highly cohesive soils, since it is easily conducted on undisturbed BOREHOLE SAMPLES, and much cheaper than a triaxial test. The cohesion of the soil (*see* COULOMB'S EQUATION) is half the unconfined compressive strength.

UNDERLINING FELT *See* SARKING.

UNDERPIN To provide a new foundation for a wall or column in an existing building without removing the superstructure.

UNDER-REINFORCED *See* BALANCED DESIGN.

UNDISTURBED SAMPLES *See* BOREHOLE SAMPLES.

UNIDIRECTIONAL MICROPHONE A microphone which picks up sound from one direction only.

UNIFORM GRAVEL or **SAND** Material retained between two adjacent sieves so that all particles are of approximately the same size.

UNIT STRESS A term generally synonymous with STRESS.

UNIT WEIGHT The density of a material, *i.e.* its weight per unit volume.

UNIVERSAL TESTING MACHINE A TESTING MACHINE capable of exerting tensile, compressive or flexural forces on a specimen under test, as opposed to a machine designed for one kind of test only.

UNIVERSE In statistical terminology, a specified group, *e.g.* of persons or concrete cylinders. A universe may have several POPULATIONS associated with it.

UPLIFT An upward force. *See also* FROST HEAVE *and* QUICKSAND.

UPRIGHTS Vertical members, usually of timber, *e.g.* the sides of a door frame.

UPSTAND BEAM A beam projecting above an adjoining slab.

UREA FORMALDEHYDE *See* FORMALDEHYDE.

USASI United States of America Standards Institute, now called American National Standards Institute Inc.

US CUSTOMARY UNITS The system of FPS units as defined by the National Bureau of Standards, Washington. There are some differences between the customary units of the US and those of the United Kingdom, Canada and Australia (*e.g.* in the GALLON and the TON).

UTIBTP Union Technique Interfédérale du Bâtiment et des Travaux Publics, Paris.

U-TIE A U-shaped WALL TIE.

U-TUBE *See* MANOMETER and Fig. 101.

U-VALUE *See* THERMAL TRANSMITTANCE.

V

VACUUM An empty space. A perfect vacuum is unobtainable on earth, and a pressure of 10^{-6} mm of mercury is considered a good vacuum.

VACUUM CONCRETE Concrete from which water is extracted with a vacuum mat before hardening occurs. The concrete has a water content adequate for placing in the formwork, but this is subsequently reduced to give a higher strength concrete, *see* ABRAMS' LAW.

VACUUM LIFTING Raising an object, *e.g.* a precast concrete panel, with a suction attachment. It allows uniform distribution of the lifting force, but the cost of the capital equipment is appreciable.

VALENCY The combining power of an atom. A hydrogen atom is univalent, and it can combine with one other hydrogen atom (to form hydrogen gas) or one chlorine atom, which is also univalent (to form hydrochloric acid). Oxygen, which is bivalent, combines with two hydrogen atoms (to form water). Carbon which has a valency of four, combines with four hydrogen atoms (to form methane). Some elements have more than one valency, *e.g.* sulphur may have a valency of 2, 4 or 6.

VALLEY The intersection between two sloping surfaces of a roof, the opposite of *hip*. A valley must itself have a slope so that it can discharge rainwater.

VALUE OF A COLOUR *See* COLOUR.

VALVE (*d*) A device for regulating the flow of a liquid or gas in a pipeline. (*b*) A *thermionic* tube used as an electronic rectifier or amplifier.

VANE ANEMOMETER *See* ANEMOMETER.

VANE TEST A SHEAR TEST for determining the strength of soil on the site. A four-bladed vane is inserted into the soil at the foot of a *borehole*. It is rotated by a rod at the surface and the force is measured when the soil shears.

VANISHING POINT *See* PROJECTION (*perspective*).

VAPOUR A gas which is at a temperature below its critical temperature, and can therefore be liquefied by a suitable increase in pressure.

VAPOUR BARRIER An airtight skin, *e.g.* of aluminium, which prevents moisture from the warm damp air in a building from passing into and condensing within a colder space. It is particularly needed in cool climates, at night and in winter, to protect the insulation from filling with condensation water or ice, and it is therefore placed on the inner,

warm face of the insulation. A barrier which satisfactorily stops the ingress of liquid water, is not necessarily sufficient to stop water vapour.

VAPOUR PRESSURE The pressure exerted by a vapour. The term is generally taken as synonymous with *saturated vapour pressure*, which is the pressure of a vapour in contact with its liquid form. The saturated vapour pressure falls as the temperature falls. *See* SATURATED AIR.

VARIANCE The square of the STANDARD DEVIATION, *i.e.* the average of the squares of the *deviations* of a number of observations of a quantity from their mean value.

VARNISH A resin dissolved in oil or spirit, which dries to a brilliant, thin, protective film. Varnish may be put on unpainted wood, put over paint to increase its gloss, or mixed with paint. The term LACQUER is usually reserved for finishes based on cellulose compounds. *See also* PAINT.

VASARI'S LIVES *Vite de' piu eccelenti architetti, pittori e scultori italiani*, published by Giorgio Vasari in 1550. It exerted great influence on architectural taste, and was highly regarded by Michelangelo. Vasari was the architect of the Uffizi palace in Florence.

VAULT (*a*) An arched masonry or concrete roof. (*b*) A room or passage with an arched masonry roof. (*c*) A room below ground, of massive construction, not necessarily with a vaulted roof. (*d*) A safe room for the storage of valuables, usually below ground, but rarely with a vaulted ceiling.

VECTOR A quantity which has magnitude as well as direction. It may be represented by a straight line drawn from a point in a given direction for a given distance. A *scalar* has magnitude, but no direction. In vector algebra, vectors are usually distinguished from scalars by the use of **bold face**.

VEHICLE The liquid part of a paint, as opposed to the pigment. Also called *medium*.

VELOCITY OF LIGHT The velocity of electromagnetic radiation, which includes light, is a universal constant. Its value in a vacuum is 299 796 kilometres per second (186 293 miles per second).

VELOCITY OF SOUND The maximum velocity of pressure waves is the same, whether the frequency is audible or not. The velocity of sound in air is 344 metres per second (1130 feet per second).

VELOCITY RATIO *See* MECHANICAL ADVANTAGE.

VENEER *See* BRICK VENEER and WOOD VENEER.

VENETIAN BLIND A window blind composed of numerous thin slats which can be raised and lowered with ease. The slats (formerly of timber, now usually of plastic) can be rotated by pulling a cord, to admit varying amounts of light and air. A Venetian blind is normally placed inside the window to protect it from the weather, and it is thus only partially effective in excluding *thermal radiation*. In air-conditioned buildings with double glazing the blinds are therefore fixed between the inner and outer panes of glass.

VENETIAN RED *See* IRON OXIDE.

VENTILATING BRICK *See* AIR BRICK.

VENTURI TUBE A constriction inserted in a line of piping, together with a MANOMETER to measure the loss of pressure over the convergent part of the constriction (Fig. 101). From this the rate of flow can be calculated by BERNOULLI'S THEOREM.

VERDIGRIS The green basic copper carbonate formed on copper roofs and statues exposed to the atmosphere. Although a corrosion product, it gives a highly esteemed *patina* if properly controlled. Verdigris is poisonous.

VERGE The edge of a sloping roof which overhangs a gable (Fig. 45).

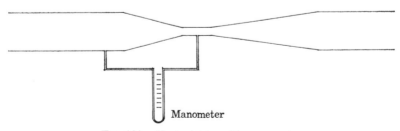

Manometer

FIG. 101. Venturi tube with manometer.

VERGE BOARD *See* BARGE BOARD.

VERMICULATION Decoration of masonry with shallow, irregular channels, resembling worm tracks; it is one form of RUSTICATION.

VERMICULITE A generic name for hydrous silicates of aluminium, magnesium and iron, which occur as minerals in plate form, and show marked exfoliation on heating. The term often implies exfoliated vermiculite, which is used for thermal insulation and fire protection, often as an aggregate in plaster or concrete.

VERMILION A brilliant red, slightly orange-coloured pigment, derived from cinnabar, which is mercuric sulphide (HgS). It is one of the traditional pigments, but is now too expensive for general use.

VERNACULAR CONSTRUCTION Construction technique traditional to the region. It often makes the best possible use of locally available materials, and takes proper advantage of the local climate. While local traditions deserve careful study, they do not invariably represent the best possible solution. Some are based on superstitions, and some have been brought by conquerors in ancient times unchanged from a previous home region.

VERNIER A device for measuring length more accurately than is possible with an ordinary scale. It consists of a subsidiary (*vernier*) scale which slides alongside the main scale. This carries one additional division, *i.e.* ten vernier divisions equal nine divisions on the main scale. By noting which division on the vernier scale is exactly in line with the main scale, the measurement can be taken to one more decimal place than is shown in the main scale. *See also* MICROMETER.

VERRAZANO NARROWS BRIDGE *See* SPAN.

VERTEX *See* CROWN.

VERTICAL In line with the direction of the gravitational forces (dead loads). The *horizontal* direction is at right angles to the vertical.

VERTICAL SASH *See* SASH WINDOW.

VESTIBULE The enclosed, or partially enclosed space in front of the main entrance of an ancient Roman house. Hence a small entrance hall in a modern building.

VIADUCT Originally a road carried on masonry arches and high piers over a deep ravine. Now applied to any bridge which has relatively short spans and tall supports, instead of a single long span. *See also* AQUEDUCT.

VIBRATED CONCRETE Concrete compacted by vibration during and after placing. Since vibration helps to place a comparatively dry concrete with satisfactory compaction (*see* ABRAMS' LAW), it increases the effective concrete strength. On the other hand, displacement of the reinforcement is a possible danger. The vibrator can be fixed to the formwork, or an internal (or poker) vibrator can be immersed in the wet concrete.

VIBRATION *See* FORCED VIBRATION and EARTHQUAKE LOADING.

VIBRATION PICKUP An instrument for measuring the velocity or acceleration of a vibration as a function of frequency.

VICAT TEST *See* INITIAL SET.

VICKERS DIAMOND HARDNESS TEST An indentation hardness test employing a diamond with a 136° pyramid. The impression of the diagonal is converted into the hardness number from a table.

VIERENDEEL GIRDER A girder (named after a Belgian engineering professor) without diagonals, so that it can be used in walls which require openings for windows or doors. All the joints are rigid, so that this structure must be solved by moment distribution or some other hyperstatic design method. *Also called an open-frame girder*.

VILLA Originally a country mansion, together with its farm buildings. Now generally a suburban house of some size, standing in its own grounds.

VINYL *See* POLYVINYL.

VIOLLET-LE-DUC'S DICTIONARY *Dictionnaire raisonné de l'architecture française*, published by Eugene-Emmanuel Viollet-le-Duc between 1854 and 1868. It developed a rational explanation of Gothic structure, and drew analogies with the newly developed iron skeleton frame. All thrusts are conducted from the ribs to the flying buttresses and the buttresses, and the thin walls can be replaced by large openings. Analyses carried out in the twentieth century with the aid of the theory of shells have shown these arguments to be unduly conservative. The Dictionary had great influence on the *Gothic Revival*, and on the restorations of medieval buildings undertaken in the nineteenth and twentieth centuries. Viollet-le-Duc himself carried out numerous restorations, and it seems likely (in the absence of proper records of the original dimensions) that this resulted, in many cases, in a coarsening of the structure which brought it in line with the new theory. *See also* FUNCTIONALISM.

VISCOSITY Internal friction (due to cohesion) in fluids, or in solids with flow characteristics. *See* FLUIDITY and RHEOLOGY.

VISCOUS FLOW Continuous deformation over a period of time. It is generally proportional to the applied stress, and it may occur at quite low stresses. *See also* CREEP, PLASTIC DEFORMATION, NEWTON, MAXWELL and KELVIN BODIES.

VISIBLE SPECTRUM The visible range of electromagnetic RADIATION, ranging in wavelength from 7600 to 3900 Å or 760 to 390 nm (violet to red light). *See also* ULTRA-VIOLET and INFRA-RED RADIATION.

VISTA A view seen through an avenue of trees, or some other narrow opening.

VISUAL ACUITY The capacity of seeing distinctly objects very close together.

VITREOUS ENAMEL Hard, impervious and weather-resistant finish, also called *porcelain enamel*, applied to steel and aluminium sheet, particularly for CURTAIN WALLS. The process consists of fusing a thin coating of glass to the metal base at temperatures above 1500°F (800°C) for steel and 1000°F (550°C) for aluminium. At these temperatures the metal and the glass combine, producing a product with the surface hardness of glass and the strength of metal. Thicker layers of vitreous enamel, usually white, are used on steel and cast iron for baths, and occasionally washbasins. The thicker layers are necessary for abrasion resistance, but they are more easily chipped by a hard blow because of the brittleness of the enamel.

VITRIOL   (*a*) *Oil of vitriol.* Concentrated sulphuric acid (H_2SO_4). (*b*) *Blue vitriol.* Copper sulphate ($CuSO_4$. $5H_2O$). (*c*) *Green vitriol.* Ferrous sulphate ($FeSO_2$.$7H_2O$). (*d*) *White vitriol.* Zinc sulphate ($ZnSO_4$.$7H_2O$).

VITRUVIUS' TEN BOOKS *De architectura*, written by Marcus Vitruvius Pollo in the first century BC. It is the only complete treatise on architecture surviving from Antiquity, and it exercised enormous influence on the early RENAISSANCE. *See also* ALBERTI and FILARETE.

VOIDS The spaces between the particles of soil or concrete aggregate, whether occupied by air, or water or both. The *voids ratio* is the ratio of voids to solids in a sample. It is closely related to POROSITY.

VOLATILE Readily evaporating at room temperature.

VOLT (V) Unit of electrical potential, or electromotive force (*emf*), named after the eighteenth century Italian physicist A. G. A. Volta. It represents the potential difference against which one JOULE of work is done in the transfer of one COULOMB.

VOLTAGE Electromotive force measured in volts.

VOLTMETER Instrument for measuring the electromotive force (or electric potential) directly, calibrated in volts.

VORTEX A whirl of fluid.

VOUSSOIR A wedge-shaped stone or brick, used in the construction of an *arch*, or a horizontally curved wall.

VULCANISATION Treatment of rubber with sulphur to cross-link the elastomer chains.

W Chemical symbol for *tungsten* (wolframium).

WAFFLE SLAB *See* RIBBED SLAB.

WAGON VAULT A BARREL VAULT.

WAINSCOT Wood-panelling of the lower part of an interior wall, usually terminating with a DADO.

WALDRAM DIAGRAM A graphical solution to the trigonometric equations for SUNLIGHT PENETRATION, developed by P. J. Waldram in 1933.

WALK-UP APARTMENT HOUSE An apartment house without an elevator; generally four storeys or less.

WALL BOARD BUILDING BOARD suitable for interior walls.

WALL PAPER Decorative printed paper, usually sold in rolls, for sticking on a plastered wall. The wall requires a sealer, to prevent it from discolouring wall paper.

WALL PLATE A horizontal piece of timber, laid flat along the top of the wall at the level of the eaves; it carries the rafters.

WALL, RETAINING *See* RETAINING WALL.

WALL TIE A piece of metal built into the bed joints across the cavity of a CAVITY WALL. It may consist of a piece of twisted galvanised steel strip, or of a bar in the shape of a hoop, a U or a Z.

WARD LEONARD CONTROL A method of speed control for large electric motors, widely used for elevators. It consists of a shunt motor which drives a variable-voltage dc generator, and this in turn drives a dc motor, whose speed is thus infinitely variable, and whose direction of driving can be reversed.

WARMTH As an acoustic quality, the LIVENESS of the bass, or fullness of bass tone relative to that of the mid-frequency tone.

WARPING Distortion from a plane surface, particularly in timber; it may be caused by careless seasoning.

WARREN TRUSS An *isostatic* truss consisting of top and bottom chords connected only by diagonals without vertical members, as distinct from a HOWE and a PRATT TRUSS. Some of the diagonals are in tension and some are in compression.

WASHER A ring placed under a nut or a bolt head. *See also* SPRING WASHER and TAPER WASHER.

WASHING SODA Sodium carbonate (Na_2CO_3).

WASTE DISPOSAL UNIT An electrically operated rubbish grinder, which is placed adjacent to or below the kitchen sink. It can usually grind up kitchen waste, but not bottles, tins or newspapers. The ground material is washed into the ordinary kitchen drain.

WASTE PIPE The pipe which discharges liquid waste into the SOIL DRAIN.

WATER–CEMENT RATIO The ratio of the amount of water, excluding that absorbed by the concrete aggregate, to the amount of cement in concrete or mortar. It has a determining influence on the strength of concrete and mortar (*see* ABRAMS' LAW and Fig. 1). *See also* MIX PROPORTIONS.

WATER, COMBINED *See* COMBINED WATER.

WATER GAUGE A U-tube MANOMETER, filled with water.

WATER-GLASS A concentrated solution of *sodium* or *potassium silicate*. It is used for waterproofing brick, stone and concrete, and as a surface hardener for concrete floors.

WATER HAMMER A sudden very high pressure in a pipe, often indicated by a loud noise, caused by stopping the flow of water too rapidly.

WATER LEVEL A simple instrument for setting out levels on a building site. It consists of a transparent tube filled with water, with the two ends held up. The level of the water in both is the same, if there are no air locks in the tube.

WATER OF CAPILLARITY *See* CAPILLARY ACTION.

WATER PAINT Any paint which can be thinned with water. The term includes oil-bound or emulsion paints, whose binder is insoluble in water, but which can be thinned with water.

WATER SEAL The seal in the TRAP of a drain, which prevents odours from the sewer from entering the building.

WATER SOFTENER *See* HARD WATER.

WATER STAIN *See* SPIRIT STAIN.

WATER TABLE The level below which the ground is saturated with water.

WATER TANK *See* HOOP TENSION.

WATERPROOFING *See* SURFACE and INTEGRAL WATERPROOFING.

WATT (W) Unit of power, named after the Scottish eighteenth century inventor of the steam engine. It is specifically used for electrical power, and 1 watt is the energy expended by a current of 1 ampere across a potential difference of 1 volt.

WATT HOUR (Wh) The unit of electrical energy. It is the energy delivered by 1 watt in 1 hour. It is customary to use kilowatt hours; $1 \text{ kWh} = 1000 \text{ Wh} = 3 \cdot 6 \times 10^6$ JOULES $= 3 \cdot 6$ MJ.

WATTLE-AND-DAUB Infilling for the walls of HALF-TIMBERED houses, traditional in some parts of Europe. It consists of branches or thin lathes (*wattles*) plastered with clay (*daub*).

WAVELENGTH *See* FREQUENCY.

WAVY GRAIN *See* FIGURE.

WAX STAIN A semi-transparent pigment dispersed in BEESWAX, thinned with turpentine.

WBT Wet-bulb temperature.

WC Water closet.

WEAR TESTS *See* ABRASION.

WEATHER-STRUCK JOINT or **WEATHER JOINT** *See* STRUCK JOINT.

WEATHERBOARD *See* CLAPBOARD *and* WOOD SIDING.

WEATHERCOCK Originally a vane in the form of a cock which turns with its head to the wind, to indicate its direction; also applied to a weather vane of any other shape.

WEATHERING *See* WEATHEROMETER.

WEATHEROMETER A machine for determining the weather-resisting properties of materials (such as paints and plastics) by cycles imitating as closely as possible natural weathering conditions. Most machines employ ultra-violet light, high or low temperatures, and moisture. The result is not as accurate as placing the samples on an exposed site for natural weathering; however, since the cycles are speeded up, the result is obtained very much quicker. Weatherometers, like abrasion testing machines, are useful for comparing a new material with a similar material of known performance. Their reliability for predicting the performance of a completely new type of material is questionable.

WEB (*a*) The vertical part of a joist. (*b*) The plate connecting the flanges of a PLATE-GIRDER. (*c*) The RIB of a concrete T-beam.

WEBB STIFFENER *See* STIFFENER.

WEDGE THEORY A theory for the stability of a retaining wall, based on the weight of the wedge of soil which would slide forward if the wall failed. It was originally proposed by C. A. Coulomb in 1776 and revised by C. F. Jenkin in 1931.

WEEPHOLE A small hole left at the base of a retaining wall, cavity wall, window or curtain wall to allow accumulated condensation or other moisture to escape.

WEIGH BATCHER A batching plant for concrete in which all materials (except the water) are weighed. *See* MIX-PROPORTIONING.

WEIGHTING STATISTICAL DATA Multiplying the data by a factor, or weight. If the data are of unequal reliability, and if the difference between their trustworthiness and importance can be properly assessed, it is proper to multiply each value by a factor which assigns to it its proper weight in relation to the others. However, weighting always lays the observer open to the suspicion that he is 'cooking' the data to arrive at a preconceived conclusion.

WEISSENHOF *See* DEUTSCHER WERKBUND.

WELDING Uniting two pieces of metal by raising the temperature of the metal surfaces to a plastic or molten condition, with or without the addition of additional welding metal, and with or without the addition of pressure. SOLDERING and BRAZING are carried out at a lower temperature. *See* ARC WELDING, FUSION WELDING and RESISTANCE WELDING.

WELLPOINT DEWATERING Draining a volume of soil to be excavated by sinking wellpoints around it, and pumping the water from them. Wellpoints are usually tubes, approximately 2 in (50 mm) diameter, which are jetted into the soil. The method is particularly suitable for soils which do not contain too much fine material, when it may be much cheaper than SHEET PILING. *See also* ELECTRO-OSMOSIS.

WELSBACH MANTLE An INCANDESCENT mantle composed of cotton impregnated with thorium and cerium oxide, invented by Welsbach in 1885 to improve the performance of *gaslight*.

WERKBUND *See* DEUTSCHER WERKBUND.

WET-AND-DRY BULB THERMO-METER A *psychrometer*, or instrument for measuring the relative humidity in the atmosphere. It consists of an ordinary thermometer (*dry-bulb thermometer*), which records air temperature (*dry-bulb temperature*) in the ordinary way; and another thermometer (*wet-bulb thermometer*) whose bulb is wrapped in a damp wick dipping into water. The latter records the *wet-bulb temperature*, which is lower because of the cooling effect of the wick due to evaporation. It depends on the RELATIVE HUMIDITY of the air, and tables are provided which give the relative humidity in terms of the wet-bulb and the dry-bulb temperature. *See also* ASSMAN PSYCHROMETER and WHIRLING PSYCHROMETER.

WET-BULB GLOBE THERMO-METER INDEX (WBGT) Criterion for determining the COMFORT ZONE evolved by Yaglou and Minard in 1957 for the control of heat casualties at military training centres in the USA. It takes into account the temperature and humidity of the air, and radiation from the sun and the terrain, and it makes some allowance for wind speed.

WET-BULB THERMOMETER *See* WET-AND-DRY-BULB THERMO-METER.

WET CONSTRUCTION Conventional construction, which relies for jointing on wet concrete, mortar or plaster, as opposed to DRY CONSTRUCTION (*see also* Fig. 94).

WET MIX A concrete mix containing too much water. To the untrained observer it may look like a better concrete than a correctly propor-tioned mix, and it is also easier to place and to finish; however, its strength is lower, and it may produce LAITANCE. *See* ABRAMS' LAW.

WETTING AGENT A substance which lowers the surface tension of liquids, and thus facilitates the wetting of solid surfaces and the penetration of liquids into capillaries.

WF Wide-flange section, used in structural steelwork.

WHEATSTONE BRIDGE An apparatus for measuring electrical resistance by the zero method, comprising two parallel resistance branches, each branch consisting of two resistances in series. It is employed in conjunction with THERMOCOUPLES and ELECTRIC RESISTANCE STRAIN GAUGES (Fig. 102).

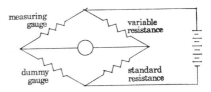

FIG. 102. Wheatstone bridge circuit for use with electric resistance strain gauges.

WHIRLING PSYCHROMETER A psychrometer in which the WET AND DRY BULB THERMOMETERS are mounted on a handle, so that they can be rotated in the air to give an approximately standardised rate of ventilation.

WHISPERING GALLERY A room shaped like a *sound mirror*, so that faint sounds can be heard across extraordinary distances. W. C. Sabine states that 'it is probably that all existing whispering galleries . . . are accidents; it is equally certain that all could have been predetermined without difficulty, and like most accidents could have been improved upon'. The absence of whispering galleries in modern buildings is due partly to a lack of interest in producing freaks, and partly to the declining popularity of domes.

WHITE ANT *See* TERMITE.

WHITE CEMENT A pure white PORTLAND CEMENT. Since the grey colour of Portland cement comes from impurities, a white cement requires raw materials of low iron content, or firing of the clinker by a reducing flame. The resulting material is much more expensive than ordinary (grey) Portland cement. It is used only for decorative surface finishes, and as the basis for the lighter COLOURED CEMENTS.

WHITE GOLD Gold alloyed with palladium, nickel or zinc.

WHITE LEAD An opaque white pigment, used extensively as an undercoat for exterior paint, and for pottery glazes. Because it is poisonous (*see* LITHOPONE) it is now rarely used for finishing coats. It consists of basic lead carbonate ($2PbCO_3.Pb(OH)_2$).

WHITE METAL A general term

covering alloys based on antimony, lead or tin, used to reduce friction.

WHITE NOISE A mixture of noise of all FREQUENCIES, in the same way as white light is a mixture of light of all frequencies.

WHITE SPIRIT A THINNER for oil paint, distilled from petroleum at 150 to 200°C. It is frequently used as a substitute for TURPENTINE.

WHITEWASH A cheap finish for external walls formed by soaking QUICKLIME in an excess of water. A binder, such as casein is sometimes added. Also called *limewash*.

WHITING Crushed chalk ($CaCO_3$), probably the cheapest white pigment. Also called *Paris white*. It is used in distemper, putty, and as an extender for other pigments.

WHITNEY STRESS BLOCK A pseudo-plastic stress distribution for the design of reinforced concrete by the ULTIMATE STRENGTH THEORY. C. S. Whitney observed in 1937 that the exact shape of the stress block (Fig. 103), is not important, provided that its area and the location of its centroid (which determines the length of the lever arm) give the correct ultimate resistance moment. The use of a rectangular stress block, which

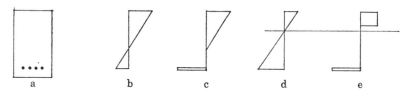

FIG. 103.   Whitney stress block: (a) reinforced concrete cross section; (b) strain distribution in elastic range; (c) stress distribution in elastic range; (d) strain distribution at ultimate load; (e) stress distribution at ultimate load.

yields simple equations, does not in itself imply fully PLASTIC STRESS DISTRIBUTION. *See also* STRESS–STRAIN DIAGRAM OF CONCRETE, and Figs. 91 and 92.

WILLIOT–MOHR DIAGRAM A graphical method for determining the deflection of pin-jointed trusses.

WIND LOAD The (positive or negative) force of the wind acting on a building. Wind applies (positive) pressure to the windward side of buildings, and (negative) suction to the leeward side (Fig. 104). The

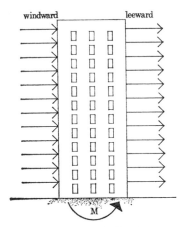

windward   leeward

FIG. 104. Wind load.

magnitude of the wind forces depends on the location of the building. They are higher in the tropics than in the temperate zone. They are higher near the sea than inland. They are higher on the top of hills, and lower when sheltered by surrounding buildings. In particular, they are higher above the ground, so that tall buildings have high wind loads not merely because they have bigger surfaces. From fluid mechanics and the density of the air,

the horizontal wind pressure (in lb/ft^2) is approximately

$$p = \frac{v^2}{400}$$

where v is the wind velocity in miles per hour; it is usually specified in a building code, but it can also be obtained from ground observation. In SI units, $p = 0{\cdot}6 \ v^2$ (approx.), where p is in N/m^2 and v is in m/sec. The increase of wind velocity with height is obtained from the empirical formula

$$\frac{v_1}{v_2} = \left(\frac{h_1}{h_2}\right)^{\frac{1}{7}}$$

See also WIND PRESSURES.

WIND PRESSURE The force exerted by the wind blowing against a building, or any portion of it. It is sometimes separated into positive pressure, exerted on a windward vertical face, and negative pressure, or *suction*, exerted on a leeward vertical face. A flat roof is invariably subject to suction, irrespective of the direction of the wind. A sloping roof also is always subject to suction if the wind blows parallel to the ridge. It it blows perpendicular to the ridge, then the windward side is subject to positive wind pressure only if the slope of the roof is more than 30° (Fig. 105). *See also* WIND LOAD.

WIND TUNNEL An apparatus for producing a steady stream of air past a MODEL for AERODYNAMIC investigations. The purpose may be to study the ventilation inside the building, and the effect which this has on its thermal properties; or the wind pressure acting on a building; or the vibrations produced by wind on a tall, flexible structure; or the eddies which may be produced, for example

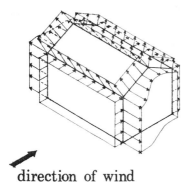

direction of wind

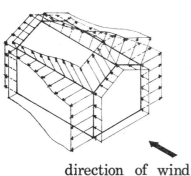

direction of wind

FIG. 105. Wind pressure.

in circular courtyards, and create an unpleasant environment. In studying the effect of wind *around* buildings it is necessary to model the buildings in the path of the wind. The large buildings need to be represented with some accuracy; but the effect of the small buildings can be represented by a roughening of the surface which produces the correct BOUNDARY LAYER. A wind tunnel which correctly models the boundary layer thus needs additional length in front of the model of the building.

WINDOW *See* CASEMENT WINDOW, SASH WINDOW, DEAD LIGHT, SKYLIGHT and LANTERN.

WINDOW EFFICIENCY RATIO *See* DAYLIGHT FACTOR.

WINDOW GLASS *See* SHEET GLASS.

WINDOW SHUTTERS *See* SHUTTERS.

WINDOW WALL An outside wall consisting largely of glass.

WINDOW WEIGHTS *See* SASH WINDOW.

WING Part of a building projecting from one side of the main body.

WIRE GAUGE *See* STANDARD WIRE GAUGE.

WIRE LATH *See* EXPANDED METAL.

WIRE MESH *See* MESH REINFORCEMENT.

WIRE ROPE *See* SUSPENSION CABLE *and* TENDON.

WIRECUT BRICK Bricks shaped by extrusion and then cut to length by a set of wires. They are less dense than *pressed bricks*, and they have no frog.

WIRED GLASS *See* SAFETY GLASS (*a*).

WITHDRAWAL LOAD The resistance of a nail to being pulled out after driving.

WITHE *See* WYTHE.

WITTKOWER, R. *See* HARMONIC PROPORTIONS.

WOOD-FRAME CONSTRUC-TION *See* FRAME CONSTRUCTION.

WOOD SIDING Wall cladding for frame building consisting of wooden boards. Called *weather boarding* in England and Australia.

WOOD VENEER Thin layer of wood of uniform thickness, used as a facing. The object may be to strengthen the wood by varying the direction of the grain (as in plywood) or to attach a decorative surface to a less attractive timber. Decorative veneers are made as thin as possible. Decorative veneers may be *sliced*, *i.e.* cut transversely, or *rotary*, *i.e.* cut on a lathe. *Matched veneers* are sliced veneers, made in successive cuts, which are used left-face and right-face, so that the figure of the wood is repeated in mirror-image in the matched pair.

WORK *See* HORSEPOWER.

WORK HARDENING *See* COLD-WORKING.

WORKABILITY OF CONCRETE The ability of freshly mixed concrete or mortar to flow, and fill the formwork without voids. Also called *consistency*. *See also* SLUMP TEST.

WORKING LOAD The normal dead, live, wind and earthquake load which the structure is required to support in service. It is generally specified in building codes. In limit design the working load is multiplied by a load factor to give the ultimate load at which the building is designed to fail.

WORKING LOAD DESIGN Design based on the working loads.

The stresses in the structure under these loads may not exceed the MAXIMUM PERMISSIBLE STRESSES. Also called ELASTIC DESIGN. The main alternative approach is called ULTIMATE STRENGTH DESIGN, PLASTIC DESIGN or LIMIT DESIGN.

WORKING PLANE The real or imaginary surface at which work is normally done, and at which consequently the illumination is specified and measured. This plane is normally horizontal, and 850 mm or 2 ft 9 in above the floor.

WORKING STRESS The MAXIMUM PERMISSIBLE STRESS under the action of the working loads. It is normally specified in the building code.

WORKING STRESS DESIGN *See* WORKING LOAD DESIGN.

WRIGHT, FRANK LLOYD *See* ORGANIC ARCHITECTURE.

WROUGHT ALUMINIUM ALLOYS Alloys which can be cold rolled, extruded, pressed or drawn. Many wrought aluminium alloys are not suitable for casting and vice versa, although there are alloys which can be used for both purposes.

WROUGHT IRON Iron with a low carbon content, less than mild steel (Fig. 71). It is one of the two traditional forms of iron, the other being *cast iron*. Steel is intermediate between the two, but prior to the invention of the *Bessemer* process it could only be produced at great expense. Wrought iron is soft, easily worked, and rusts less than steel, but it is more expensive. The term 'wrought iron' now usually denotes a very mild steel.

WYTHE (*a*) One leaf of a cavity wall. (*b*) A half-brick wall. Also spelt *withe*.

X

X-RAYS Electromagnetic RADIA-TION of very short wavelength, also called *Röntgen rays* after W. K. von Röntgen, the German physicist who discovered them in 1895. They have the capacity to penetrate materials opaque to light, and are employed in *radiography* to produce photographic pictures of certain invisible features. Apart from the important medical applications, they can be used for non-destructive testing, *e.g.* to show up defects in WELDING. *See also* GAMMA RAYS.

in 1943. It is based on the observation that concrete slabs fail following the formation of a number of *large* cracks, just sufficient to turn the hyperstatic slab into a mechanism. Their failing loads can therefore be derived from considerations of LIMIT DESIGN. The theory is admitted by the building codes of Scandinavian countries and Australia.

YIELD POINT The lowest stress at which STRAIN increases without increase in STRESS (Fig. 93). Only a few materials (including structural steel) exhibit a marked yield point which delineates the boundary between the elastic and the plastic state. For other materials the transition from elastic to plastic behaviour is gradual, and a PROOF STRESS is defined as an artificial boundary.

YIELD STRESS The stress of a material at the YIELD POINT.

YOUNG'S MODULUS *See* MODULUS OF ELASTICITY.

Y

YAGLOU, C. P. *See* EFFECTIVE TEMPERATURE.

YEAR RINGS *See* GROWTH RINGS.

YIELD-LINE THEORY A theory for the ultimate strength of reinforced concrete slabs, proposed by the Danish engineer K. W. Johansen

Z

ZANDMAN, F. *See* PHOTOSTRESS METHOD.

ZENITH The highest point in the sky, immediately overhead at the time of an observation. Its altitude is 90°. *See also* HORIZON.

ZIGGURAT A tower, built in ancient Assyria and Babylonia as a temple, in which each storey is smaller than that below it.

ZINC A white metallic element, which is highly resistant to atmospheric corrosion, and is consequently used to protect steel (*see* GALVANISING, SHERARDISING). It is one of the constituents of BRASS. Its chemical symbol is Zn, its atomic number is 30, its valency is 2, its atomic weight is 65·38, its specific gravity is 7·14, and its melting point is 419·4°C.

ZINC WHITE *See* CHINESE WHITE.

Zn Chemical symbol for *zinc*.

Z-SECTION A metal section shaped ⌐_.

Z-TIE *See* WALL TIE.

A Survey of the Literature of Architectural Science

1. GENERAL ABSTRACTING PERIODICALS

(a) *Architectural Index*, Sausalito, California. Annual index of American architectural magazines, published by The Architectural Index.

(b) *Engineering Index*, New York. Annual index of engineering periodicals on a world-wide basis, published by the Engineering Societies Library.

(c) *Art Index*, New York. Quarterly index, which includes the major architectural magazines, published by H. W. Wilson Co.

(d) *Canadian Building Abstracts*, Ottawa. Quarterly abstracts of Canadian literature, published by the Division of Building Research, NRC.

(e) *Building Science Abstracts*, London. Monthly abstracts of periodicals (and occasionally books) received in the library of the British Building Research Station. World-wide coverage with a scientific bias.

(f) *Current Information in the Construction Industry, Department of the Environment* (formerly the *Library Bulletin of the Ministry of Public Building and Works*, London). Fortnightly abstracts of books, pamphlets and periodicals received in the Ministry's library. World-wide coverage with a constructional bias.

(g) *Library Bulletin of the Royal Institute of British Architects*, London. Quarterly abstracts of books, pamphlets and periodicals received in the RIBA library. World-wide coverage with an architectural bias.

(h) *RIBA Annual Review of Periodical Articles*, London. World-wide coverage with an architectural bias.

(i) *Review* section of the *Architectural Science Review*, Sydney. (*See* Appendix A, Section 3(c)).

Note: (i) The *Avery Index of Architectural Periodicals*, published by H. K. Hall, Boston (Mass) contains the periodicals catalogue compiled over the years at the Avery Library of Architecture (Columbia University, New York) in 12 volumes. Supplementary volumes have been published in 1965, 1966, 1967 and 1968.

(ii) The *CIB* (*see* Appendix C, Section 1(a)) has asked one building research organisation in each country to compile abstracts of the local literature on building research. These are published at irregular intervals on sheets of paper which can be cut into slips measuring 105 by 75 mm to form a *card index*; the abstract appears on one side in the language of the country of origin, and on the other in English. At present no abstracts are published for the USA, the United Kingdom, Australia, or the USSR; but most Continental countries participate. Although the service is primarily intended for CIB members, several building research organisations are willing to make their abstracts available on request.

2. ABSTRACTING PERIODICALS DEALING WITH SPECIFIC ASPECTS OF ARCHITECTURAL SCIENCE

(a) *Housing and Planning References*, Washington, D.C. Published in alternate months by the Housing and Home Finance Agency.
(b) *Applied Mechanics Reviews*, New York. Published monthly by the American Society of Mechanical Engineers.
(c) *Papers and Books on Cement and Concrete*, published in the *Magazine of Concrete Research* (*see* Appendix A, Section 5(p)).
(d) *Plastics Abstracts*, Welwyn, Herts, England. Published weekly by Plastics Investigations.
(e) *Chemical Abstracts*, Washington. Published weekly by the American Chemical Society.
(f) *Fire Research Abstracts and Reviews*, Washington, D.C. Published three times annually by the Division of Engineering, National Research Council.
(g) *Thermal Abstracts*, Bracknell, Berks, England. Published in alternate months by the Heating and Ventilating Research Association.
(h) *Acoustics Abstracts*, Brentwood, England. Published in alternate months by the Multi-Science Publishing Co.
(i) *Abstracts* published in *Lighting Research and Technology* (*see* Appendix A, Section 5(q)).

3. ARCHITECTURAL SCIENCE PERIODICALS

(a) *Building Research*, Washington. Published quarterly by the Building Research Institute.
(b) *Building Science*, Oxford, England. Published quarterly by Pergamon Press.
(c) *Architectural Science Review*, Australia. Published quarterly by Research Publications, Melbourne.
(d) *Building Forum*, Sydney. Published quarterly by the Building Science Forum of Australia.
(e) *Build International*, London. Published in alternate months by Applied Science.

(f) *Building Research and Practice*, Rotterdam. Published in alternate months in English by CIB (*see* Appendix C, Section 1(a)).

4. SELECTED ARCHITECTURAL AND BUILDING PERIODICALS

(a) *Architectural Forum*, New York. Monthly.
(b) *Architectural Record*, New York. Monthly.
(c) *House and Home*, New York. Monthly.
(d) *Journal of the American Institute of Architects*, Washington. Monthly.
(e) *Progressive Architecture*, New York. Monthly. (Recently ceased publication.)
(f) *Architects' Journal*, London. Weekly.
(g) *Architect and Building News*, London. Weekly.
(h) *Building*, London. Weekly.
(i) *Architectural Design*, London. Monthly.
(j) *Architectural Review*, London. Monthly.
(k) *Journal of the Royal Institute of British Architects*, London. Monthly.
(l) *Concrete Quarterly*, London. Magazine on concrete architecture.
(m) *Modular Quarterly*, London. Magazine of the Modular Society.
(n) *Industrialised Building*, London. Monthly.
(o) *System Building and Design*, London. Monthly.
(p) *Architecture in Australia*, Sydney. Quarterly magazine of the Royal Australian Institute of Architects.
(q) *Architecture Française*, Paris. (French, with English summary.) Alternate months.
(r) *Architecture Aujourd'hui*, Boulogne, France. (French, with English summary.) Alternate months.
(s) *Bauwelt*, Berlin. (German). Weekly.
(t) *Werk*, Winterthur, Switzerland. (German, with English summary.) Monthly.
(u) *Casabella Continuita*, Milan. (Italian, with English summary.) Monthly.
(v) *Domus*, Milan. (Italian, with English summary.) Monthly.
(w) *Prefabricare*, Milan. (Italian, with English summary.) Alternate months.
(x) *Informes de la Construcción*, Madrid. (Spanish, with English summary.) Ten issues annually.
(y) *Architektura SSSR*, Moscow. (Russian, with English list of contents.) Monthly.
(z) *Japan Architect*, Tokyo. (English.) Monthly.

5. SELECTED ENGINEERING PERIODICALS

(a) *Engineering News-Record*, New York. Weekly.
(b) *Civil Engineering*, New York. Published monthly by the American Society of Civil Engineers.
(c) *Journal of the Structural Division, American Society of Civil Engineers*, New York. Monthly.
(d) *Journal of the American Concrete Institute*, Detroit. Monthly.
(e) *Fire Technology*, Boston. Quarterly.
(f) *Illuminating Engineering*, New York. Monthly magazine of the Illuminating Engineering Society, New York.
(g) *Journal of the Acoustical Society of America*, New York. Alternate months.
(h) *ASHRAE Journal*, New York. Monthly. (*See* Appendix C, Section 3(b).)
(i) *Heating, Piping and Air Conditioning*, New York. Monthly.
(j) *The Engineer*, London. Weekly.
(k) *Engineering*, London. Weekly.
(l) *Civil Engineering and Public Works Review*, London. Monthly.
(m) *Insulation*, London. Alternate months.
(n) *The Structural Engineer*, London. Monthly magazine of the Institution of Structural Engineers.
(o) *Concrete*, London. Monthly magazine of the Concrete Society.
(p) *Magazine of Concrete Research*, London. Quarterly.
(q) *Lighting Research and Technology*, London. Quarterly magazine of the Illuminating Engineering Society, London.
(r) *Journal of the Institution of Heating and Ventilating Engineers*, London. Monthly.
(s) *Journal of the Royal Society of Health*, London. Monthly.
(t) *Acier, Stahl, Steel*, Brussels. (English.) Monthly.
(u) *Acustica*, Stuttgart, Germany. (English, French or German, with English summaries.) Alternate months.
(v) *Bulletin of the International Association of Shell Structures*, Madrid. (English.) Quarterly.
(w) *Indian Concrete Journal*, Bombay. (English.) Monthly.
(x) *Transactions of the Architectural Institute of Japan*, Tokyo. (Papers on engineering topics, in English or Japanese.) Monthly.
(y) *Journal of the Institution of Engineers, Australia*, Sydney. Monthly.

6. BIBLIOGRAPHIES

Bibliographies are rarely printed as separate books. Standard textbooks contain references to the literature on their subject, and these are frequently the best sources of bibliographical information.

In addition, many libraries compile typed or duplicated lists of books and papers from periodicals for the use of their readers, and they are often willing to make them available to others, although not necessarily free of charge. These include:

Engineering Societies Library, New York.
US Department of Housing and Urban Development, Washington, D.C.
American Concrete Institute, Detroit.
Educational Resources Information Center, Madison, Wisconsin.
Division of Building Research, National Research Council, Ottawa, Canada.
Building Research Station, Garston, Herts, England.
Royal Institute of British Architects, London.
Cement and Concrete Association, London.
National Swedish Institute for Building Research, Stockholm (some in English).
Department of Architectural Science, University of Sydney, Australia.

The remainder of Appendix A contains, of necessity, a personal selection of the books available.

7. DICTIONARIES

(a) H. H. Saylor: *Dictionary of Architecture*. Wiley, New York, 1952.
(b) M. S. Briggs: *Everyman's Concise Encyclopaedia of Architecture*. Dutton, New York, 1959.
(c) D. Ware and B. Beatty: *A Short Dictionary of Architecture*. Allen and Unwin, London, 1953.
(d) J. Fleming, H. Honour and N. Pevsner: *The Penguin Dictionary of Architecture*. Penguin Books, Harmondsworth (England), 1966.
(e) J. Harris and J. Lever: *Illustrated Glossary of Architecture 850–1830*. Faber, London, 1966.
(f) H. Hatje: *Encyclopaedia of Modern Architecture*. Thames and Hudson, London, 1963.
(g) A. E. Burke, J. R. Dalzell and G. Townsend: *Architectural and Building Trades Dictionary*. American Technical Society, Chicago, 1955.
(h) J. S. Scott: *A Dictionary of Building*. Penguin Publications, Harmondsworth, 1964.
(i) J. S. Scott: *A Dictionary of Civil Engineering*. Penguin Publications, Harmondsworth, 1965.

(j) J. Moreau: *Dictionnaire Technique Américain–Français de Construction*. Dunod, Paris, 1960.

(k) H. Bücksch: *Dictionary of Civil Engineering and Construction Machinery and Equipment*. French–English and English–French Edition. Éditions Eyrolles, Paris, 1962.
H. Bücksch: *Dictionary of Civil Engineering and Construction Machinery and Equipment*. German–English and English–German Edition. Bauverlag, Wiesbaden, 1964.

(l) C. J. van Mansum: *Dictionary of Building Construction in Four Languages*. Elsevier, Amsterdam, 1968. The dictionary is alphabetically arranged in English, with translations into French, Dutch and German. There are alphabetical indices in the other languages.

(m) R. Walther: *Dictionary of Mechanics, Strength of Materials and Materials*. English–German and German–English. Pergamon, Oxford (England), 1965.

(n) A. Schlomann: *Baukonstruktionen*. R. Oldenbourg, Munich, 1919. In spite of its age, this remains a useful work. It gives the words in German, English, French, Russian, Italian and Spanish, generally with an illustration. There is an alphabetical index.

8. HANDBOOKS

(a) F. S. Merritt: *Building Construction Handbook*. McGraw-Hill, New York, 1965.

(b) N. W. Kay: *The Modern Building Encyclopaedia*. Odhams Press, London, 1959.

(c) C. Hornbostel: *Materials for Architecture*. Reinhold, New York, 1961.

(d) I. E. Morris: *Handbook of Structural Design*. Reinhold, New York, 1963.

(e) American Institute of Timber Construction: *Timber Construction Manual*. Wiley, New York, 1966.

(f) British Steel Producers Conference: *Steel Designers' Manual*. Crosby Lockwood, London, 1966.

(g) W. S. Lalonde and M. F. Janes: *Concrete Engineering Handbook*. McGraw-Hill, New York, 1961.

(h) G. H. Tryon: *Fire Protection Handbook*. National Fire Protection Association, Boston, 1962.

(i) American Society of Heating, Refrigerating and Air Conditioning Engineers. *ASHRAE Guide and Data Book*. Volume 1: *Fundamentals and Equipment*. Volume 2: *Applications*. Each volume published in alternate years. The Society, New York, 1969 and 1970.

(j) C. M. Harris: *Handbook of Noise Control.* McGraw-Hill, New York, 1957.

(k) Tufts College Institute of Applied and Experimental Psychology: *Handbook of Human Engineering Data.* US Naval Training Service Center, Port Washington (N.Y.), 1960.

9. BOOKS ON BUILDING CONSTRUCTION, STRUCTURES AND MATERIALS

(See also Appendix A, Section 8)

(a) Building Research Station: *Principles of Modern Building.* HM Stationery Office, London, 1959 and 1961. 2 *volumes.*

(b) W. C. Huntington: *Building Construction.* Wiley, New York, 1963.

(c) H. Parker, C. M. Gay and J. W. MacGuire: *Materials and Methods of Architectural Construction.* Wiley, New York, 1958.

(d) D. Rosenthal: *Introduction to Properties of Materials.* Van Nostrand, Princeton (N.J.), 1964.

(e) J. E. Gordon: *The New Science of Strong Materials, or Why You Don't Fall Through the Floor.* Penguin Publications, Harmondsworth (England), 1968.

(f) L. A. Ragsdale and E. A. Rayham: *Building Materials Practice.* Arnold, London, 1964.

(g) L. E. Akers: *Particle Board and Hardboard.* Pergamon Press, Oxford (England), 1966.

(h) I. Skeist: *Plastics in Building.* Reinhold, New York, 1966.

(i) R. McGrath and A. C. Frost: *Glass in Architecture and Decoration.* Architectural Press, London, 1961.

(j) W. F. Cassie: *Fundamental Foundations.* Applied Science, London, 1968.

(k) E. Torroja: *Philosophy of Structures.* University of California Press, Berkeley, 1958.

(l) H. J. Cowan: *Architectural Structures.* Elsevier, New York, 1970.

(m) A. J. S. Pippard and J. F. Baker: *The Analysis of Engineering Structures.* Arnold, London, 1968.

(n) A. S. Hall and R. W. Woodhead: *Frame Analysis.* Wiley, New York, 1967.

(o) F. B. Johnson: *Designing, Engineering and Constructing Masonry Products.* Gulf Publishing Co., Houston, 1969.

(p) B. Bresler and T. Y. Lin: *Design of Steel Structures.* Wiley, New York, 1960.

(q) L. S. Beedle: *Plastic Design of Steel Frames.* Wiley, New York, 1965.

(r) P. M. Ferguson: *Reinforced Concrete Fundamentals.* Wiley, New York, 1965.

(s) P. Collins: *Concrete, a Vision of a New Architecture.* Faber, London, 1959.

(t) J. G. Wilson: *Exposed Concrete Finishes.* CR Books, London, 1962 and 1964. 2 *volumes.*

(u) B. Lewicki: *Building with Large Prefabricates.* Elsevier, Amsterdam, 1966.

(v) CIB: *Towards Industrialised Building.* Elsevier, Amsterdam, 1966.

(w) G. S. Ramaswamy: *Design and Construction of Concrete Shells.* McGraw-Hill, New York, 1968.

(x) A. M. Haas: *Thin Concrete Shells.* Wiley, New York, 1962 and 1967. 2 *volumes.*

(y) F. Otto: *Tensile Structures.* MIT Press, Cambridge (Mass), 1967 and 1969. 2 *volumes.*

(z) R. M. Davies: *Space Structures.* Blackwell, Oxford (England), 1967.

(aa) National Research Council of Canada: *Wind Effects on Buildings and Structures.* University of Toronto Press, Toronto, 1968. 2 *volumes.*

(bb) E. Mackey: *The Design of High Buildings.* Hong Kong University Press, Hong Kong, 1963.

(cc) IASS: *Proceedings of the Symposium on Tower-Shaped Steel and Reinforced Concrete Structures.* International Association for Shell Structures, Madrid, 1966.

(dd) RILEM: *Symposium on the Observation of Structures.* Laboratório Nacional de Engenharia Civil, Lisbon, 1955. 2 *volumes.*

(ee) T. H. McKaig: *Building Failures.* McGraw-Hill, New York, 1962.

10. BOOKS ON ENVIRONMENTAL SCIENCE
(See also Appendix A, Section 8)

(a) O. G. Edholm: *The Biology of Work.* Weidenfeld and Nicolson, London, 1967.

(b) R. Sommer: *Personal Space.* Prentice-Hall, Englewood Cliffs (N.J.), 1969.

(c) R. H. Day: *Human Perception.* Wiley, Sydney, 1969.

(d) J. F. Young: *Cybernetics.* Iliffe, London, 1969.

(e) R. L. Gregory: *Eye and Brain—The Psychology of Seeing.* Weidenfeld and Nicolson, London, 1966.

(f) H. C. Weston: *Sight, Light and Work.* H. K. Lewis, London, 1962.

(g) J. A. Lynes: *Principles of Natural Lighting.* Applied Science, London, 1968.

(h) R. G. Hopkinson, P. Petherbridge and J. Longmore: *Daylighting*. Heinemann, London, 1966.

(i) P. Moon: *The Scientific Basis of Illuminating Engineering*. Dover, New York, 1961.

(j) H. Hewitt and A. S. Vause: *Lamps and Lighting*. Arnold, London, 1966.

(k) L. L. Beranek: *Music, Acoustics and Architecture*, Wiley, New York, 1962.

(l) A. Lawrence: *Architectural Acoustics*, Applied Science, London, 1970.

(m) P. H. Parkin and H. R. Humphreys: *Acoustics, Noise and Buildings*. Faber, London, 1958.

(n) P. H. Parkin, H. J. Purkis and W. E. Scholes: *Field Measurements of Sound Insulation between Dwellings*. HM Stationery Office, London, 1960.

(o) B. Givoni: *Man, Climate and Architecture*. Applied Science, London, 1969.

(p) L. H. Newburgh: *Physiology of Heat Regulation and the Science of Clothing*. Hafner, New York, 1968.

(q) J. F. van Straaten: *Thermal Performance of Buildings*. Applied Science, London, 1967.

(r) T. Bedford: *Basic Principles of Ventilation and Heating*. H. K. Lewis, London, 1964.

(s) B. Y. Kinzey and H. M. Sharp: *Environmental Technologies in Architecture*. Prentice-Hall, Englewood Cliffs, 1963.

(t) Carrier Air Conditioning Company: *Handbook of Air Conditioning System Design*. McGraw-Hill, New York, 1965.

(u) P. Jay and J. Hemsley: *Electrical Services in Buildings*. Applied Science, London, 1968.

(v) R. R. Adler: *Vertical Transportation in Buildings*. Elsevier, New York, 1970.

(w) L. S. Nielsen: *Standard Plumbing Engineering Design*. McGraw-Hill, New York, 1963.

11. BOOKS ON BUILDING ECONOMICS, SYSTEMS ANALYSIS, COMPUTERS AND MODEL ANALYSIS

(a) A. B. Handler: *Systems Approach to Architecture*. Elsevier, New York, 1970.

(b) RIBA: *The Co-ordination of Dimensions for Building*. Royal Institute of British Architects. London, 1965.

(c) P. A. Stone: *Building Economy*. Pergamon Press, Oxford (England), 1966.

(d) P. A. Stone: *Building Design Evaluation: Costs-in-Use*. Spon, London, 1967.

(e) R. Clements and D. Parkes: *Manual of Maintenance—Building and Building Services*. Business Publications, London, 1965.

(f) J. M. Antill and R. W. Woodhead: *Critical Path Methods in Construction Practice*. Wiley, New York, 1965.

(g) A. Huitson and J. Keen: *Essentials of Quality Control*. Heinemann, London, 1965.

(h) D. Campion: *Computers in Architectural Design*. Applied Science, London, 1968.

(i) G. N. Harper: *Computer Applications in Architecture and Engineering*. McGraw-Hill, New York, 1968.

(j) H. J. Cowan, J. S. Gero, G. D. Ding and R. W. Muncey: *Models in Architecture*. Applied Science, London, 1968.

12. BOOKS ON THE HISTORY OF ARCHITECTURAL SCIENCE

(a) H. J. Cowan: *An Historical Outline of Architectural Science*. Elsevier, Amsterdam, 1966.

(b) S. Timoshenko: *History of the Strength of Materials*. McGraw-Hill, New York, 1953.

(c) H. Straub: *A History of Civil Engineering*. Leonard Hill, London, 1952.

(d) R. Banham: *The Architecture of the Well-Tempered Environment*. Architectural Press, London, 1969.

(e) R. B. White: *Prefabrication—A History of its Development in Great Britain*. HM Stationery Office, London, 1965.

(f) C. W. Condit: *The Chicago School of Architecture*. University of Chicago Press, Chicago, 1964.

(g) N. Pevsner: *The Sources of Modern Architecture*. Thames and Hudson, London, 1968.

(h) J. Joedicke: *A History of Modern Architecture*. Architectural Press, London, 1962.

(i) Banister Fletcher: *A History of Architecture on the Comparative Method*. Athlone Press, London, 1961.

(j) R. Wittkower: *Architectural Principles in the Age of Humanism*. Tiranti, London, 1963.

(k) Vitruvius: *The Ten Books of Architecture*. (Translated by M. H. Morgan.) Dover, New York, 1960.

(l) L. B. Alberti: *The Ten Books of Architecture*. (Translated by G. Leoni.) Tiranti, London, 1955. Facsimile of an edition published in London in 1755.

(m) A. Palladio: *The Four Books of Architecture*. (Translated by G. Leoni.) Dover, New York, 1965. Facsimile of the edition by Isaac Ware, London, 1738.

(n) H. Wotton: *The Elements of Architecture*. University Press of Virginia, Charlottesville, 1968. Facsimile of the First Edition, London, 1624.

APPENDIX B

Information Processing

1. INTRODUCTION

It probably remains true that in a small library a person who knows its contents thoroughly can find useful information more quickly without any system than a skilled librarian (who does not know its contents) using the most up-to-date information retrieval system. The use of a system is no substitute for the expert knowledge which enables the reader to distinguish the significant from the irrelevant, since any method must err either on the side of retrieving all items which could conceivably have a bearing on the subject, or else of deleting some publications whose relevance is doubtful.

Classification systems for library books are not easily altered, because the adoption of a new system necessitates reclassification of all material in the permanent collection, and some antiquated systems remain in use. However, this restriction does not apply to systems for retrieving technical literature from periodicals, trade catalogues, etc., which generally cease to be of any but historical interest within ten years. Thus it is not uncommon to start a new information retrieval system at intervals of, say, five years, which facilitates the introduction of new and more efficient systems. It is better to search several catalogues than to burden the latest catalogue with obsolete material; only recent material need be examined, but older references are available if required.

Although a central information retrieval system is evidently desirable, it has been achieved only in a few scientific fields. No general abstracting system has been devised for building, architecture, architectural engineering or architectural science. Moreover, none appear satsifactory for all types of literature.

In 1950 the *International Council for Building Documentation* (*CIBD*) was set up under the auspices of the Economic Commission for Europe. This later became a joint committee (*International Building Classification Committee—IBCC*) of the *International Council for Building Research, Studies and Documentation* (*CIB*) and of the *International Federation for Documentation* (*FID*). It established two parallel systems of classification, SfB and ABC, which will be described.

312

2. THE SfB SYSTEM

This classification was devised in Sweden by the *Samarbetskommittén for Byggnadsfragor* (Co-ordinating Committee for Building) from whose initials it takes its name. It consists of three basic tables which form the co-ordinates of the system. The first describes the *functional element*, the second the *construction* and the third the *material*.

SfB Table 1: Functional Elements

(1)	*External Elements: General*
(11)	Ground: General
(12)	Drainage: General
(13)	Retaining structures: General
(14)	Roads and pavings: General
(15)	Garden: General
(16)	Foundations: General
(17)	Piles: General
(18)	Footings: General
(19)	Other substructures: General
(2)	*Primary Elements: General*
(20)	Accessories: Structural fixings
(21)	Walls: General
(22)	Partitions: General
(23)	Floors, structural, galleries: General
(24)	Stairs and ramps: General
(25)	Ceilings, suspended: General
(26)	Roofs, structural, flat, balconies: General
(27)	Roofs, structural, pitched
(28)	Elements above roof: General
(3)	*Secondary Elements: General*
(30)	Secondary elements, accessories, ironmongery: General
(31)	Secondary elements, windows: General
(32)	Secondary elements, doors: General
(33)	Secondary elements, floors, grilles, etc.: General
(34)	Secondary elements, stairs, handrails, etc.: General
(35)	Secondary elements, screens, louvres: General
(36)	Secondary elements, roofs, flat, pavement lights, etc.: General
(37)	Secondary elements, roof lights and traps, etc.: General
(38)	Secondary elements, roof eaves, verges, gutters, rails, etc.: General
(4)	*Finishes*
(40)	Finishes, accessories
(41)	Finishes, external: General
(42)	Finishes, internal: General
(43)	Finishes, floor: General

313

(44)	Finishes, stair: General
(45)	Finishes, sills, skirtings, cover strips
(46)	Finishes, flat roofs
(47)	Finishes, roof: General
(48)	Finishes, roof, flashings
(5)	*Services Installations: Sanitation, Heating, Ventilation: General*
(50)	Installations, accessories: General
(51)	Installations, refuse disposal: General
(52)	Installations, drainage, sanitation: General
(53)	Installations, water, hot and cold: General
(54)	Installations, gas, compressed air, steam, refrigeration, etc.: General
(56)	Installations, heating: General
(57)	Installations, ventilation, air conditioning: General
(6)	*Service Installations: Electrical and Mechanical: General*
(60)	Electrical accessories: General
(63)	Installations, electrical, lighting and power: General
(64)	Installations, communications, radio, TV: General
(66)	Installations, mechanical, lifts, escalators, etc.: General
(68)	Installations, special, lightning conductors, fire-fighting, etc.: General
(7)	*General Spaces, Fixtures and Equipment*
(70)	Accessories: General spaces
(71)	Entrances, fixtures and equipment: General
(72)	Rooms, fixtures and equipment: General
(73)	Kitchens, fixtures and equipment: General
(74)	Cloakrooms, bathrooms and lavatories, fixtures and equipment: General
(75)	Laundries, fixtures and equipment: General
(76)	Stores, cupboards, fixtures and equipment: General
(77)	Plant, fixtures and equipment, boiler rooms, garages: General
(78)	External, fixtures and equipment: General
(8)	*Special spaces, Fixtures and Equipment*
	This is the place for items special to one building type
(9)	*Building Types*

SfB Table II: Construction

Exact terms used for headings and sub-divisions will differ according to the application of the tables for filing or specifications and quantities, etc.

A	*Used for Special Purposes*	
B	*Used for Special Purposes*	
C	*Used for Special Purposes*	
	Operations	Products
D		*Materials (Generally Aggregates, Binders, Mortars, Bitumen, Chemicals, etc.)*
E	*Concreting*	*Concrete Mass, Reinforcement*
F	*Brickwork, Masonry*	*Bricks and Blocks*
G	*Erection*	*Structural Units*
H	*Assembly*	*Sections and Bars (other than I, J and R)*
I	*Pipework*	*Tubes and Pipes*
J	*Wire work*	*Wires and Mesh*
K	*Insulating*	*Insulating Products*
L	*Coating (Asphalting, Membranes, etc.)*	*Foils and Felts, etc.*
M	*Sheet Working*	*Thin Plain Sheets (Sheet Metal, etc.)*
N	*Roofing, Siding, etc.*	*Tiles and Sheets (overlapping, etc.)*
P	*Plastering*	*Plasters*
Q	*Acoustic Treatment*	*Acoustic Products*
R	*Glazing, Sheeting*	*Sheets (Rigid), (Plaster board, Plywood, Glass, etc.)*
S	*Tiling*	*Tiles (Butt jointing)*
T	*Flooring (Special)*	*Flooring Products (other than E, H and S)*
U	*Facing (Special)*	*Claddings and Facings (other than N and S)*
V	*Painting*	*Paints, Varnishes*
X	*Installation of*	*Assemblies, Fixtures and Equipment*
Y	*Used for Special Purposes*	
Z	*Used for Special Purposes*	

SfB Table III: Materials

a   *Used for Special Purposes*

b   *Used for Special Purposes*

c   *Used for Special Purposes*

d–o   *Material in Components, e.g., Bricks, Blocks, Pipes, etc.*

d   *Metal*
d1   Cast iron
d2   Wrought iron, steel (also galvanised, enamelled, etc.)
d3   Steel alloy, stainless steel
d4   Aluminium and aluminium alloy
d5   Copper
d6   Copper alloy, bronze, brass
d7   Zinc
d8   Lead

e   *Stone, Natural*
e1   Granite and igneous rock
e2   Marble
e3   Limestone
e4   Sandstone
e5   Slate
e8   Asbestos

f   *Concrete, Artificial Stone (using Lime or Cement in Solid Blocks)*
f1   Sand lime concrete
f2   Cement concrete
f3   Terrazzo
f4   Lightweight concrete

g   *Clayware, Ceramics*
g1   Cob, pise, adobe
g2   Heavy burnt clay
g3   Terracotta, faience, earthenware
g4   Semi-vitreous ware, stoneware
g5   Vitreous ware, porcelain
g6   Fire-resistant ware, refractory ware
g7   Chemically resistant ware

h   *Other Mineral Materials (in Solid Blocks and Sheets)*
h1   Asbestos cement
h2   Gypsum
h3   Magnesia

i	*Wood*
i1	Timber (including rot-proofed)
i2	Softwood
i3	Hardwood
i4	Laminated wood, plywood
j	*Organic Fibre (in Boards, etc.)*
j1	Wood fibre, wood particles
j2	Pulp
j3	Organic fabric
j4	Corrugated paper
k	*Cork and other Organic Material (in Solid Products, Slabs, Sheets)*
k1	Cork
k3	Wood wool cement
k4	Reeds, straw
k5	Peat
m	*Felted Material, etc.*
m1	Mineral wool including glass fibre
m2	Seaweed
m3	Organic wool (wadding, wool, hair)
n	*Plastic, Composition, Linoleum, Rubber*
n1	Asphalt
n2	Impregnated fibre and felt
n3	Bituminous fabric
n4	Linoleum
n5	Rubber
n6	Plastic
n8	'Thermoplastic' Composition
o	*Glass*
p–s	*Materials in Formless Products*
p	*Loose Fill, Aggregate*
p1	Stone aggregate, sand, gravel, shingle, crushed granite, shale
p2	Crushed clay bricks, concrete, lightweight concrete, etc.
p3	Granulated slag, clinker (lightweight aggregate)
p4	Ash
p5	Shavings (including seaweed, wood wool, paper wool)
p6	Powder
p7	Organic wool, mineral wool
p8	Plastic
p9	Fluid, gas

q	Mortar (with Lime or Cement) and Mass Concrete
q1	Lime
q2	Cement
q4	Mortar with lime and cement, concrete mass (including coarse stuff for rendering, screeding)
q5	Terrazzo
q6	Lightweight concrete

r	Gypsum, Plasters, Magnesite
r1	Clay, mortar, fire-resistant mortar, chemically resistant mortar
r2	Gypsum
r3	Magnesia
r4	Synthetic resins

s	Bituminous Material, Tar, Asphalt, etc.
s1	Bitumen, pitch, tar
s2	Bitumen solution and emulsion
s4	Mastic asphalt
s5	Asphalt, bitumen macadam and tar macadam

t–w	Agents, Chemicals, etc.

t	Fixing Material, Adhesives, Mastics and Jointing Compounds
t1	Welding Material
t2	Soldering Material
t3	Adhesives
t4	Putty, mastics, jointing material, etc.

u	Protective Materials
u1	Means against corrosion
u2	Admixtures for cement and concrete, etc.
u3	Means for rot proofing and insect attack prevention
u4	Means for fire proofing
u5	Means for treating of floors, wax, etc.
u6	Water repellent finish
u8	Surface means for diffusion proofing

v	Oils, Varnishes, Paints
v1	Stopping, putty, paint fillers
v2	Pigment, stains
v3	Vehicle, oil
v4	Varnishes
v6	Paints, oil, emulsion, water, etc.

w	Chemicals
w1	Rust removing agents
w2	Solvents, thinners, drying agents, emulsifying agents

| w3 | Water glass |
| w4 | Polishing agents |

| *x* | *Used for Special Purposes* |

| *y* | *Used for Special Purposes* |

| *z* | *Used for Special Purposes* |

By combination of the three tables, any part of a building or construction process can be described with reasonable accuracy, and the terms can be, and have been, translated into different languages.

The SfB system is particularly well suited to trade catalogues, and to articles and pamphlets describing practical building operations. It has been widely adopted in Europe, including the United Kingdom, and also in Australia, for this purpose. In America the *AIA* (American Institute of Architects) *Standard Filing System* and the *Uniform System for Construction Specifications, Data Filing and Cost Accounting* (also published by AIA) continue to be used.

3. THE ABC SYSTEM

The SfB system is too much oriented towards practical problems to be useful for classifying research papers, and it therefore co-exists with the *Abridged Building Classification* (*ABC*) which is a selection from the *Universal Decimal Classification* (*UDC*) made by the International Building Classification Committee (IBCC).

The decimal classification system was devised by Melvil Dewey, a US librarian, in 1876, and many public libraries still use the original Dewey system because of the cost of changing to a more modern one. Partly because it was written a hundred years ago, this has many weaknesses, especially in its classification of science and technology. The revised version is known as the *Universal Decimal Classification* and the first complete English abridged edition was published by the British Standards Institution in 1948.

Outline of the Main Divisions of the Universal Decimal Classification

0   **GENERALITIES**
00   Prolegomena. Fundamentals of knowledge and culture
01   Bibliography. Catalogues
02   Libraries. Librarianship
03   Encyclopaedias. Dictionaries. Reference books
04   Essays. Pamphlets, offprints, brochures and the like
05   Periodicals. Reviews
06   Corporate bodies. Institutions. Associations. Congresses. Exhibitions. Museums
07   Newspapers. Journalism
08   Polygraphies. Collective works
09   Manuscripts. Rare and remarkable works. Curiosa

1   *PHILOSOPHY. METAPHYSICS. PSYCHOLOGY. LOGIC. ETHICS AND MORALS*
11   Metaphysics
13   Metaphysics of spiritual life. Occultism
14   Philosophical systems
15   Psychology
16   Logic. Theory of knowledge. Logical method
17   Ethics. Moral science. Convention
18   Aesthetics

2   *RELIGION. THEOLOGY*
21   Natural theology
22   Holy Scripture. The Bible
23   Dogmatic theology
24   The religious life. Practical theology
25   Pastoral theology
26   The Christian church in general
27   General history of the Christian church
28   Christian churches or worshipping bodies
29   Non-Christian religions. Comparative religion

3   *SOCIAL SCIENCES. ECONOMICS. LAW. GOVERNMENT. EDUCATION*
30   General sociology. Sociography
31   Statistics
32   Political science. Politics. Current affairs
33   Political and social economy. Economics
34   Jurisprudence. Law. Legislation
35   Public administration. Military science. Defence
36   Social relief and welfare. Insurance
37   Education
38   Trade. Commerce. Communication and transport
39   Ethnography. Custom and tradition. Folklore

320

83	Germanic literature: German, Dutch and Scandinavian
84	Romance literature. French literature
85	Italian literature. Roumanian literature
86	Spanish literature. Portuguese literature
87	Classical, Latin and Greek literature
88	Slavonic literature. Baltic literature
89	Oriental, African and other literature

9	*GEOGRAPHY. BIOGRAPHY. HISTORY*
91	Geography, exploration and travel
92	Biography. Genealogy. Heraldry
93	History in general. Sources. Ancient history
94	Mediaeval and modern history
940	History of Europe
950	History of Asia
960	History of Africa
970	History of North America
980	History of South America
990	History of Oceania, Australasia and Polar regions

For use in building, the Universal Decimal Classification is too detailed in Sections 1 (Philosophy), 2 (Religion), 4 (Philology), and 8 (Literature), but insufficiently subdivided in Sections 6 (Technology) and 7 (Arts), particularly in subsections 62 (Engineering), 69 (Building) and 72 (Architecture). The ABC Classification, which runs to only 70 pages, deletes the inessential parts of UDC.

The process is taken further by many librarians and research workers. Thus building economics, which would probably be put into 69 in a general library, may be placed under 33 (Economics) in a specialist building science library.

In addition, existing UDC divisions can be further divided. A shell roof, for example, is 624.074.4 (6 technology, 62 engineering, 624 civil engineering, 624.07 structural elements, 624.074 tridimensional structures). This can be subdivided into saddle shells 624.074.43, hyperbolic paraboloids 624.074.431, etc.

The process of subdivision can be carried on indefinitely, but this leads to an increasing likelihood of misclassification. Even within the existing abridged classification one person might consider that an article on reinforced concrete hyperbolic paraboloid roofs is primarily concerned with geometry (513), another with shell structures (624.074.4), and yet another with shell roofs (69.024.4); there are still further possibilities, such as reinforced concrete slabs (666.982), reinforced concrete products (691.328) and reinforced concrete construction

(693.55). While the SfB system is too insensitive for scientific publications, a decimal system, because of its very versatility, is liable to misinterpretation.

If one single research worker operates a personal card index, he is likely to classify hyperbolic paraboloids always under the one classification. However, different research workers are likely to choose different numbers to suit their special interests, and an untrained librarian relying on an index might use one classification now and another later on. This suggests the use of multiple classification, which is the basis of keyword indexing.

4. THE KEYWORD SYSTEM

The large proportion of European (including British) documents relating to building now come with a dual pre-classification in both the SfB and ABC system. This is a great convenience for filing, although it does not entirely overcome the problem of mis-classification. If the interest of the person who made the pre-classification differs from that of the eventual user, then the document will not be found, unless it has been re-classified.

This problem can only be overcome by multiple classification. However, if more than one tag is given to an item, then it can no longer be put into a box or file together with all other related items. The SfB and ABC systems have the great advantage that related items are to be found in adjacent files, and that the right material might thus be found by 'browsing'. Since this is no longer possible in a multiple classification system, there is no merit in maintaining the subject sequence, and the alphabetical listing of the words of the index of the ABC classification may be used instead. The use of ordinary words makes the task of classification and recovery much easier, particularly to the non-specialist who immediately recognises 'shell roof' but not '69.024.4'.

It is, however, necessary to restrict the number of words if the recovery system is to be satisfactory. Thus *keywords* are definite words listed in a standard list, or *thesaurus*; no others may be used for classifying information. The most widely used list of keywords is contained in the *Thesaurus of Engineering Terms* published by the Engineers Joint Council, which contains about 100,000 terms.

Many American and some European periodical articles are now pre-classified by keywords. It is, of course, necessary to have as many copies of the index card as there are keywords, plus one for an author index. Multiple indexing gives greater certainty of retrieving all relevant material in an index, but the time taken for retrieval is longer by hand. However, the process can be computerised if the operation is sufficiently large to warrant it.

REFERENCES

1. *Abridged Building Classification (ABC) for Architects, Builders and Civil Engineers.* Second Edition. Bouwcentrum, Rotterdam, 1955. Supplement, 1965.
2. *Recent Developments in Building Classification.* International Building Classification Committee, Bouwcentrum, Rotterdam, 1959.
3. *SfB/UDC Building Filing Manual—Recommendations for Standard Practice in Pre-classification and Filing.* RIBA Technical Information Service, London, 1961.
4. L. M. Giertz: *A Survey of Activities of the International Building Classification Committee,* 1959–1962. Danish Centre for Documentation, Copenhagen, 1962.
5. R. Mølgaard-Hansen: *Bibliography on Building Documentation,* 1938–1962. IBCC Report No. 6. The Danish Institute of Building Research, Copenhagen, 1962.
6. B. Agard Evans: *Building Classification Practices.* CIB Report No. 6. International Council for Building Research Studies and Documentation, Rotterdam, 1966.
7. *Nonconventional Technical Information Systems in Current Use, No. 3.* National Science Foundation, Washington, 1962.
8. *Thesaurus of Engineering Terms.* Engineers Joint Council, New York, 1964.
9. *Bibliography on Filing, Classification and Indexing Systems, and Thesauri for Engineering Offices and Libraries.* ESL Bibliography No. 15. Engineering Societies Library, New York, 1966.
10. D. P. Delany and H. H. Neville: Information Retrieval at the (British) Building Research Station. *Build International,* Vol. 2 (May, 1969), pp. 6–9.

Organisations Which Contribute to Architectural Science

1. INTERNATIONAL ORGANISATIONS

(a) CIB—Conseil International du Bâtiment pour la Recherche l'Étude et la Documentation—International Council for Building Research Studies and Documentation. Bouwcentrum, Weena 700, Rotterdam, Holland.

(b) UIA—Union Internationale des Architectes—International Union of Architects. 4 Impasse d'Antin, Paris 8, France.

(c) World Federation of Engineering Organisations, 5 Savoy Place, London, WC2, England.

(d) RILEM—Réunion Internationale des Laboratories d'Essais et des Recherches sur les Matériaux et les Constructions—International Union of Testing and Research Laboratories for Materials and Structures. 12 Rue Brancion, Paris 15, France.

(e) IABSE—International Association for Bridge and Structural Engineering. Eidgenössische Technische Hochschule, Zürich, Switzerland.

(f) IASS—International Association for Shell Structures. Alfonso XII, 3, Madrid 7, Spain.

(g) FIP—Fédération Internationale la Précontrainte—International Federation for Prestressing. Cement and Concrete Association, 52 Grosvenor Gardens, London, SW1, England.

(h) International Society of Soil Mechanics and Foundation Engineering. Institution of Civil Engineers, Great George Street, London, SW1, England.

(i) CIE—Commission Internationale de l'Éclairage—International Commission on Illumination. 57 Rue Cuvier, Paris 5, France.

(j) FIBTP—Fédération Internationale du Bâtiment et des Travaux Publics—International Federation of Building and Public Works. 9 Rue La Pérouse, Paris 16, France.

(k) IFHP—International Federation for Housing and Planning. 43 Wassenaarseweg, Den Haag, Holland.

(l) CINVA—Centro Interamericano de Vivienda y Planeamiento—Inter-American Housing and Planning Center. Apartado Aéreo 6209, Bogotá, Columbia.

(m) ISO—International Organisation for Standardisation. 1 Rue de Varembé, Genève, Switzerland.

(n) FID—Fédération Internationale de Documentation—International Federation for Documentation. 7 Hofweg, Den Haag, Holland.

2. BUILDING RESEARCH ORGANISATIONS

(a) Building Research Advisory Board, Division of Engineering, National Research Council, 2101 Constitution Avenue, Washington D.C. 20418, USA.

(b) Building Research Institute (now a part of the BRAB), 2101 Constitution Avenue, Washington, D.C. 20418, USA.

(c) National Bureau of Standards. Connecticut Avenue and Van Ness Street NW, Washington, D.C. 20234, USA.

(d) Division of Building Research, National Research Council, Ottawa, Ont., Canada.

(e) Building Research Establishment, Watford, WD2 7JR, England.

(f) Division of Building Research, CSIRO, Graham Road, Highett, Victoria 3190, Australia.

(g) Commonwealth Experimental Building Station, PO Box 30, Chatswood, NSW 2067, Australia.

(h) Central Building Research Institute, Roorkee, UP, India.

(i) Structural Engineering Research Centre, Roorkee, UP, India.

(j) National Building Research Institute, PO Box 395, Pretoria, South Africa.

(k) Centre Scientifique et Technique du Bâtiment, 4 Avenue du Recteur Poincaré, Paris 16, France.

(l) Centre Expérimental de Recherches et d'Études du Bâtiment et des Travaux Publics (Experimental Station for Study and Research on Building and Civil Engineering). 12 Rue Brancion, Paris 15, France.

(m) Dokumentationsstelle für Bautechnik in der Fraunhofer-Gesellschaft (Documentation Centre of Building Technics in the Fraunhofer-Gesellschaft). Siberburgstrasse 119A, 7000 Stuttgart, German Federal Republic.

(n) Bundesanstalt für Materialsprüfung (Federal Institute of Testing Materials). Unter den Eichen 87, 1000 Berlin-Dahlem, German Federal Republic.

(o) Deutsche Bauakademie (German Academy of Building). Hannoversche Strasse 30, Berlin N4, German Democratic Republic.

(p) Institut TNO voor Bouwmaterialen en Bouwconstructies (Institute TNO for Building Materials and Building Constructions). PO Box 49, Delft, Holland.

(q) Stichting Bouwcentrum (Foundation Bouwcentrum). 700 Weena, Rotterdam, Holland.

(r) Statens Byggeforskningsinstitut (Danish National Institute for Building Research). 20 Borgergade, København K, Denmark.

(s) Norges Byggforskningsinstitutt (Norwegian Building Research Institute). 1 Forskningsveien, Blindern, Norway.

(t) Statens Institut för Byggnadsforskning (National Swedish Institute for Building Research). Linnegatan 64, Stockholm Ö, Sweden.

(u) Research Department, Gosstroi USSR, Prospekt Marksa 43, Moscow K-25, Russia.

(v) Instytut Techniki Budowlanej (Building Research Institute). 2 Wawelska, Warszawa, Poland.

(w) Vyzkumny Ústav Vystavby a Architektury (Research Institute for Building and Architecture). 3 Letenská, Praha 1, Czechoslovakia.

(x) Hungarian Institute of Building Science, ÉTI. 2–4 Beloiannisz u., Budapest 5, Hungary.

(y) Institutul Central de Studii, Cercetari Stiintifice şi Proiectare în Constructii, Architectura şi Sistematizare (Central Institute for Studies, Scientific Research and Design in Building, Architecture and Planning). 53–55 Str. Niculae Filipescu, Bucuresti 13, Roumania.

(z) Jugoslovenski Gradjevinski Centar (Yugoslav Building Centre). 21 Bozidara Adzije, Beograd, Yugoslavia.

(aa) Associazione Italiana per gli Studi dei Problemi dell'Abitazione e per le Applicazioni della Ricerca Scientifica dell'Edilizia (Italian Association for the Study of Housing Problems and the Application of the Scientific Research to Building Construction). 15 Via Lariana, Roma, Italy.

(bb) Instituto Eduardo Torroja da la Construcción y del Cemento (Building and Cement Research Institute Eduardo Torroja). Apartado 19.002, Costillares (Chamartin), Madrid 16, Spain.

(cc) Laboratório Nacional de Engenharia Civil (National Civil Engineering Laboratory). Avenida do Brasil, Lisboa, Portugal.

(dd) Building Research Station. Technion City, Haifa, Israel.

(ee) Asian Regional Institute for School Building Research, PO Box 1368, Colombo 7, Ceylon.

(ff) Regional Housing Centre. Bandung, Indonesia.

(gg) Kenchiku Kenkyujo (Building Research Institute). 4-chome, Hyakunin-cho, Shinjuku-ku, Tokyo, Japan.

(hh) Instituto del Cemento Portland Argentino (Argentine Institute of Portland Cement). Calle San Martin 1137, Buenos Aires, Argentina.

(jj) Centro Regional de Construcciones Escolares para América Latina (Regional School Building Center for Latin America). Apartado-Postal 41-518, México 10, D.F., Mexico.

3. INSTITUTES OF ARCHITECTS AND ENGINEERS

(a) American Institute of Architects. 1735 New York Avenue NW, Washington, D.C. 20006, USA.

(b) American Society of Civil Engineers; American Society of Mechanical Engineers; Institute of Electrical and Electronics Engineers; American Institute of Chemical Engineers; American Society of Heating, Refrigerating and Air-Conditioning Engineers; Illuminating Engineering Society; all at United Engineering Center, 345 East 47th Street, New York, N.Y. 10017, USA.

(c) American Academy of Environmental Engineers. PO Box 9728, Washington, D.C. 20016, USA.

(d) American Society for Engineering Education, 1345 Connecticut Avenue NW, Washington, D.C. 20036, USA.

(e) Royal Architectural Institute of Canada, 151 Slater Street, Ottawa 4, Ont., Canada.

(f) Engineering Institute of Canada, 2050 Mansfield Street, Montreal 2, P.Q., Canada.

(g) Royal Institute of British Architects, 66 Portland Place, London, W1N 4AD, England.

(h) Institution of Civil Engineers, Great George Street, London, SW1, England.

(i) Institution of Structural Engineers, 11 Upper Belgrave Street, London, SW1, England.

(j) Institution of Mechanical Engineers, 1 Birdcage Walk, London, SW1, England.

(k) Institution of Electrical Engineers, Savoy Place, London, WC2, England.

(l) Institution of Heating and Ventilating Engineers, 19 Cadogan Square, London, SW1, England.

(m) The Illuminating Engineering Society, York House, Westminster Bridge Road, London, SE1, England.

(n) Royal Society of Health, 90 Buckingham Palace Road, London, SW1, England.

(o) Royal Australian Institute of Architects, 2a Mugga Way, Canberra, ACT 2603, Australia.

(p) Institution of Engineers, Australia, 157 Gloucester Street, Sydney, NSW 2000, Australia.

(q) New Zealand Institute of Architects, PO Box 438, Wellington, C1, New Zealand.

(r) New Zealand Institution of Engineers, PO Box 12241, Wellington, New Zealand.

(s) Indian Institute of Architects, Dr Dadabhai Naoroji Road, Fort, Bombay 1, India.

(t) Institute of Engineers India, 8 Gokhali Road, Calcutta, India.

(u) Institute of South African Architects, PO Box 7322, Johannesburg, South Africa.

(v) South African Society of Civil Engineers, PO Box 1183, Johannesburg, South Africa.

(w) Conseil Supérieur de l'Ordre des Architectes. 10 Rue Portallis, Paris 16, France.

(x) Société des Ingénieurs Civils de France, 19 Rue Blanche, Paris 9, France.

(y) Bund Deutscher Architekten, Rosenburgweg 14, 5300 Bonn, German Federal Republic.

(z) Verein Deutscher Ingenieure, Prinz-Georg-Strasse 77–79, 4 Düsseldorf 10, German Federal Republic.

(aa) Academy of Architecture of the USSR, Pushkinskaya Oulitsa 36, Moscow, Russia.

(bb) The All-Union Council of Scientific and Engineering Societies, Academy of Sciences of the USSR, Leninsky Prospekt, Moscow, Russia.

(cc) The Architectural Society of the People's Republic of China, Pai Wang Chuang, West District, Peking, China.

(dd) The Chinese Society of Civil Engineering, Cheh Kung Chuan Street No. 19, West District, Peking, China.

(ee) Nippon Kenchiku Gakkai (Architectural Institute of Japan), 3-1 Ginza-Nishi, Chuo-ku, Tokyo, Japan.

(ff) Dobuku Gakkai (Japan Society of Civil Engineers), 1-Chome, Yotsuya, Shinjuku-ku, Tokyo, Japan.

(gg) Federación Nacional de Colegios de Arquitectos de la Republica Mexicana, Avenida Veracruz 24, México, D.F., Mexico.

(hh) Associación de Ingenieros y Arquitectos de Mexico, de Alvarado 58, México, D.F., Mexico.

(ii) Instituto de Arquitectos do Brasil, Avenida Rio Branco 185–8, S/817, Rio de Janeiro, Brazil.

(jj) Instituto Nacional de Tecnologia, Avenida Venezuela 81, Rio de Janeiro, Brazil.

(kk) Sociedad Central de Arquitectos, Montevideo 942, Buenos Aires, Argentina.

(ll) Centro Argentino de Ingenieros, Cerrito 1250, Buenos Aires, Argentina.

4. STANDARDS ORGANISATIONS

(a) International Organisation for Standardisation, 1 Rue de Verembé, Genève 20, Switzerland.

(b) American National Standards Institute Inc., 1430 Broadway, New York, N.Y. 10018, USA.
(*Note*: In the USA some of the most important building codes and standards are drafted by specialist institutes, such as the American Concrete Institute).

(c) American Society for Testing Materials, 1916 Race Street, Philadelphia, Pa 19104, USA.

(d) Canadian Standards Association, 178 Rexdale Boulevard, Rexdale 603, Ont., Canada.

(e) British Standards Institution, 101–113 Pentonville Road, London, N1, England.

(f) Standards Association of Australia, 80 Arthur Street, North, Sydney 2060, Australia.

(g) Standards Association of New Zealand, Private Bag, GPO, Wellington, New Zealand.

(h) Indian Standards Institution, 19 University Road, Civil Lines, Delhi 8, India.

(i) South African Bureau of Standards, Private Bag 191, Pretoria, South Africa.

(j) Association Française de Normalisation, Tour Europe 92, Courbevoie, France.

(k) Deutscher Normenausschuss, Burggrafenstrasse 4/7, 1 Berlin 30, German Federal Republic.

(l) Komitet Standartove, Mer i Izmeritel nyh Briborov pri Sovete, Ministrov USSR, Leninsky Prospekt 96, Moscow M-49, Russia.

(m) Japanese Standards Association, 1–24 Akasaka 4, Minato-ku, Tokyo, Japan.

5. UNIVERSITY DEPARTMENTS

This section is selective and limited to specialised departments in English-speaking institutions. Contributions to architectural science came from individual research workers in most major universities which have schools of architecture and engineering.

(a) Program in Building Systems Design, School of Architecture and Environmental Design, State University of New York in Buffalo. Buffalo, N.Y. 14214, USA.

(b) Department of Architectural Technology, School of Architecture, 405 Avery Hall, Columbia University. New York, N.Y. 10027, USA.

(c) Program in Architectural Science, College of Architecture, Art and Planning, Cornell University. Ithaca, N.Y. 14850, USA.

(d) Department of Architectural Engineering, Pennsylvania State University. University Park, Pa 16802, USA.

(e) Architectural Research Laboratory, College of Architecture and Design, University of Michigan. Ann Arbor, Mich. 48104, USA.

(f) Small Homes Council, University of Illinois. Urbana, Ill. 61801, USA.

(g) Environmental Systems Studies, College of Architecture, Virginia Polytechnic Institute. Blacksburg, Va 24061, USA.

(h) Program in Construction Systems, School of Engineering, Auburn University. Auburn, Ala, 36830, USA.

(i) Division of Building Research, School of Architecture, University of Southern California. Los Angeles, Cal. 9007, USA.

(j) Faculté de l'Aménagement, Université de Montréal, Montréal 101, P.Q., Canada.

(k) Department of Design, University of Waterloo. Waterloo, Ont., Canada.

(l) Bartlett School of Architecture, University of London. University College, Gower Street, London, WC1, England.

(m) Department of Building, University of Aston. Gosta Green, Birmingham 4, England.

(n) Department of Building Science, University of Liverpool, Liverpool, England.

(o) Department of Building, University of Manchester. Institute of Science and Technology, Manchester 1, England.

(p) Department of Building Science, University of Sheffield. Sheffield, England.

(q) Department of Architectural Science, University of Edinburgh. Edinburgh, Scotland.

(r) Department of Building Science, Strathclyde University. Glasgow, C1, Scotland.
(s) Department of Architectural Science, University of Sydney. Sydney, NSW 2006, Australia.
(t) Department of Building Science, Faculty of Architecture, University of Witwatersrand. Johannesburg, South Africa.

An *International Directory of Postgraduate Teaching Institutions in Building Science* will be published by the University of Sydney, Australia, in 1974.

APPENDIX D

Mathematical Tables

The following tables are included since they are not readily available. Tables of common logarithms and of circular functions in terms of degrees of angle are omitted, because they may be found in any book containing mathematical tables.

Mathematical Tables follow

1. POWERS OF NUMBERS

n	n^{-1}	n^2	n^3	$(n)^{1/2}$	$(10n)^{1/2}$	$(n)^{1/3}$	$(10n)^{1/3}$	Diff	$(100n)^{1/3}$
1	1·0	1	1	1·0	3·162 3	1·0	2·154 4		4·641 6
2	0·5	4	8	1·414 2	4·472 1	1·259 9	2·714 4		5·848 0
3	0·333 333	9	27	1·732 1	5·477 2	1·442 2	3·107 2		6·694 3
4	0·25	16	64	2·0	6·324 6	1·587 4	3·420 0		7·368 1
5	0·2	25	125	2·236 1	7·071 1	1·710 0	3·684 0		7·937 0
6	0·166 667	36	216	2·449 5	7·746 0	1·817 1	3·914 9		8·434 3
7	0·142 857	49	343	2·645 8	8·366 6	1·912 9	4·121 3		8·879 0
8	0·125	64	512	2·828 4	8·944 3	2·0	4·308 9		9·283 2
9	0·111 111	81	729	3·0	9·486 8	2·080 1	4·481 4		9·654 9
10	0·1	100	1 000	3·162 3	10·0	2·154 4	4·641 6	1 498	10·0
11	0·090 909	121	1 331	3·316 6	10·488	2·224 0	4·791 4	1 410	10·323
12	0·083 333	144	1 728	3·464 1	10·954	2·289 4	4·932 4	1 334	10·627
13	0·076 923	169	2 197	3·605 6	11·402	2·351 3	5·065 8	1 267	10·914
14	0·071 429	196	2 744	3·741 7	11·832	2·410 1	5·192 5	1 208	11·187
15	0·066 667	225	3 375	3·873 0	12·247	2·466 2	5·313 3	1 155	11·447
16	0·062 5	256	4 096	4·0	12·649	2·519 8	5·428 8	1 109	11·696
17	0·058 824	289	4 913	4·123 1	13·038	2·571 3	5·539 7	1 065	11·935
18	0·055 556	324	5 832	4·242 6	13·416	2·620 7	5·646 2	1 027	12·164
19	0·052 632	361	6 859	4·358 9	13·784	2·668 4	5·748 9	991	12·386
20	0·05	400	8 000	4·472 1	14·142	2·714 4	5·848 0	959	12·599
21	0·047 619	441	9 261	4·582 6	14·491	2·758 9	5·943 9	929	12·806
22	0·045 455	484	10 648	4·690 4	14·832	2·802 0	6·036 8	901	13·006
23	0·043 478	529	12 167	4·795 8	15·166	2·843 9	6·126 9	876	13·200
24	0·041 667	576	13 824	4·899 0	15·492	2·884 5	6·214 5	851	13·389
25	0·04	625	15 625	5·0	15·811	2·924 0	6·299 6	829	13·572

Along with the preceding column, this table of differences may be used for interpolation.

n		n²	n³						
26	0·038 462	676	17 576	5·099 0	16·125	2·962 5	6·382 5	808	13·751
27	0·037 037	729	19 683	5·196 2	16·432	3·0	6·463 3	788	13·925
28	0·035 714	784	21 952	5·291 5	16·733	3·036 6	6·542 1	770	14·095
29	0·034 483	841	24 389	5·385 2	17·029	3·072 3	6·619 1	752	14·260
30	0·033 333	900	27 000	5·477 2	17·321	3·107 2	6·694 3	736	14·422
31	0·032 258	961	29 791	5·567 8	17·607	3·141 4	6·767 9	720	14·581
32	0·031 25	1 024	32 768	5·656 9	17·889	3·174 8	6·839 9	705	14·736
33	0·030 303	1 089	35 937	5·744 6	18·166	3·207 5	6·910 4	691	14·888
34	0·029 412	1 156	39 304	5·831 0	18·439	3·239 6	6·979 5	688	15·037
35	0·028 571	1 225	42 875	5·916 1	18·708	3·271 1	7·047 3	665	15·183
36	0·027 778	1 296	46 656	6·0	18·974	3·301 9	7·113 8	653	15·326
37	0·027 027	1 369	50 653	6·082 8	19·235	3·332 2	7·179 1	641	15·467
38	0·026 316	1 444	54 872	6·164 4	19·494	3·362 0	7·243 2	629	15·605
39	0·025 641	1 521	59 319	6·245 0	19·748	3·391 2	7·300 1	620	15·741
40	0·025	1 600	64 000	6·324 6	20·0	3·420 0	7·368 1	609	15·874
41	0·024 390	1 681	68 921	6·403 1	20·248	3·448 2	7·429 0	599	16·005
42	0·023 810	1 764	74 088	6·480 7	20·494	3·476 0	7·488 9	589	16·134
43	0·023 256	1 849	79 507	6·557 4	20·736	3·503 4	7·547 8	581	16·261
44	0·022 727	1 936	85 184	6·633 2	20·976	3·530 3	7·605 9	572	16·386
45	0·022 222	2 025	91 125	6·708 2	21·213	3·556 9	7·663 1	563	16·510
46	0·021 739	2 116	97 336	6·782 3	21·448	3·583 0	7·719 4	556	16·631
47	0·021 277	2 209	103 823	6·855 7	21·679	3·608 8	7·775 0	547	16·751
48	0·020 833	2 304	110 592	6·928 2	21·909	3·634 2	7·829 7	540	16·869
49	0·020 408	2 401	117 649	7·0	22·136	3·659 3	7·883 7	533	16·985
50	0·02	2 500	125 000	7·071 1	22·361	3·684 0	7·937 0	526	17·100

1. POWERS OF NUMBERS—contd.

n	n^{-1}	n^2	n^3	$(n)^{1/2}$	$(10n)^{1/2}$	$(n)^{1/3}$	$(10n)^{1/3}$	Diff	$(100n)^{1/3}$
51	0·019 608	2 601	132 651	7·141 4	22·593	3·708 4	7·989 6	519	17·213
52	0·019 231	2 704	140 608	7·211 1	22·804	3·732 5	8·041 5	512	17·325
53	0·018 868	2 809	148 877	7·280 1	23·022	3·756 3	8·092 7	506	17·435
54	0·018 519	2 916	157 464	7·348 5	23·238	3·779 8	8·143 3	499	17·544
55	0·018 182	3 025	166 375	7·416 2	23·452	3·803 0	8·193 2	494	17·652
56	0·017 857	3 136	175 616	7·483 3	23·664	3·825 9	8·242 6	487	17·758
57	0·017 544	3 249	185 193	7·549 8	23·875	3·848 5	8·291 3	483	17·863
58	0·017 241	3 364	195 112	7·615 8	24·083	3·870 9	8·339 6	476	17·967
59	0·016 949	3 481	205 379	7·681 1	24·290	3·893 0	8·387 2	471	18·070
60	0·016 667	3 600	216 000	7·746 0	24·495	3·914 9	8·434 3	466	18·171
61	0·016 393	3 721	226 981	7·810 2	24·698	3·936 5	8·480 9	461	18·272
62	0·016 129	3 844	238 328	7·874 0	24·900	3·957 9	8·527 0	456	18·371
63	0·015 873	3 969	250 047	7·937 3	25·100	3·979 1	8·572 6	451	18·469
64	0·015 625	4 096	262 144	8·0	25·298	4·0	8·617 7	447	18·566
65	0·015 385	4 225	274 625	8·062 3	25·495	4·020 7	8·662 4	442	18·663
66	0·015 152	4 356	287 496	8·124 0	25·690	4·041 2	8·706 6	437	18·758
67	0·014 925	4 489	300 763	8·185 4	25·884	4·061 5	8·750 3	434	18·852
68	0·014 706	4 624	314 432	8·246 2	26·077	4·081 7	8·793 7	429	18·945
69	0·014 493	4 761	328 509	8·306 6	26·268	4·101 6	8·836 6	424	19·038
70	0·014 286	4 900	343 000	8·366 6	26·458	4·121 3	8·879 0	421	19·129
71	0·014 085	5 041	357 911	8·426 1	26·646	4·140 8	8·921 1	417	19·220
72	0·013 889	5 184	373 248	8·485 3	26·833	4·160 2	8·962 8	413	19·310
73	0·013 699	5 329	389 017	8·544 0	27·019	4·179 3	9·004 1	409	19·399
74	0·013 514	5 476	405 224	8·602 3	27·203	4·198 3	9·045 0	406	19·487
75	0·013 333	5 625	421 875	8·660 3	27·386	4·217 2	9·085 6	402	19·574

336

n	$1/n$	n^2	n^3	$\sqrt{n}$	$\sqrt{10n}$	$\sqrt[3]{n}$	$\sqrt[3]{10n}$		$\sqrt[3]{100n}$
76	0·013 158	5 776	438 976	8·717 8	27·568	4·235 8	9·125 8	399	19·661
77	0·012 987	5 929	456 533	8·775 0	27·749	4·254 3	9·165 7	395	19·747
78	0·012 821	6 084	474 552	8·831 8	27·928	4·272 7	9·205 2	391	19·832
79	0·012 658	6 241	493 039	8·888 2	28·107	4·290 8	9·244 3	389	19·916
80	0·012 5	6 400	512 000	8·944 3	28·284	4·308 9	9·283 2	385	20·0
81	0·012 346	6 561	531 441	9·0	28·460	4·326 7	9·321 7	382	20·083
82	0·012 195	6 724	551 368	9·055 4	28·636	4·344 5	9·359 9	379	20·165
83	0·012 048	6 889	571 787	9·110 4	28·810	4·362 1	9·397 8	376	20·247
84	0·011 905	7 056	592 704	9·165 2	28·983	4·379 5	9·435 4	373	20·328
85	0·011 765	7 225	614 125	9·219 5	29·155	4·396 8	9·472 7	370	20·408
86	0·011 628	7 396	636 056	9·273 6	29·326	4·414 0	9·509 7	367	20·488
87	0·011 494	7 569	658 503	9·327 4	29·496	4·431 0	9·546 4	364	20·567
88	0·011 364	7 744	681 472	9·380 8	29·665	4·448 0	9·582 8	362	20·646
89	0·011 236	7 921	704 969	9·434 0	29·833	4·464 7	9·619 0	359	20·724
90	0·011 111	8 100	729 000	9·486 8	30·0	4·481 4	9·654 9	357	20·801
91	0·010 989	8 281	753 571	9·539 4	30·166	4·497 9	9·690 5	354	20·878
92	0·010 870	8 464	778 688	9·591 7	30·332	4·514 4	9·725 9	351	20·954
93	0·010 753	8 649	804 357	9·643 7	30·496	4·530 7	9·761 0	349	21·029
94	0·010 638	8 836	830 584	9·695 4	30·659	4·546 8	9·795 9	346	21·105
95	0·010 526	9 025	857 375	9·746 8	30·822	4·562 9	9·830 5	343	21·179
96	0·010 417	9 216	884 736	9·798 0	30·984	4·578 9	9·864 8	342	21·253
97	0·010 309	9 409	912 673	9·848 9	31·145	4·594 7	9·899 0	339	21·327
98	0·010 204	9 604	941 192	9·899 5	31·305	4·610 4	9·932 9	337	21·400
99	0·010 101	9 801	970 299	9·949 9	31·464	4·626 1	9·966 6	334	21·472
100	0·01	10 000	1 000 000	10·0	31·623	4·641 6	10·0		21·544

2. CIRCULAR FUNCTIONS IN TERMS OF RADIANS

Degree	Radian	Sine	Tangent	Cosine	Logarithm of Sine	Logarithm of Tangent	Logarithm of Cosine
5·73	0·10	0·099 8	0·100 3	0·995 0	$\bar{2}$·999 3	$\bar{1}$·001 5	$\bar{1}$·997 8
6·30	0·11	0·109 8	0·110 4	0·994 0	$\bar{1}$·040 5	$\bar{1}$·043 1	$\bar{1}$·997 4
6·88	0·12	0·119 7	0·120 6	0·992 8	$\bar{1}$·078 1	$\bar{1}$·081 3	$\bar{1}$·996 9
7·45	0·13	0·129 6	0·130 7	0·991 6	$\bar{1}$·112 7	$\bar{1}$·116 4	$\bar{1}$·996 3
8·02	0·14	0·139 5	0·140 9	0·990 2	$\bar{1}$·144 7	$\bar{1}$·149 0	$\bar{1}$·995 7
8·59	0·15	0·149 4	0·151 1	0·988 8	$\bar{1}$·174 5	$\bar{1}$·179 4	$\bar{1}$·995 1
9·17	0·16	0·159 3	0·161 4	0·987 2	$\bar{1}$·202 3	$\bar{1}$·207 8	$\bar{1}$·994 4
9·74	0·17	0·169 2	0·171 7	0·985 6	$\bar{1}$·228 4	$\bar{1}$·234 7	$\bar{1}$·993 7
10·31	0·18	0·179 0	0·182 0	0·983 9	$\bar{1}$·252 9	$\bar{1}$·260 0	$\bar{1}$·992 9
10·89	0·19	0·188 9	0·192 3	0·982 0	$\bar{1}$·276 1	$\bar{1}$·284 0	$\bar{1}$·992 1
11·46	0·20	0·198 7	0·202 7	0·980 1	$\bar{1}$·298 1	$\bar{1}$·306 9	$\bar{1}$·991 3
12·03	0·21	0·208 5	0·213 1	0·978 0	$\bar{1}$·319 0	$\bar{1}$·328 7	$\bar{1}$·990 4
12·61	0·22	0·218 2	0·223 6	0·975 9	$\bar{1}$·338 9	$\bar{1}$·349 5	$\bar{1}$·989 4
13·18	0·23	0·228 0	0·234 1	0·973 7	$\bar{1}$·357 9	$\bar{1}$·369 5	$\bar{1}$·988 4
13·75	0·24	0·237 7	0·244 7	0·971 3	$\bar{1}$·376 0	$\bar{1}$·388 7	$\bar{1}$·987 4
14·32	0·25	0·247 4	0·255 3	0·968 9	$\bar{1}$·393 4	$\bar{1}$·407 1	$\bar{1}$·986 3
14·90	0·26	0·257 1	0·266 0	0·966 4	$\bar{1}$·410 1	$\bar{1}$·424 9	$\bar{1}$·985 2
15·47	0·27	0·266 7	0·276 8	0·963 8	$\bar{1}$·426 1	$\bar{1}$·442 1	$\bar{1}$·984 0
16·04	0·28	0·276 4	0·287 6	0·961 1	$\bar{1}$·441 5	$\bar{1}$·458 7	$\bar{1}$·982 7
16·62	0·29	0·285 9	0·298 4	0·958 3	$\bar{1}$·456 3	$\bar{1}$·474 8	$\bar{1}$·981 5
17·19	0·30	0·295 5	0·309 3	0·955 3	$\bar{1}$·470 6	$\bar{1}$·490 4	$\bar{1}$·980 2
17·76	0·31	0·305 1	0·320 3	0·952 3	$\bar{1}$·484 4	$\bar{1}$·505 6	$\bar{1}$·978 8
18·33	0·32	0·314 6	0·331 4	0·949 2	$\bar{1}$·497 7	$\bar{1}$·520 3	$\bar{1}$·977 4
18·91	0·33	0·324 0	0·342 5	0·946 0	$\bar{1}$·510 6	$\bar{1}$·534 7	$\bar{1}$·975 9
19·48	0·34	0·333 5	0·353 7	0·942 8	$\bar{1}$·522 1	$\bar{1}$·549 7	$\bar{1}$·974 4

20·63	0·36	0·352 3	0·376 4	0·935 9	1·546 9	1·575 7	1·971 2
21·20	0·37	0·361 6	0·387 9	0·932 3	1·558 2	1·588 7	1·969 6
21·77	0·38	0·370 9	0·399 4	0·928 7	1·569 3	1·601 4	1·967 9
22·34	0·39	0·380 2	0·411 1	0·924 9	1·580 0	1·613 9	1·966 1
22·92	0·40	0·389 4	0·422 8	0·921 1	1·590 4	1·626 1	1·964 3
23·49	0·41	0·398 6	0·434 6	0·917 1	1·600 5	1·638 1	1·962 4
24·06	0·42	0·407 8	0·446 6	0·913 1	1·610 4	1·649 9	1·960 5
24·64	0·43	0·416 9	0·458 6	0·909 0	1·620 0	1·661 5	1·958 6
25·21	0·44	0·425 9	0·470 8	0·904 8	1·629 3	1·672 8	1·956 5
25·78	0·45	0·435 0	0·483 1	0·900 5	1·638 5	1·684 0	1·954 5
26·36	0·46	0·443 9	0·495 4	0·896 1	1·647 3	1·695 0	1·952 3
26·93	0·47	0·452 9	0·508 0	0·891 6	1·656 0	1·705 8	1·950 2
27·50	0·48	0·461 8	0·520 6	0·887 0	1·664 4	1·716 5	1·947 9
28·07	0·49	0·470 6	0·533 4	0·882 3	1·672 7	1·727 0	1·945 6
28·65	0·50	0·479 4	0·546 3	0·877 6	1·680 7	1·737 4	1·943 3
29·22	0·51	0·488 2	0·559 4	0·872 7	1·688 6	1·747 7	1·940 9
29·79	0·52	0·496 9	0·572 6	0·867 8	1·696 2	1·757 8	1·938 4
30·37	0·53	0·505 5	0·585 9	0·862 8	1·703 7	1·767 8	1·935 9
30·94	0·54	0·514 1	0·599 4	0·857 7	1·711 1	1·777 7	1·933 3
31·51	0·55	0·522 7	0·613 1	0·852 5	1·718 2	1·787 5	1·930 7
32·09	0·56	0·531 2	0·626 9	0·847 3	1·725 3	1·797 2	1·928 0
32·66	0·57	0·539 6	0·641 0	0·841 9	1·732 1	1·806 8	1·925 3
33·23	0·58	0·548 0	0·655 2	0·836 5	1·738 8	1·816 4	1·922 4
33·80	0·59	0·556 4	0·669 6	0·830 9	1·745 4	1·825 8	1·919 6
34·37	0·60	0·564 6	0·684 1	0·825 3	1·751 8	1·835 1	1·916 6

Degree	Radian	Sine	Tangent	Cosine	Logarithm of Sine	Logarithm of Tangent	Logarithm of Cosine
34·38	0·60	0·564 6	0·684 1	0·825 3	$\bar{1}$·751 8	$\bar{1}$·835 1	$\bar{1}$·916 6
34·95	0·61	0·572 9	0·698 9	0·819 6	$\bar{1}$·758 1	$\bar{1}$·844 4	$\bar{1}$·913 6
35·52	0·62	0·581 0	0·713 9	0·813 9	$\bar{1}$·764 2	$\bar{1}$·853 6	$\bar{1}$·910 6
36·10	0·63	0·589 1	0·729 1	0·808 0	$\bar{1}$·770 2	$\bar{1}$·862 8	$\bar{1}$·907 4
36·67	0·64	0·597 2	0·744 5	0·802 1	$\bar{1}$·776 1	$\bar{1}$·871 9	$\bar{1}$·904 2
37·24	0·65	0·605 2	0·760 2	0·796 1	$\bar{1}$·781 9	$\bar{1}$·880 9	$\bar{1}$·901 0
37·82	0·66	0·613 1	0·776 1	0·790 0	$\bar{1}$·787 5	$\bar{1}$·889 9	$\bar{1}$·897 6
38·39	0·67	0·621 0	0·792 3	0·783 8	$\bar{1}$·793 1	$\bar{1}$·898 9	$\bar{1}$·894 2
38·96	0·68	0·628 8	0·808 7	0·777 6	$\bar{1}$·798 5	$\bar{1}$·907 8	$\bar{1}$·890 7
39·53	0·69	0·636 5	0·825 3	0·771 2	$\bar{1}$·803 8	$\bar{1}$·916 6	$\bar{1}$·887 2
40·11	0·70	0·644 2	0·842 3	0·764 8	$\bar{1}$·809 0	$\bar{1}$·925 5	$\bar{1}$·883 6
40·68	0·71	0·651 8	0·859 5	0·758 4	$\bar{1}$·814 1	$\bar{1}$·934 3	$\bar{1}$·879 9
41·25	0·72	0·659 4	0·877 1	0·751 8	$\bar{1}$·819 1	$\bar{1}$·943 0	$\bar{1}$·876 1
41·83	0·73	0·666 9	0·894 9	0·745 2	$\bar{1}$·824 0	$\bar{1}$·951 8	$\bar{1}$·872 3
42·40	0·74	0·674 3	0·913 1	0·738 5	$\bar{1}$·828 8	$\bar{1}$·960 5	$\bar{1}$·868 3
42·97	0·75	0·681 6	0·931 6	0·731 7	$\bar{1}$·833 6	$\bar{1}$·969 2	$\bar{1}$·864 3
43·54	0·76	0·688 9	0·950 5	0·724 8	$\bar{1}$·838 2	$\bar{1}$·977 9	$\bar{1}$·860 2
44·12	0·77	0·696 1	0·969 7	0·717 9	$\bar{1}$·842 7	$\bar{1}$·986 6	$\bar{1}$·856 1
44·69	0·78	0·703 3	0·989 3	0·710 9	$\bar{1}$·847 1	$\bar{1}$·995 3	$\bar{1}$·851 8
45·26	0·79	0·710 4	1·009 2	0·703 8	$\bar{1}$·851 5	0·004 0	$\bar{1}$·847 5
45·84	0·80	0·717 4	1·029 6	0·696 7	$\bar{1}$·855 7	0·012 7	$\bar{1}$·843 0
46·41	0·81	0·724 3	1·050 5	0·689 5	$\bar{1}$·859 9	0·021 4	$\bar{1}$·838 5
46·98	0·82	0·731 1	1·071 7	0·682 2	$\bar{1}$·864 0	0·030 1	$\bar{1}$·833 9
47·56	0·83	0·737 9	1·093 4	0·674 9	$\bar{1}$·868 0	0·038 8	$\bar{1}$·829 2

$\bar{1}$·814 5	0·065 0	$\bar{1}$·879 6	0·652 4	1·161 6	0·757 8	0·86	49·27
$\bar{1}$·809 4	0·073 8	$\bar{1}$·883 3	0·644 8	1·185 3	0·764 3	0·87	49·85
$\bar{1}$·804 2	0·082 7	$\bar{1}$·886 9	0·637 2	1·209 7	0·770 7	0·88	50·42
$\bar{1}$·798 9	0·091 5	$\bar{1}$·890 5	0·629 4	1·234 6	0·777 1	0·89	50·99
$\bar{1}$·793 5	0·100 4	$\bar{1}$·893 9	0·621 6	1·260 2	0·783 3	0·90	51·57
$\bar{1}$·788 0	0·109 4	$\bar{1}$·897 4	0·613 7	1·286 4	0·789 5	0·91	52·14
$\bar{1}$·782 3	0·118 4	$\bar{1}$·900 7	0·605 8	1·313 3	0·795 6	0·92	52·71
$\bar{1}$·776 6	0·127 4	$\bar{1}$·904 0	0·597 8	1·340 9	0·801 6	0·93	53·29
$\bar{1}$·770 7	0·136 5	$\bar{1}$·907 2	0·589 8	1·369 2	0·807 6	0·94	53·86
$\bar{1}$·764 7	0·145 6	$\bar{1}$·910 3	0·581 7	1·398 4	0·813 4	0·95	54·43
$\bar{1}$·758 5	0·154 8	$\bar{1}$·913 4	0·573 5	1·428 4	0·819 2	0·96	55·00
$\bar{1}$·752 3	0·164 1	$\bar{1}$·916 4	0·565 3	1·459 2	0·824 9	0·97	55·58
$\bar{1}$·745 9	0·173 5	$\bar{1}$·919 3	0·557 0	1·490 9	0·830 5	0·98	56·15
$\bar{1}$·739 3	0·182 9	$\bar{1}$·922 2	0·548 7	1·523 7	0·836 0	0·99	56·72
$\bar{1}$·732 6	0·192 4	$\bar{1}$·925 0	0·540 3	1·557 4	0·841 5	1·00	57·30
$\bar{1}$·725 8	0·202 0	$\bar{1}$·927 8	0·531 9	1·592 2	0·846 8	1·01	57·87
$\bar{1}$·718 8	0·211 7	$\bar{1}$·930 5	0·523 4	1·628 1	0·852 1	1·02	58·44
$\bar{1}$·711 7	0·221 5	$\bar{1}$·933 1	0·514 8	1·665 3	0·857 3	1·03	59·01
$\bar{1}$·704 3	0·231 4	$\bar{1}$·935 7	0·506 2	1·703 6	0·862 4	1·04	59·59
$\bar{1}$·696 9	0·241 4	$\bar{1}$·938 2	0·497 6	1·743 3	0·867 4	1·05	60·16
$\bar{1}$·689 2	0·251 5	$\bar{1}$·940 7	0·488 9	1·784 4	0·872 4	1·06	60·73
$\bar{1}$·681 4	0·261 8	$\bar{1}$·943 1	0·480 1	1·827 0	0·877 2	1·07	61·31
$\bar{1}$·673 3	0·272 1	$\bar{1}$·945 4	0·471 3	1·871 2	0·882 0	1·08	61·88
$\bar{1}$·665 1	0·282 6	$\bar{1}$·947 7	0·462 5	1·917 1	0·886 6	1·09	62·45
$\bar{1}$·656 7	0·293 3	$\bar{1}$·950 0	0·453 6	1·964 8	0·891 2	1·10	63·03

341

2. CIRCULAR FUNCTIONS IN TERMS OF RADIANS—*contd.*

Degree	Radian	Sine	Tangent	Cosine	Logarithm of Sine	Logarithm of Tangent	Logarithm of Cosine
63·03	1·10	0·891 2	1·965	0·453 6	1̄·950 0	0·293 3	1̄·656 7
63·60	1·11	0·895 7	2·014	0·444 7	1̄·952 2	0·304 1	1̄·648 0
64·17	1·12	0·900 1	2·066	0·435 7	1̄·954 3	0·315 1	1̄·639 2
64·74	1·13	0·904 4	2·120	0·426 7	1̄·956 4	0·326 3	1̄·630 1
65·32	1·14	0·908 6	2·176	0·417 6	1̄·958 4	0·337 6	1̄·620 8
65·89	1·15	0·912 8	2·234	0·408 5	1̄·960 4	0·349 2	1̄·611 2
66·46	1·16	0·916 8	2·296	0·399 3	1̄·962 3	0·360 9	1̄·601 3
67·04	1·17	0·920 8	2·360	0·390 2	1̄·964 1	0·372 9	1̄·591 2
67·61	1·18	0·924 6	2·427	0·380 9	1̄·966 0	0·385 1	1̄·580 8
68·18	1·19	0·928 4	2·498	0·371 7	1̄·967 7	0·397 6	1̄·570 1
68·75	1·20	0·932 0	2·572	0·362 4	1̄·969 4	0·410 3	1̄·559 1
69·33	1·21	0·935 6	2·650	0·353 0	1̄·971 1	0·423 3	1̄·547 8
69·90	1·22	0·939 1	2·733	0·343 6	1̄·972 7	0·436 6	1̄·536 1
70·47	1·23	0·942 5	2·820	0·334 2	1̄·974 3	0·450 2	1̄·524 1
71·05	1·24	0·945 8	2·912	0·324 8	1̄·975 8	0·464 2	1̄·511 6
71·62	1·25	0·949 0	3·010	0·315 3	1̄·977 3	0·478 5	1̄·498 8
72·19	1·26	0·952 1	3·113	0·305 8	1̄·978 7	0·493 2	1̄·485 5
72·77	1·27	0·955 1	3·224	0·296 3	1̄·980 0	0·508 3	1̄·471 7
73·34	1·28	0·958 0	3·341	0·286 7	1̄·981 4	0·523 9	1̄·457 5
73·91	1·29	0·960 8	3·467	0·277 1	1̄·982 6	0·540 0	1̄·442 7
74·48	1·30	0·963 6	3·602	0·267 5	1̄·983 9	0·556 6	1̄·427 3
75·06	1·31	0·966 2	3·747	0·257 9	1̄·985 1	0·573 7	1̄·411 4
75·63	1·32	0·968 7	3·903	0·248 2	1̄·986 2	0·591 4	1̄·394 8
76·20	1·33	0·971 2	4·072	0·238 5	1̄·987 3	0·609 8	1̄·377 4

$\overline{1}$·320 6	0·669 6	$\overline{1}$·990 3	0·209 2	4·674	0·977 9	1·36	77·92
$\overline{1}$·299 8	0·691 4	$\overline{1}$·991 2	0·199 5	4·913	0·979 9	1·37	78·50
$\overline{1}$·277 9	0·714 1	$\overline{1}$·992 0	0·189 6	5·177	0·981 9	1·38	79·07
$\overline{1}$·254 8	0·738 0	$\overline{1}$·992 9	0·179 8	5·471	0·983 7	1·39	79·64
$\overline{1}$·230 4	0·763 3	$\overline{1}$·993 6	0·170 0	5·798	0·985 4	1·40	80·21
$\overline{1}$·204 4	0·790 0	$\overline{1}$·994 4	0·160 1	6·165	0·987 1	1·41	80·79
$\overline{1}$·176 7	0·818 3	$\overline{1}$·995 0	0·150 2	6·581	0·988 6	1·42	81·36
$\overline{1}$·147 2	0·848 5	$\overline{1}$·995 7	0·140 3	7·055	0·990 1	1·43	81·93
$\overline{1}$·115 4	0·881 9	$\overline{1}$·996 3	0·130 4	7·619	0·991 5	1·44	82·51
$\overline{1}$·081 0	0·915 8	$\overline{1}$·996 8	0·120 5	8·238	0·992 7	1·45	83·08
$\overline{1}$·043 6	0·953 7	$\overline{1}$·997 3	0·110 6	8·989	0·993 9	1·46	83·65
$\overline{1}$·002 7	0·995 1	$\overline{1}$·997 8	0·100 6	9·887	0·994 9	1·47	84·23
$\overline{2}$·957 5	1·040 7	$\overline{1}$·998 2	0·090 7	10·984	0·995 9	1·48	84·80
$\overline{2}$·906 9	1·091 7	$\overline{1}$·998 6	0·080 7	12·350	0·996 7	1·49	85·37
$\overline{2}$·849 7	1·149 3	$\overline{1}$·998 9	0·070 7	14·101	0·997 5	1·50	85·94
$\overline{2}$·783 6	1·215 6	$\overline{1}$·999 2	0·060 8	16·428	0·998 2	1·51	86·52
$\overline{2}$·705 6	1·293 8	$\overline{1}$·999 4	0·050 8	19·670	0·998 7	1·52	87·09
$\overline{2}$·610 5	1·389 1	$\overline{1}$·999 6	0·040 8	24·499	0·999 2	1·53	87·66
$\overline{2}$·479 5	1·520 3	$\overline{1}$·999 8	0·030 2	33·138	0·999 5	1·54	88·24
$\overline{2}$·317 8	1·682 1	$\overline{1}$·999 9	0·020 8	48·101	0·999 8	1·55	88·81
$\overline{2}$·033 3	1·966 7	0·000 0	0·010 8	92·623	0·999 9	1·56	89·38
$\overline{3}$·901 1	3·098 9	0·000 0	0·008 0	1 255·866	1·000 0	1·57	89·95
$-\infty$	$+\infty$	0·000 0	0·000 0	∞	1·000 0	$\pi/2$	90·00

343

3. EXPONENTIAL AND HYPERBOLIC FUNCTIONS

x	e^x	e^{-x}	$\cosh x$	$\sinh x$	$\tanh x$	$\log \cosh x$	$\log \sinh x$
0·1	1·105 2	0·904 8	1·005 0	0·100 2	0·099 7	0·002 2	$\bar{1}$·000 7
0·2	1·221 4	0·818 7	1·020 1	0·201 3	0·197 4	0·008 6	$\bar{1}$·303 9
0·3	1·349 9	0·740 8	1·045 3	0·304 5	0·291 3	0·019 3	$\bar{1}$·483 6
0·4	1·491 8	0·670 3	1·081 1	0·410 8	0·379 9	0·033 9	$\bar{1}$·613 6
0·5	1·648 7	0·606 5	1·127 6	0·521 1	0·462 1	0·052 2	$\bar{1}$·716 9
0·6	1·822 1	0·548 8	1·185 5	0·636 7	0·537 0	0·073 9	$\bar{1}$·803 9
0·7	2·013 8	0·496 6	1·255 2	0·758 6	0·604 4	0·098 7	$\bar{1}$·880 0
0·8	2·225 5	0·449 3	1·337 4	0·888 1	0·664 0	0·126 3	$\bar{1}$·948 5
0·9	2·459 6	0·406 6	1·433 1	1·026 5	0·716 3	0·156 3	0·011 4
1·0	2·718 3	0·367 9	1·543 1	1·175 2	0·761 6	0·188 4	0·070 1
1·1	3·004 2	0·332 9	1·668 5	1·335 7	0·800 5	0·222 3	0·125 7
1·2	3·320 1	0·301 2	1·810 7	1·509 5	0·833 7	0·257 8	0·178 8
1·3	3·669 3	0·272 5	1·970 9	1·698 4	0·861 7	0·294 7	0·230 0
1·4	4·055 2	0·246 6	2·150 9	1·904 3	0·885 4	0·332 6	0·279 7
1·5	4·481 7	0·223 1	2·352 4	2·129 3	0·905 1	0·371 5	0·328 2
1·6	4·953 0	0·201 9	2·577 5	2·375 6	0·921 7	0·411 2	0·375 8
1·7	5·473 9	0·182 7	2·828 3	2·645 6	0·935 4	0·451 5	0·422 5
1·8	6·049 6	0·165 3	3·107 5	2·942 2	0·946 8	0·492 4	0·468 7
1·9	6·685 9	0·149 6	3·417 7	3·268 2	0·956 3	0·533 7	0·514 3
2·0	7·389 1	0·135 3	3·762 2	3·626 9	0·964 0	0·575 4	0·559 5
2·1	8·166 2	0·122 5	4·144 3	4·021 9	0·970 4	0·617 5	0·604 4
2·2	9·025 1	0·110 8	4·567 9	4·457 1	0·975 8	0·659 7	0·649 1
2·3	9·974 2	0·100 3	5·037 2	4·937 0	0·980 1	0·702 2	0·693 5
2·4	11·023 2	0·090 7	5·557 0	5·466 2	0·983 7	0·744 8	0·737 7
2·5	12·182 5	0·082 1	6·132 3	6·050 2	0·986 6	0·787 6	0·781 8

x							
2·6	0·825 7	0·830 5	0·989 0	6·694 7	6·769 0	0·074 3	13·463 8
2·7	0·869 6	0·873 5	0·991 0	7·406 3	7·473 5	0·067 2	14·879 7
2·8	0·913 4	0·916 6	0·992 6	8·191 9	8·252 7	0·060 8	16·444 6
2·9	0·957 1	0·959 7	0·994 0	9·059 6	9·114 6	0·055 0	18·174 1
3·0	1·000 8	1·002 9	0·995 1	10·018	10·068	0·049 8	20·085 5
3·1	1·044 4	1·046 2	0·995 9	11·076	11·122	0·045 0	22·198 0
3·2	1·088 0	1·089 4	0·996 7	12·246	12·287	0·040 8	24·532 5
3·3	1·131 6	1·132 7	0·997 3	13·538	13·575	0·036 9	27·112 6
3·4	1·175 1	1·176 1	0·997 8	14·965	14·999	0·033 4	29·964 1
3·5	1·218 6	1·219 4	0·998 2	16·543	16·573	0·030 2	33·115 5
3·6	1·262 1	1·262 8	0·998 5	18·285	18·313	0·027 3	36·598 2
3·7	1·305 6	1·306 1	0·998 8	20·211	20·236	0·024 7	40·447 3
3·8	1·349 1	1·349 5	0·999 0	22·339	22·362	0·022 4	44·701 2
3·9	1·392 5	1·392 9	0·999 2	24·691	24·711	0·020 2	49·402 4
4·0	1·436 0	1·436 3	0·999 3	27·290	27·308	0·018 3	54·598 2
4·1	1·479 5	1·479 7	0·999 5	30·162	30·178	0·016 6	60·340 3
4·2	1·522 9	1·523 1	0·999 6	33·336	33·351	0·015 0	66·686 3
4·3	1·566 4	1·566 5	0·999 6	36·843	36·857	0·013 6	73·699 8
4·4	1·609 8	1·609 9	0·999 7	40·719	40·732	0·012 3	81·450 9
4·5	1·653 2	1·653 3	0·999 7	45·003	45·014	0·011 1	90·017 1
4·6	1·696 7	1·696 8	0·999 8	49·737	49·747	0·010 1	99·484 3
4·7	1·740 1	1·740 2	0·999 8	54·969	54·978	0·009 1	109·947 2
4·8	1·783 6	1·783 6	0·999 9	60·751	60·759	0·008 2	121·510 4
4·9	1·827 0	1·827 0	0·999 9	67·141	67·149	0·007 4	134·289 8
5·0	1·870 4	1·870 5	0·999 9	74·203	74·210	0·006 7	148·413 2

The Periodic Table of the Chemical Elements

Group 0	Group I a	Group I b	Group II a	Group II b	Group III a	Group III b	Group IV a	Group IV b	Group V a	Group V b	Group VI a	Group VI b	Group VII a	Group VII b	Group VIII
	1 H														
2 He	3 Li		4 Be		5 B		6 C		7 N		8 O		9 F		
10 Ne	11 Na		12 Mg		13 Al		14 Si		15 P		16 S		17 Cl		
18 A	19 K		20 Ca		21 Sc		22 Ti		23 V		24 Cr		25 Mn		26 Fe 27 Co 28 Ni
		29 Cu		30 Zn		31 Ga		32 Ge		33 As		34 Se		35 Br	
36 Kr	37 Rb		38 Sr		39 Y		40 Zr		41 Nb		42 Mo		43 Ma		44 Ru 45 Rh 46 Pd
		47 Ag		48 Cd		49 In		50 Sn		51 Sb		52 Te		53 I	
54 Xe	55 Cs		56 Ba		57 La		72 Hf		73 Ta		74 W		75 Re		76 Os 77 Ir 78 Pt
		79 Au		80 Hg		81 Tl		82 Pb		83 Bi		84 Po		85 At	
86 Rn	87 Fr		88 Ra		89 Ac		90 Th		91 Pa		92 U		93 Np		94 Pu 95 Am 96 Cm
		97 Bk		98 Cf		99 E		100 Fm		101 Mv		102 No			

Rare-earth elements (box, in Group III):

57 La 58 Ce 59 Pr 60 Nd 61 Il 62 Sm 63 Eu 64 Gd 65 Tb

66 Dy 67 Ho 68 Er 69 Tm 70 Yb 71 Lu

The elements, shown by their chemical symbol, are arranged in ascending order of their atomic number, which is equal to the total number of planetary electrons in an atom of the element, or to the net positive charge on the nucleus. With a few exceptions, this is also the order of increasing atomic weight.

Apart from Group 0, which is chemically inert, electropositive (metallic) properties are most marked towards the bottom left-hand corner of the table, and electro-negative properties towards the top right-hand corner. Elements in the same vertical line have a similar chemical character.

The rare-earth elements, Nos. 57–71, are regarded as occupying a single place in the periodic system.

APPENDIX F

Mass per Unit Volume of Common Materials

Note: In the metric SI system, the force due to a given mass is not numerically equal to it, since 1 kgf = 9·81 N (*see* Appendix G).

	lb/ft^3	*kg/m*^{3}
Water		
Fresh water at 4°C	62·425	1000
Fresh water at 100°C	59·8	958
Sea water	64	1020
Freshly fallen snow	8	130
Ice	57	916
Other liquids		
Alcohol	50	800
Olive oil	57	910
Crude petroleum	55	880
Petrol (gasoline)	43–48	690–770
Mineral lubricating oil	57	910
Mercury	847	13 560
Minerals		
Loose earth	80	1300
Packed earth	90–110	1450–1750
Undisturbed clay	120–140	1900–2250
Sand	90–100	1450–1600
Gravel	100–120	1600–1900
Limestone or sandstone	140–160	2250–2550
Dolomite	180	2880
Marble	170	2700
Quartz or flint	165	2650
Shale or slate	170–185	2700–2950
Granite	165	2650
Natural pumice	40	640
Soapstone or talc	170	2720
Gypsum or alabaster	160	2550

	lb/ft^3	kg/m^3
Brick and concrete		
Brickwork	110–140	1750–2250
Cellular concrete	50–60	800–950
Concrete made with lightweight aggregate	90–110	1450–1750
Normal concrete	130–155	2050–2500
Metals		
Magnesium	110	1740
Pure aluminium	168	2690
Cast aluminium	160	2560
Wrought iron	480	7700
Cast iron	450	7200
Steel	490	7850
Copper	560	8960
Brass	510–530	8200–8500
Bronze	540–550	8700–8800
Lead	710	11 340
Timber		
Balsa wood	7	110
Western red cedar (air dried)	23	370
Douglas fir (air dried)	33	530
Eucalypts (air dried) up to	69	1100
Eucalypts (green) up to	80	1280
Miscellaneous materials		
Cork	15	240
Raw rubber	58	930
Vulcanised rubber	75	1200
Plastics	70–100	1100–1600
Glass	155–195	2500–3100
Cement	80–95	1300–1500
Ash	40–45	640–720

APPENDIX G

Metric Conversion Tables

1. BRITISH TO METRIC UNITS

Length
1 mile (m)	= 1·609 kilometres (km)
1 engineer's chain	= 30·48 metres (m)
1 chain	= 20·117 m
1 yard (yd)	= 0·914 m
1 foot (ft)	= 0·3048 m
1 inch (in)	= 25·40 millimetres (mm)

Area
1 square mile	= 2·590 square kilometres
1 acre	= 4047 square metres
1 square chain	= 404·7 m²
1 square yard	= 0·836 m²
1 square foot	= 929 square centimetres
1 square inch	= 6·452 cm²
Note: 1 acre	= 10 square chains

Volume
1 cubic yard	= 0·7646 cubic metre
1 cubic foot	= 28·32 cubic decimetres
1 board foot of timber	= 2·360 dm³
1 cubic inch	= 16·39 cubic centimetres
Note: 1 board foot	= 1 square foot × 1 inch

Capacity
1 Imperial gallon	= 4·546 litres (l)
1 US gal	= 3·785 l
Notes: 1 Imperial gallon	= 1·200 94 US gallons
1 litre	= 1 dm³

Second Moment of Area
1 ft⁴	= 86·3 dm⁴
1 in⁴	= 41·62 cm⁴

Velocity
  1 mile per hour (mph)   = 1·609 km/h
  1 ft/s   = 0·3048 m/s

Mass
  1 long ton = 2240 pounds (lb) = 1016 kilograms (kg)
                                  = 1·016 tonnes
  1 short ton = 2000 lb   = 0·907 tonnes
  1 lb   = 0·4536 kg
  1 ounce   = 28·35 g

Mass per Unit Area
  1 lb/ft^2 (psf)   = 4·882 kg/m^2
  1 lb/in^2 (psi)   = 703 kg/m^2

Mass per Unit Volume
  1 lb/ft^3 = 16·019 kg/m^3
  1 lb/in^3 = 27·68 g/cm^3

Force
  1 lbf   = 0·4536 kgf
                                  = 4·448 newtons (N)

Moment of Force
  1 lbf ft   = 1·356 N m
  1 lbf in   = 0·1130 N m

Pressure and Stress
  1 lbf/ft^2 (psf)   = 47·88 N/m^2 = 47·88 pacsal (Pa)
                                  = 4·882 kgf/m^2
  1 lbf/in^2 (psi)   = 68·95 millibar (mb)
                                  = 6895 N/m^2 = 6895 Pa
                                  = 0·0703 kgf/cm^2
  1 inch of Mercury   = 33·86 mb = 3386 N/m^2

Power
  1 horse-power (hp)   = 1·014 metric horse-power
                                  = 0·746 kilowatt (kW)

Energy, Work and Heat
  1 ft lbf   = 1·356 joules (J)
  1 British thermal unit (BThU)   = 1·055 kJ
                                  = 0·252 kilogram calorie (kcal)
  1 horsepower-hour (hp h)   = 2·685 megajoules (MJ)
                                  = 0·746 kWh

Temperature
 °F (Fahrenheit)   = 1·8°C + 32 (Celsius or Centigrade)
 °R (Rankine)   = 1·8°K (Kelvin)
                            = °F + 459·67

Heat Flow Rate
 1 BThU/h   = 0·2931 W
 1 ton of refrigeration   = 3517 W

Thermal Capacity per Unit Volume
 1 BThU/ft^3/°F   = 67·07 kJ/m^3/°C

Thermal Conductance
 1 BThU/ft^2/h/°F   = 5·678 W/m^2/°C

Thermal Conductivity
 1 BThU ft/ft^2/h/°F   = 1·731 W/m/°C
 1 BThU in/ft^2/h/°F   = 0·1442 W/m/°C
 1 BThU in/ft^2/s/°F   = 519·2 W/m/°C

Thermal Resistivity
 1 ft^2 h °F/BThU in   = 6·93 m°C/W
 1 ft^2 h °F/BThU ft   = 0·577 m°C/W

Illumination
 1 foot-candle = 1 lumen/ft^2 (lm/ft^2)
                            = 10·76 lux = 10·76 lm/m^2

Luminance
 1 foot-lambert   = 3·426 candela per square metre
                              (cd/m^2)
 1 cd/ft^2 = 10·76 cd/m^2
 1 cd/in^2 = 1550 cd/m^2

2. METRIC TO BRITISH UNITS

Length
 1 km   = 0·6214 miles
 1 m   = 1·094 yd
 1 cm   = 0·3937 in
 1 mm   = 0·039 37 in
 Note: 1 μm (micrometre or
                   micron) = 10^{-6} m
 1 nm (nanometre or
              millimicron) = 10^{-9} m
 1 Å (Ångström)   = 10^{-10} m

Area
1 km^2	= 247·1 acres
	= 0·3861 square mile
1 hectare	= 2·471 acres
1 m^2	= 1·196 yd^2
1 cm^2	= 0·155 in^2

Volume
1 m^3	= 1·308 yd^3
1 dm^3	= 0·035 31 ft^3
1 cm^3	= 0·0610 in^3
1 litre	= 0·2200 Imperial gal
	= 0·2642 US gal

Velocity
1 km/h	= 0·621 mile/h
1 m/s	= 3·280 ft/s

Mass
1 tonne	= 0·984 long tons (2240 lb)
	= 1·102 short tons (2000 lb)
1 kg	= 2·205 lb
1 g	= 0·035 27 ounces
	= 15·43 grains

Mass per Unit Area
1 kg/m^2	= 0·2048 psf
1 kg/cm^2	= 14·22 psi

Mass per Unit Volume
1 kg/m^3	= 0·062 43 lb/ft^3

Force
1 N	= 0·2248 lbf
1 kgf	= 2·205 lbf

Note: 1 kgf = 9·81 N
1 dyne = 10^{-5} N

Moment of Force
1 N m	= 0·738 lbf ft
1 kgf m	= 7·233 lbf ft

Pressure and Stress
    1 kN/m² = 1 kPa
                                        = 20·89 psf
                                        = 0·1450 psi
                                        = 0·2953 inches of mercury
    Note: 1 millibar   = 100 N/m²
        1 kgf/cm²   = 98·1 kN/m²
    1 mm of mercury   = 1 torr = 133·3 N/m²
    1 atmosphere   = 101·3 kN/m²
    1 pascal (Pa)   = 1 N/m²

Power
    1 kW   = 1·341 hp
    1 metric horsepower   = 0·986 hp

Energy, Work and Heat
    1 J   = 0·738 lbf ft
    1 k cal   = 3·968 BThU
    1 kgf m   = 7·233 lbf ft
    Note: 1 kgf m   = 9·81 J
        1 erg   = 10^{-7} J
        1 W s   = 1 J
        1 Wh   = 3·60 kJ
        1 k cal   = 4·186 kJ

Temperature
    °C   = (°F − 32)/1·8
    °K   = °R/1·8

Heat Flow Rate
    1 kW   = 3410 BThU/h
                                        = 0·284 tons of refrigeration
    Note: 1 k cal/s   = 4·187 kW = 4·187 kJ/s
        1 k cal/h   = 1·163 W

Thermal Capacity per Unit Volume
    1 J/cm³/°C   = 14·90 BThU/ft³/°F
    Note: 1 k cal/m³/°C   = 0·004 187 J/cm³/°C

Thermal Conductance
    1 W/m²/°C   = 0·176 BThU/ft²/h/°F
    Note: 1 k cal/m²/h/°C   = 1·163 W/m²/°C

Thermal Conductivity
    1 W/m/°C   = 0·577 BThU ft/ft²/h/°F

Thermal Resistivity
  1 m°C/W $= 1{\cdot}731$ ft^2 h °F/BThU ft

Illumination
  1 lm/m^2 $= 0{\cdot}0929$ lm/ft^2

Luminance
  1 cd/m^2 $= 0{\cdot}0929$ cd/ft^2

This dictionary contains about 4500 entries ranging in length from one line to $1\frac{1}{2}$ pages. While Professor Cowan has concentrated on the scientific aspects of architecture, he has also included the most frequently encountered terms from neighbouring fields, such as fine art, the history of architecture, the craft traditions of the building industry, structural, mechanical and electrical engineering, materials science, physics and chemistry. The book will probably be most helpful to architecture students, it will also be of considerable value to students of engineering, practising architects, practising engineers and other professional persons engaged in the building industry, and as a general reference book.

The coverage is probably much wider than that of any other dictionary published, and the comparison of different meanings for the same word in neighbouring fields should be particularly helpful. There is an extensive system of cross-references, so that the reader who cannot remember the exact term he is looking for, but knows a related word, should have no difficulty in finding the information he is seeking. Since it is the work of one author the treatment is consistent throughout.

The appendices give the current addresses of important international organisations in architectural science, a review of the literature on the subject, and technical data which are not readily obtainable from trade publications.

ISBN 0 470 18070 6